Contents

Acknowledgements

When this Fourth Edition is published, it will be almost fifteen years since we discussed the original idea for the Reader and began work on it. That idea had a lot to do with our being members of the enthusiastic course team for a new BA Communication Studies Degree at Sunderland Polytechnic. We now have other people to thank, including the undergraduate and postgraduate students we have taught at Liverpool and at Trondheim Universities and our colleagues at these institutions. Both of us found the experience of being members of the Communication and Cultural Studies panel of the Council for National Academic Awards during a formative phase in the development of the field particularly helpful. We thank other members and officers of that panel for the forum they provided and we also acknowledge the extremely useful comments on past editions made by staff teaching at universities and colleges in Britain and, more recently, in the USA. Finally, the publication of the Fourth Edition is a good time to register how much we have enjoyed working with editorial staff at Edward Arnold, in what has become a quite closely collaborative venture. In particular we are grateful for the initial encouragement of Sarah Barrett, the continuing support of Christopher Wheeler, the efficient editorial eye of Helen Tuschling and the enthusiasm and judgement of Lesley Riddle, who has steered us through our more recent revisions and re-structurings.

The publisher wishes to thank the following for permission to use their copyright material:

Addison-Wesley Publishing Company for the extract from *Person Perception*, © 1970 by A.H. Hastorf, J. Schneider and J. Polefka; George Allen & Unwin, Hemel Hempstead and Basic Books, New York for an extract from *The Interpretation of Dreams*; Edward Arnold (Publishers) Ltd for 'Documentary Meanings and the Discourse of Interpretation' by John Corner and Kay Richardson from John Corner (ed.) *Documentary and the Mass Media*, 1986; Association for Educational Communications and Technology, Washington for George Gerbner's 'Basic Generalized Graphic Model of Communication' from 'On content-analysis and critical research in mass communications' in *Audio-Visual Communication Review* (AVCR) Vol. 6, No. 3, Spring 1958, pp. 85–108; Blackwell Publishers for the extract from Anthony Giddens, *Modernity and Self-Identity*, Polity: Oxford, 1991, pp. 14–27; Cambridge University Press for the extract from Deborah Tannen, *Talking Voices: Repetition, Dialogue and Imagery in Conversational Discourse*, 1989, pp. 176–94; Indiana University Press for the extract from Bill Nichols's *Ideology and the Image*; Hroar Klempe for 'Music, text and image in commercials for Coca-Cola' © Hroar Klempe 1993; Lawrence Erlbaum Associates, Inc. for 'Captured on Videotape: Camcorders and the Personalization of Television' by Lawrence J. Vale from Julia Dobrow (ed.) *Social and Cultural Aspects of VCR Use*, 1990; Lexington Books for the extract from *A Perspective on Social Communications* by Stuart J. Sigman, Lexington, Mass.: Lexington Books, D.C. Heath and Company, copyright © 1987, D.C. Heath and Company; Methuen & Co Ltd for the extract from *Watching Dallas: Soap Opera and the Melodramatic Imagination* by Ien Ang; MIT Press, Mass. for Colin Cherry's 'What is Communication?' from *On Human Communication*; Mouton de Gruyter, A Division of Walter de Gruyter & Co. for the extract from William T. Scott's *The Possibility of Communication*, 1990, pp. 126–31; Trevor Pateman for 'Impossible Discourse' from *Language, Truth and Politics*; Penguin Books Ltd for 'Verbal and Non-Verbal Communication' from Michael Argyle's *The Psychology of Interpersonal Behaviour* (Pelican Original 3/e. 1978) © Michael Argyle, 1967, 1972, 1978; Penguin Books Ltd (Allen Lane, the Penguin Press, 1969 Pelican Books, 1971) © Erving Goffman 1959 and Doubleday & Co. Inc., NY © Erving Goffman 1958 for the extract from *The Presentation of Self in Everyday Life*; Routledge and Kegan Paul for 'Social Class, Language and Socialization' from Basil Bernstein's *Class, Codes and Control* Vol 1 and for John Ellis's 'Broadcast TV as Sound and Image' from *Visible Fictions* and for the extract from *Learning to Write* by Gunther Kress; Sage Publications Ltd for 'Talk, Identity and Performance: The Tony Blackburn Show', by Graham Brand and Paddy Scannell, reprinted with permission from Paddy Scannell (ed.) *Broadcast Talk*, © 1991, Sage Publications Ltd; St Martin's Press, Inc. and Macmillan, London and Basingstoke for 'Beyond Alienation: An Integrational Approach to Women and Language' from *Feminism and Linguistic Theory* by Deborah Cameron; Universe Books, New York and Hutchinson (Publishers) Ltd, London for Jean Aitchison's 'Defining Language' from *The Articulate Mammal*; University of Toronto Press and the Open University, Milton Keynes for the extract from *Visualizing Deviance: A Study of News Organization* by Richard Ericson, Patricia Baranek and Janet Chan, 1987 reprinted by permission of University of Toronto Press; The William Alanson White Psychiatric Foundation, Inc. for 'Mass Communication and Para-Social Interaction' by Donald Horton and Richard Wohl from *Psychiatry*, Vol. 19, 1956, pp. 215–29, reprinted by special permission of and © renewed 1984 by The William Alanson White Psychiatric Foundation, Inc.

Every effort has been made to trace copyright holders of material produced in this book. Any rights not acknowledged here will be acknowledged in subsequent printings if notice is given to the publishers.

Introduction

This is the fourth edition of *Communication Studies*. Like its predecessors, it is addressed to the student reader and is designed for use on undergraduate and postgraduate courses. Currently, these courses either form part of specialist degrees in Communication or are devised as options within broader programmes. There is also a strong Communication Studies element in schools and colleges. Here, the book should be of help to teachers in preparing a syllabus.

The grouping together of complementary and often convergent lines of enquiry in the humanities and social sciences to make a new category within which to organize teaching and research – Communication Studies – is now some twenty years old as a feature of British Higher Education and is older still in its separate development within the North American system. It thus needs less explanation and defence than it did when the first edition of the book appeared in 1980. Nevertheless, a few brief comments on the character of the field as it has become established and on shifts in its emphases over the last decade might be helpful.

Perhaps it is worth observing first of all that in the opinion of the editors and in the practice of the majority of institutions offering courses, the term Communication Studies does not indicate a new discipline. This is not a point about its status but about its character – it is essentially a way of organizing and relating ideas, methods of study and topics drawn from several disciplines and thus might usefully be described as a 'field of study' (Environmental Studies, Urban Studies and Women's Studies are other, cognate examples). Such a gathering and interconnecting activity inevitably produces work of a distinctive character, but this distinctiveness is grounded in a basic familiarity with a number of discipline-based ideas and methods. These circumstances have sometimes led to discussion as to whether Communication Studies is *interdisciplinary* (operating 'across and between' disciplines) or *multidisciplinary* (recognizing a range of distinctive disciplinary inputs). Such discussion can quickly become irritatingly abstract but it would be appropriate to note that most degree courses in Britain begin with a strongly multidisciplinary programme, and then move towards a final year in which a more interdisciplinary approach is encouraged, though options may also allow for further discipline specialization.

Within this broad pattern, the present volume is intended principally to further 'integrated study' within the earlier part of a programme, the part when introductory specialist studies are concurrently being undertaken but when there is a need for reading and discussion which will encourage thinking across a broad range of issues and topics. At its best, such work can light up large areas of the whole field and generate an early sense of conceptual and substantive interconnections along with an informed enthusiasm for the 'communication' focus.

At root, Communication Studies is about how human meanings are made through the production and reception of various types of sign. It is about visual and verbal sign systems and the technologies used to articulate, record and convey them. It is also about the political and social character of signifying activities. Section I of this book explores matters of definition and approach in detail and the subsequent sections present the reader with what we hope are examples of clear and stimulating inquiry into the varieties and social conditions of meaning-making.

Studying communication is vulnerable to two main dangers as an academic enterprise – those of 'bittiness' and of excessive formalism. Bittiness often follows from the desire on the part of some communication specialists to make connections just about everywhere on the map of human knowledge ('communication' can be a disastrously capacious category with which to work!). Bits are all that this type of intellectual promiscuity has time to offer. Formalism results from academics developing a concern with the nature of signs and sign forms which is so exclusive as to obscure from consideration the specific historical and social circumstances within which signification is shaped and has its consequences. There is more about these two problems in Section I but we very much hope to have produced here a collection in which continuity, depth of treatment and a sense of social contexts and functions are apparent alongside the formal variety. In practice, most institutions teaching in the area achieve cohesion and an appropriate scholarly thoroughness for their courses by radically selecting from the range of possible topics and approaches. This book engages with some of the more common topics, ideas, issues addressed, though it often seeks to do so by providing a fresh perspective or an original line of commentary.

In revising the volume for this edition we have taken account of a number of changes and developments in the field. The emphasis on the reception side of communicative activity, an emphasis which developed strongly in the late '80s, is given further prominence in this edition. A return to certain fundamental questions about the relationship of the media to group and individual perception is also signalled, suggesting a need to re-evaluate the terms in which the psychological dimension of media processes can best be analysed. There is also, in some pieces, a direct concern with the conditions of modern, or indeed post-modern, private and public life, and the character and function of communicative systems within these conditions.

Communication Studies as a way of organizing degree-level work has so far been a success. It has attracted able and enthusiastic students and promoted not only topics of study but also forms of study which have often contrasted strongly with the insular, canonical and unreflective teaching which can still sometimes be found in many traditional, single-discipline subjects. It has also informed research in the Humanities and Social Sciences, encouraging the emergence of new ideas, methods of analysis and topics.

Apart from the factors mentioned above, our selection of items has been guided by three general criteria:

First of all, we have tried to select pieces which raise questions and promote discussion in a clear and accessible way. We have therefore chosen items which have worked well in our own teaching.

Secondly, we have favoured items which indicate or explore connections between different academic areas and methods rather than those which work exclusively within the terms of one established perspective.

Thirdly, we have included examples of studies widely judged as seminal or which appear to us to have a 'classic' quality which merits re-publication here alongside more recent studies. Extracts from work by Bernstein, Goffman, Argyle, Lippmann and Lang and Lang are indicative of this category.

In preparing a fourth edition we have revised all the editorial material, updated the lists of recommended further reading, removed ten items and added eleven. We have retained the four-section structure to allow a broad classification of item without fragmentation of overall design. We continue to think that this structure usefully foregrounds connections and contrasts across the items and gives scope for different sequences of reading and use.

Where items have included references to other parts of the larger work from which the item is taken, these have been 'silently' deleted. An ellipsis in square brackets [. . .] indicates that a more substantial part of the original has been omitted.

John Corner
Jeremy Hawthorn

Section I

Communication: definitions and approaches

If you were to go out into a busy town centre and to stop passers-by, asking them each to give you an example of something describable as 'communication', you would be sure to collect an interesting variety of responses (and this is to exclude the perhaps less polite ones!). Radio and television would almost certainly be mentioned several times, with newspapers not far behind. Perhaps the telephone might top the list though. Then there would probably be mentions of letter-writing, of conversation, of public speaking and even of your own questioning. Less frequently, examples such as advertising hoardings and traffic signals would turn up. A few people might interpret your question in terms of an established plural usage of the word, to mean something to do with transport, and cite rail or air services as examples. Even with these excluded, it's pretty clear that a long list could be collected quite soon. You will notice too, that some of the suggested examples refer primarily to means of communication (e.g. the telephone and radio) while others refer to types of communicative act (e.g. letter-writing and public speaking).

Two pertinent questions might then be posed for intending students of communication. First, what is to be gained by grouping together these various message-based systems and activities so as to constitute the subject matter of a field of study? And then, how on earth is it possible to develop a unitary set of concepts and methods by which to pursue such study?

Briefly, our answer to the first question is that an understanding of the complexity of human communication processes is considerably advanced the more we are aware of the different sign-systems through which they work, the different technologies which can be used in support of them and the different kinds of context in which they occur. It certainly isn't true that everything which we might choose to call 'communication' fits neatly together to make one grand, unified phenomenon, but interconnections across different types of behavioural, verbal and visual processes are both extensive and illuminating. Moreover, as the centrality of meaning-making activities is increasingly recognised in the study of a whole range of political, social and psychological matters, it becomes appropriate to have a formal grouping of studies which can connect developments in related lines of inquiry. Communication Studies, then, makes sense as a way of organizing and developing knowledge.

Our answer to the second question is that it is not possible to get very far in this venture with any single, guaranteed set of concepts and methods. Research and teaching programmes in the area of Communication Studies need to proceed by reference to a variety of ideas and procedures used in Arts and Social Science disciplines. Since hardly any of these is uncontroversial and many have been the subject of intensive academic debate, communications students should quickly acquire a useful critical alertness (and an accompanying sense of challenge) concerning *how* academic inquiries seek to find things out and explain them.

But don't we need a reasonably tight working definition of 'communication' to guide our studies? We think not, although as Colin Cherry points out in the first item of this section, attempts at such definitions can promote useful discussion. However, most of the activities and processes we are interested in can be recognized according to the kinds of criteria which were used by our hypothetical passers-by. It is only when we have, in a sense, pushed beyond the broad category of 'communication' itself to a description and analysis of specific communicative practices and interactions that definitional precision becomes essential. Putting theoretical emphasis on categoric definition, in a search for a set of universal principles and exclusive defining features, seems an unpromising venture. It also risks making of 'communication' some mystical higher unity, contemplation of whose essential or ideal characteristics is more interesting than looking in detail at examples of what are commonly regarded as 'communicative' activity. In fact, over-use of the term can disguise the need for concepts other than 'communication' to be employed in order adequately to tackle this latter kind of investigation.

So we would largely dissociate ourselves from attempts at developing some 'general theory of communication'. This is not to deny the value of work which tries to discern general features but to note that a theory so devised would have to operate at such a rarefied altitude above actual kinds and means (let alone specific instances) of human interaction as to severely limit its usefulness. This would be largely confined to the level of broad classification. Secondly, and as a result, 'general theory' would have to omit virtually all consideration of communicational contexts – the social settings and the social relationships (including relationships of inequality and of power). These always serve to shape communicative forms and practices at the same time as such practices are one important way in which social settings and relationships themselves are constituted and changed.

As we suggested in the Introduction, a reduced awareness of communicational contexts, following from a desire to achieve the closest possible analysis of communicative forms, is already one of the main limitations with which work in Communication Studies has been confronted. This has perhaps given a measure of academic support to the current fashion for talking about the modern world as characterised by 'problems of communication' (e.g. in industry, in social welfare, in education and in relations with the Third World) as if the circumstances referred to were kinds of technical hitch, resolvable through better (more professional?) kinds of communication, without regard for such factors as material conditions and differing economic interests.

It is important then, to regard 'contexts' as something other than mere background. Certainly, as Communication Studies students, we have to declare our academic interest in specific kinds of phenomena over and above others, but a recognition of the fundamental ways in which, as Cherry puts it, 'communication is a social affair', of the ways in which communicative forms and social circumstances inter-relate, is needed too. In fact, at the end of his piece, Cherry refers to a branch of linguistics known as pragmatics which is precisely concerned to relate sign-systems both to users and to contexts of use. Pragmatics has an increasingly important function in communication research and further reference to it will appear later in the book.

All the pieces in this section have been chosen because they offer clear, preliminary ideas about the characteristics of communicative behaviour and about the analytic categories through which we can begin to explore it in a coherent way. For instance, they address questions of communicative convention – the codes or rules by which signs are selected and combined in such a way as to be meaningful to an addressee. These conventions can be as 'tight' as those regulating our word order when writing an English sentence or as relatively 'loose' as those which govern the tones of voice, facial mannerisms and general behaviour associated with 'politeness' in our culture (where a number of options and permutations exist). Or we might contrast the international visual conventions of traffic and road signs with the (again, international, but much looser) conventions at work in television 'soap opera'. Quite frequently, communication analysis involves studying 'tight' and 'loose' conventions in combination.

'Successful' use of communicative conventions suggests that meaning has effectively been transmitted, but this way of putting it radically over-simplifies what is in fact a complex interaction of intentions, sign-forms and interpretive activity, one which will involve different levels of consciousness. Communication seen as a business of 'sending things' may be an adequate perspective for the postal services but it will serve only to distort the study of most human communication and to mask important variables. To give an example, we might ask how questions of intention figure in respect of what an addressee 'makes of' (a revealing phrase in this context) some remark of ours. If we say we did not intend such a meaning, does that close the matter? Or if the addressee claims in response that, nevertheless, this is what our comment really means, does that close it from the opposite direction? Matters of setting and circumstances may quickly enter the picture here but we can easily imagine situations in which we would accept personal responsibility for an 'unintended meaning' (perhaps regarding it as a result of our careless choice of words) and also situations in which it was seen to be the addressee's problem for 'taking it the wrong way' or 'reading more into it than was intended' (again, revealing phrases). Were there to be two addressees, they might well differ in their interpretations as to what was meant, adding a further complication. Straight away, any direct analogy with the postal system or an electrical circuit looks unhelpful.

In both this and the other sections of the book, links between communicative conventions, cultural contexts of use and modes of participant understanding will be regularly explored. In examining these links, some writers have made use of the term *ideology* in order to probe the relations between communication, consciousness and power. As noted above, most communication takes place within settings of inequality, quite often framed by the broader patterns of economic inequality. The play of dominant economic and political interests in securing terms of public communication which appear neutral but which act to reinforce particular prejudices, discriminations and evaluations can be regarded as an ideological dimension to signification. This dimension often becomes 'locked into' the meaning of words and images and, as such, it permeates many aspects of culture. Its presence (and perhaps unconscious duplication) has given rise to many charges of 'biased' communication (as in accusations of sexism, racism or of social class partiality). This is particularly so where the media are concerned, though the matter is by no means confined to mass communication and it will be dealt with more thoroughly in all three subsequent sections.

Two other terms which can be helpful right from the start are *denotation* and *connotation*. These will be given more discussion later too, but the distinction is essentially between on the one hand, the literal meanings of words or those things which images primarily depict and, on the other hand, the associative meanings which words and images have gathered around them within specific cultural contexts. For instance, 'enterprise', 'gay' and even 'defence' are all words which, in different ways, generated powerful political associations in the 1980s, taking them well beyond their core meanings (in the case of 'gay', what might have started out as a new connotation has effectively become a new, alternative denotation). Likewise, images of European cities in advertisements for expensive cars are clearly designed to connote cosmopolitanism, sophistication, style and wealth rather than to provide denotative information about, say, Rome or Berlin. Connotative meanings are very important at the ideological level of communicative process referred to above.

Distinguishing between two 'levels' like this is often illuminating but you should be wary of assuming a tidy division every time.

Colin Cherry develops a preliminary discussion of the relationship between language systems and social values in his look at the problems of adequately defining communication. In the course of his piece, he attempts to relate together a number of the key terms (e.g. code, sign and rule) and to show by examples how their use might help us to think more incisively about the ways in which communication works.

His stress on communication as a 'sharing of rules' keeps our attention on those cultural conventions – the symbol system of a language, for example – which we noted above to be the primary means by which meaning is produced. It is necessary, of course, to see such 'sharing' as always occurring under the terms imposed by the broader schemes of social relationship within which individuals live, feel and think rather than posing it as somehow a natural equality in the human race. Cherry was one of the first British academics to pursue an interest in communication as an organizing concept for research. By citing and discussing a number of definitions he also indicates something of the range of phenomena which communication analysis finds itself considering.

In our second extract, we have reproduced a model of communication devised by **George Gerbner**, a distinguished American media researcher. Again, like Cherry, Gerbner has a basic understanding of general communication processes in mind. The early history of Communication Studies, particularly in the USA, was littered with various diagrammatic representations – most of them failing to make their bold, graphic depictions of process and relationship shed much light on real-world instances. We think that Gerbner's model, which concentrates only on a few features, has a more suggestive power than most and, as our accompanying note indicates, we think it encourages an investigation of several important features of signifying activity and perception.

Jean Aitchison's piece brings us in closer to the study of that most important element in communication – language. Her main aim is to consider how human language differs from the communicative behaviour of other animals. The discussion is organized around the question 'can animals talk?' and in the course of answering this she lists some of the 'design features' which we might want to use to decide whether a given piece of communication could usefully be said to employ a 'language'. Her comments on certain experiments carried out with bees and with dolphins bring out the importance of displacement ('the ability to refer to things far removed in time and place') and of arbitrariness (the lack of any *necessary* connection between conventional signs [e.g. words] and the things they refer to) in human language use. These, together with other characteristics which she mentions, will be treated in more detail in a later section.

As we noted earlier, central to many arguments about communication through language or through images or gesture, are questions of intention. In an enterprising use of diagrammatic form, **W.T. Scott** explores the permutations which can possibly occur between sender and receiver in respect of communicative intentions. In working towards a more tightly theorized account of intentionality, Scott wishes to distinguish between behaviour which is communicative and that which is informative. He also wants to investigate how intention is related both to the practices of deceit and to the perception of deceitful behaviour by receivers. His itemized discussion of the variables draws on a wide range of examples and offers a clear insight into the interactive complexity of what we routinely 'take for granted' in everyday life.

The following extract also analyses a particular aspect of communication within a broad scheme of classification. It is taken from the work of **Michael Argyle**, a social psychologist with a distinguished record in the study of interpersonal behaviour. Argyle produces a helpfully illustrated inventory of human communicative actions, one in which a clear account is given of non-verbal behaviours. Although the rules at work here are not so precise as many linguistic ones, Argyle draws our attention to the often highly convention-bound character of what goes on through this important 'channel' of interaction. Rather disappointingly, Argyle seems unwilling to ask *why* the cross-cultural differences which he so concisely documents actually occur, perhaps being too prompt in accepting ideas of innate rather than culturally-formed communicative dispositions and competences. Nevertheless, as with all the pieces in the section, the final effect is to provide some basic knowledge and to raise questions and problems in an accessible manner and with plenty of examples.

Our final item, too, concentrates on visual rather than verbal signification, this time raising questions about the use of the medium of photography to represent features of the world of appearances. Through photographic representation, meanings are generated which go far beyond the particular features thus depicted (see the discussion of denotation and connotation earlier). **Bill Nichols** applies the ideas of semiology – a theory and method of sign analysis which is particularly concerned with the relations between signs – to a number of photographic examples. His aim is to identify the processes of meaning-making at work. These include the perception of compositional form but go right 'up' to questions about the kinds of implicit evaluations and systems of social understanding which the images support. Whereas Aitchison talks of the arbitrary character of language's *symbolic* signs (no necessary relation between a word and the thing represented), Nichols engages with the *iconic* character of much visual depiction (the image resembles that which it represents). In photography, this iconicity is joined by what has been referred to as an *indexical* factor (the image is actually caused by light waves refracted from the represented object entering the camera lens, thus linking object and image in a direct, causal relationship).* The special communicative power of photography (and of film and video too) stems from this distinctive iconic-indexical mix, which gives evidential values to the image produced. Nichols would not, however, agree with the well-known comment that 'the camera never lies' and he explores the particular configurations of codes (including the verbal codes of captions) which photography can use at different levels of its communicative organisation. He refers both to the 'horizontal' axis of sign *combination* and to the 'vertical' axis of sign *selection*. These can more easily be demonstrated in respect of language (where, at the level of the sentence, they roughly equal the rules of syntax and the available options for word choice). Nevertheless, as Nichols shows, they can be put to revealing use in 'de-naturalizing' the meanings of photography, and especially the meanings of those images which have been produced in order to persuade.

Further reading

Dance, F.E. 1970: The concept of communication. *Journal of Communication* 20, 201–10.
A short but helpful discussion of some basic definitional issues.

Fiske, J. 1990: *Introduction to communication studies* (second edition) London: Routledge.
A short and well-illustrated introduction both to communication as a topic of study and to the available methods for conducting research and analysis. Fiske believes semiotics to offer a central, unifying perspective in communication research and he gives examples of its application within a number of different kinds of verbal and visual inquiry. Each chapter ends with useful suggestions for further work.

Gerbner, George 1956: Towards a general model of communication. *Audio Visual Communication Review* IV, 3.
A clearly written and very suggestive argument about the general characteristics of communication processes.

Hinde, R.A. (ed.) 1972: *Non-verbal communication*. Cambridge University Press.
A fine book, containing articles which are relevant to several strands of inquiry into communication. It is recommended here because it contains a number of pieces which discuss the principles both of human and animal communication. Other useful articles trace the importance of non-verbal behaviour in children, in the mentally ill, in acting and in the characteristic postures and gestures depicted in Western art.

O'Sullivan, T., Hartley, J., Saunders, D. and Fiske, J. 1983: *Key concepts in communication*. London: Methuen.
A useful guide to the terminology of the area which manages to give potted information without too much superficiality or loss of critical perspective. Well-referenced.

* The terms 'index', 'icon' and 'symbol' were used by the pioneer theorist Charles S. Peirce (1839–1914) to differentiate sign-types. They have been widely influential though frequently subject to confusion, and even Peirce's own usage was not consistent. There are many secondary expositions; for example J. Fiske 1990 above.

1

What is communication?

Colin Cherry

From Cherry, Colin 1957: *On human communication.* Cambridge, Mass: MIT Press, 3–9.

Communication is essentially a social affair. Man has evolved a host of different systems of communication which render his social life possible – social life not in the sense of living in packs for hunting or for making war, but in a sense unknown to animals. Most prominent among all these systems of communication is, of course, human speech and language. Human language is not to be equated with the sign systems of animals, for man is not restricted to calling his young, or suggesting mating, or shouting cries of danger; he can with his remarkable faculties of speech give utterance to almost any thought. Like animals, we too have our inborn instinctive cries of alarm, pain, etcetera: we say *Oh!, Ah!;* we have smiles, groans, and tears; we blush, shiver, yawn, and frown.[1] A hen can set her chicks scurrying up to her, by clucking – communication established by a releaser mechanism – *but human language is vastly more than a complicated system of clucking.*

The development of language reflects back upon thought; for with language thoughts may become organized, new thoughts evolved. Self-awareness and the sense of social responsibility have arisen as a result of organized thoughts. Systems of ethics and law have been built up. Man has become self-conscious, responsible, a social creature.

Inasmuch as the words we use disclose the true nature of things, as truth is to each one of us, the various words relating to personal communication are most revealing. The very word 'communicate' means 'share', and inasmuch as you and I are communicating at the moment, we are one. Not so much a union as a unity. Inasmuch as we agree, we say that we *are of one mind,* or, again, that we understand *one another.* This one another is the unity. A group of people, a society, a culture, I would define as 'people in communication'. They may be thought of as 'sharing rules' of language, custom, of habit; but who wrote these rules? These have evolved out of those people themselves – rules of conformity. Inasmuch as that conformity is the greater or the less, so is the unity. The degree of communication, the sharing, the conformity, is a measure of one-mindedness. After all, what we share, we cannot each have as our own possession, and no single person in this world has ever been born and bred in utter isolation. 'No man is an island, entire of itself.'[2]

Speech and writing are by no means our only system of communication. Social intercourse is greatly strengthened by habits of gesture – little movements of the hands and face. With nods, smiles, frowns, handshakes, kisses, fist shakes, and other gestures we can convey most subtle understanding.[3] We also have economic systems for trafficking not in ideas but in material goods and services; the tokens of communication are coins, bonds, letters of credit, and so on. We have conventions of dress, rules of the road, social formalities, and good manners; we have rules of membership and function in businesses, institutions, and families. But life in the modern world is coming to depend more and more upon 'technical' means of communication, telephone and telegraph, radio and printing. Without such technical aids the modern city-state could not exist one week, for it is only by means of them that trade and business can proceed; that goods and services can be distributed where needed; that railways can run on a schedule; that law and order are maintained; that education is possible. Communication renders true social life practicable, for communication means organization. Communications have enabled the social unit to grow, from the village to the town, to the modern city-state, until today we see organized systems of mutual dependence grown to cover whole hemispheres (McDougall, 1927).

The development of human language was a tremendous step in evolution; its power for organizing thoughts, and the resulting growth of social organization of all kinds, has given man, wars or no wars, street accidents or no street accidents, vastly increased potential for survival.

As a start, let us now take a few of the concepts and notions to do with communication, and discuss them briefly, not in any formal scientific sense, but in the language of the market place. A few dictionary definitions may serve as a starting point for our discursive approach here; later we shall see that such definitions are not at variance with those more restricted definitions used in scientific analysis. The following have been drawn from the *Concise Oxford English Dictionary*[4]

> *Communication,* n. Act of impairing (esp. news); information given; intercourse; . . . (Military, Pl.) connexion between base and front.
>
> *Message,* n. Oral or written communication sent by one person to another.
>
> *Information,* n. Informing, telling; thing told, knowledge, items of knowledge, news, (on, about) . . .
>
> *Signal,* n., v.t. & i. Preconcerted or intelligible sign conveying information . . . at a distance. . . .
>
> *Intelligence,* n. . . . understanding, sagacity . . . information, news.
>
> *News,* n. pl. Tidings, new information. . . .
>
> *Knowledge,* n. . . . familiarity gained by experience, person's range of information. . . .
>
> *Belief,* n. Trust or confidence (*in*); . . . acceptance as true or existing (of any fact, statement, etc.;. . .). . .
>
> *Organism,* n. Organized body with connected interdependent parts sharing common life . . .; whole with interdependent parts compared to living being.
>
> *System,* n. Complex whole, set of connected things or parts, organized body of material or immaterial things . . .; method, organization, considered principles of procedure , (principle of) classification;. . .

Such dictionary definitions are the 'common usages' of words; scientific usage frequently needs to be more restricted but should not violate common sense –

an accusation often mistakenly levelled against scientific words by the layman.

The most frequent use of the words listed above is in connection with *human* communication, as the dictionary suggests. The word 'communication' calls to mind most readily the sending or receipt of a letter, or a conversation between two friends; some may think of newspapers issued daily from a central office to thousands of subscribers, or of radio broadcasting; others may think of telephones, linking one speaker and one listener. There are systems too which come to mind only to specialists; for instance, ornithologists and entomologists may think of flocking and swarming, or of the incredible precision with which flight manoeuvres are made by certain birds, or the homing of pigeons – problems which have been extensively studied, yet are still so imperfectly understood. Again, physiologists may consider the communicative function of the nervous system, coordinating the actions of all the parts of an integrated animal. At the other end of the scale, the anthropologist and sociologist are greatly interested in the communication between large groups of people, societies and races, by virtue of their cultures, their economic and religious systems, their laws, languages, and ethical codes. Examples of 'communication systems' are endless and varied.

When 'members' or 'elements' are in communication with one another, they are associating, cooperating, forming an 'organization', or sometimes an 'organism'. Communication is a social function. That old cliché, 'a whole is more than the sum of the parts', expresses a truth; the whole, the organization or organism, possesses a structure which is describable as a set of *rules*, and this structure, the rules, may remain unchanged as the individual members or elements are changed. By the possession of this structure the whole organization may be better adapted or better fitted for some goal-seeking activity. Communication means a *sharing* of elements of behaviour or modes of life, by the existence of sets of rules. This word *rule* will be discussed later.

Perhaps we may be permitted to comment upon a definition of communication, as given by a leading psychologist (Stevens, 1950): '*Communication is the discriminatory response of an organism to a stimulus.*[5] The same writer emphasizes that a definition broad enough to embrace all that the word 'communication' means to different people may risk finding itself dissipated in generalities. We would agree; such definitions or descriptions serve as little more than foci for discussion. But there are two points we wish to make concerning this psychologist's definition. First, as we shall view it in our present context, communication is not the response itself but is essentially the *relationship* set up by the transmission of stimuli and the evocation of responses. Second, it will be well to expand somewhat upon the notion of a stimulus; we shall need to distinguish between human language and the communicative signs of animals, between languages, codes, and logical sign systems, at least.

The study of the signs used in communication, and of the rules operating upon them and upon their users, forms the core of the study of communication. There is no communication without a system of signs – but there are many kinds of 'signs'. Let us refer again to the *Concise Oxford English Dictionary*:

> *Sign,* n. . . . written mark conventionally used for word or phrase, symbol, thing used as representation of something . . . presumptive evidence or indication or suggestion or symptom *of* or *that,* distinctive mark, token, guarantee, password . . . portent . . .; natural or conventional motion or gesture used instead of words to convey information. . . .
>
> *Language,* n. A vocabulary and way of using it . . .
>
> *Code,* n., and v.t. Systematic collection of statutes, body of laws so arranged as to avoid inconsistency and overlapping; . . . set of rules on any subject; prevalent morality of a society or class . . .; system of mil. or nav. signals. . . .
>
> *Symbol,* n. . . . Thing regarded by general consent as naturally typifying or representing or recalling something by possession of analogous qualities or by association in fact or thought. . . .

We shall use the word *sign* for any physical event used in communication – human, animal, or machine – avoiding the term *symbol,* which is best reserved for the Crown, the Cross, Uncle Sam, the olive branch, the Devil, Father Time, and others 'naturally typifying or representing or recalling . . . by association in fact or thought', religious and cultural symbols interpretable only in specified historical contexts. The term *language* will be used in the sense of human language, 'a vocabulary [of signs] and way of using it'; as a set of signs and rules such as we use in everyday speech and conversation, in a highly flexible and mostly illogical way. On the other hand, we shall refer to the strictly formalized systems of signs and rules, such as those of mathematics and logic, as *language systems* or *sign-systems.*

The term *code* has a strictly technical usage which we shall adopt here. Messages can be coded *after* they are already expressed by means of signs (e.g. letters of the English alphabet); then a code is an agreed transformation, usually one to one and reversible, by which messages may be converted from one set of signs to another. Morse codes, semaphore, and the deaf-and-dumb code represent typical examples. In our terminology then, we distinguish sharply between *language,* which is developed organically over long periods of time, and *codes,* which are invented for some specific purpose and follow explicit rules.

Apart from our natural languages (English, French, Italian, etc.), we have many examples of *systems* of signs and rules, which are mostly of a very inflexible kind. A pack of playing cards represents a set of signs, and the rules of the game ensure communication and patterned behaviour among the players. Every motorist in Britain is given a book of rules of the road called the *Highway Code,* and adherence to these signs and rules is supposed to produce concerted, patterned behaviour on British roads. There are endless examples of such simple sign systems. A society has a structure, definite sets of relationships between individuals, which is not formless and haphazard but organized. Hierarchies may exist and be recognized, in a family, a business, an institution, a factory, or an army – functional relationships which decide to a great extent the patterned flow of communication. The communication and the structure are subject to sets of rules, rules of conduct, authoritarian dictates, systems of law; and the structures may be highly complex and varied in form. A 'code' of ethics is more like a language, having developed organically; it is a set of guiding rules concerning 'ought situations', generally accepted, whereby people in a society associate together and have social coherence. Such codes are different in the various societies of the world, though there is an overlap of varying degrees.

2

A generalized graphic model of communication

George Gerbner

From Gerbner, George 1956: Towards a general model of communication. *Audio-Visual Communication Review* IV.3, 1956.

Editors' note on Gerbner's model of communication (following page)

Gerbner's model is an attempt to depict diagrammatically some of the issues of perception and representation which must be taken into account in any study of communicational activity as a dynamic, social process.

The flow is from right to left in the diagram.

An event (E) is perceived by someone (M). The event-as-perceived (E^1) is the product of perceptual *activity* and thus the mediations and transformations of particular selective and contextual factors introduce the difference between E and E^1.

The vertical arm of the model shows the *representation* of the event ('statement about event') by the perceiver to be partly a product of the available meaning systems (e.g. print, speech, photograph, film) and of the particular conventions of use of such systems (here it is important to stress the social and historical contingency of these conventions). These formal elements (S) combine with event-related elements (E).

Finally, the lower horizontal arm shows this representation, the statement about the event (SE), being perceived (heard, read, viewed) by a second person (M^2). This perceptual activity will involve, as it did in the earlier case, a transformation such that the difference SE/SE^1 will occur.

Clearly, all academic models must be regarded as depictions of arguments and ideas rather than somehow being straightforward illustrations of the 'way things are'. Indeed, one of the strengths of Gerbner's model is that it draws our attention to just this point. As such, models invite not only our understanding but also our critical assessment of their adequacy.

The processes which Gerbner here treats graphically are nearly all the subject of theoretical debates, some of which you will find discussed elsewhere in this Reader.

The relation of language to reality and to thought, the nature of different forms of representation, the problems which follow from considering 'form' as separate from 'content', the extent to which the perception and representation of events in, say, the natural sciences involve factors fundamentally different from those in, say, sports journalism – questions such as these should be borne in mind when discussing the model.

Finally, it is worth noting that the model carries implications for the study of mass communication, since it suggests the difficulty if not the impossibility of achieving 'neutrality' and 'objectivity' in the relaying of events through the media. By emphasizing the way in which 'messages' are both *selections* from real world events and *constructions* made from the material of whatever medium is used (language, photography, video etc.) Gerbner provides a good base from which to consider the problems of reporting. In fact, the version of the model reprinted here has been used in a brief and helpful article on the professional goals and ethics of journalists (Halloran, 1969).

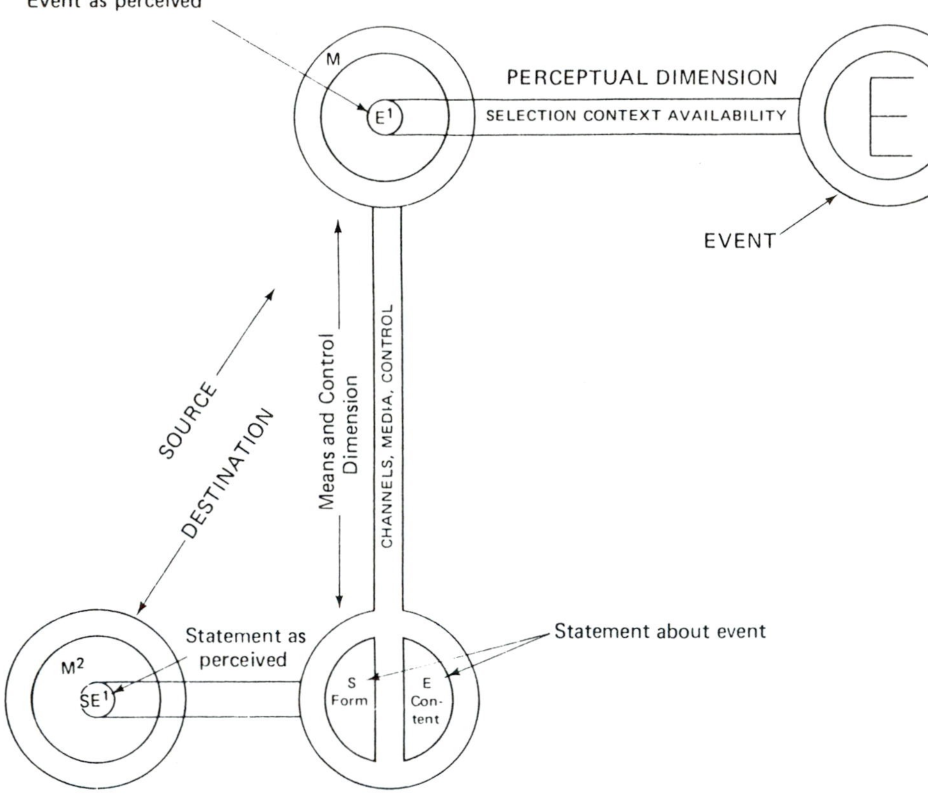

Figure 2.1 Basic generalized graphic model of communication

3

Defining language

Jean Aitchison

From Aitchison, J. 1976: *The articulate mammal.* London: Hutchinson, 36–43.

A useful first step might be to attempt to define 'language'. This is not as easy as it sounds. Most of the definitions found in elementary textbooks are too wide. For example: 'The faculty of language consists in man's ability to make noises with the vocal organs and marks on paper or some other material, by means of which groups of people "speaking the same language" are able to interact and cooperate as a group' (Robins, 1971, 12). This definition, if one ignores the word 'man' and the phrase involving 'marks on paper', might equally well apply to a pack of wolves howling in chorus.

Perhaps the most promising approach is that suggested by the linguist Charles Hockett. In a series of articles stretching over ten years he has attempted to itemize out the various 'design features' which characterize language. For example: '*Interchangeability*: Adult members of any speech community are interchangeably transmitters and receivers of linguistic signals'; '*Complete Feedback*: The transmitter of a linguistic signal himself receives the message' (Hockett, 1963, 9). Of course, such an approach is not perfect. A list of features may even be misleading, since it represents a random set of observations which do not cohere in any obvious way. To use this list to define language is like trying to define a man by noting that he has two arms, two legs, a head, a belly button, he bleeds if you scratch him, and shrieks if you tread on his toe. But in spite of this, a definition of language based on design features or 'essential characteristics' seems to be the most useful proposed so far.

But how many characteristics should be considered? Two? Ten? A hundred? The number of design features Hockett considers important has changed over the years. The longest list contains sixteen (Hockett, 1963), though perhaps most people would consider that eight features capture the essential nature of language: *use of the vocal-auditory channel, arbitrariness, semanticity, cultural transmission, duality, displacement, structure-dependence* and *creativity*.

Let us discuss each of these features in turn, and see whether it is present in animal communication. If any animal naturally possesses all the design features of human language, then clearly that animal can talk.

The use of the *vocal-auditory channel* is perhaps the most obvious characteristic of language. Sounds are made with the vocal organs, and a hearing mechanism receives them – a phenomenon which is neither rare nor particularly surprising.

The use of sound is widespread as a means of animal communication. One obvious advantage is that messages can be sent or received in the dark or in a dense forest. Not all sound signals are vocal – woodpeckers tap on wood, and rattlesnakes have a rattle apparatus on their tail. But vocal-auditory signals are common and are used by birds, cows, apes and foxes, to name just a few. The advantages of this method of producing the sound are that it leaves the body free to carry on other activities at the same time, and also requires relatively little physical energy. But this design feature is clearly neither unique to humans, nor all-important, since language can be transferred without loss to visual symbols (as in deaf-and-dumb language, or writing) and to tactile symbols (as in Braille). Patients who have had their vocal cords removed, and communicate mainly by writing, have not lost their language ability. It follows that this characteristic is of little use in an attempt to distinguish animal from human communication. So let us proceed to the second feature, arbitrariness.

Arbitrariness means that human languages use neutral symbols. There is no connection between the word 'dog' and the four-legged animal it symbolizes. It can equally be called UN CHIEN (French), EIN HUND (German), or CANIS (Latin). GÜL (Turkish) and RHODON (Greek) are equally satisfactory names for a 'rose'. As Juliet notes:

> What's in a name? that which we call a rose
> By any other name would smell as sweet.
>
> Shakespeare

Onomatopoeic words such as CUCKOO, POP, BANG, SLURP and SQUISH are exceptions to this. But there are relatively few of these in any language. On the other hand, it is normal for animals to have a strong link between the message they are sending and the signal they use to convey it. A crab which wishes to convey extreme aggression will extend a large claw. A less angry crab will merely raise a leg: 'Extending a major chaliped is more effective than raising a single ambulatory leg in causing the second crab to retreat or duck back into its shell' (Marshall, 1970). However, arbitrary symbols are not unique to man. Gulls, for example, sometimes indicate aggression by turning away from their opponent and uprooting beakfuls of grass. So we are forced to conclude that arbitrariness cannot be regarded as a critical distinction between human and animal communication.

Semanticity, the third suggested test for language ability, is the use of symbols to 'mean' or refer to objects and actions. To a human, a CHAIR 'means' a four-legged contraption you can sit on. Humans can generalize by applying this name to all types of chair, not just one in particular. Furthermore, semanticity applies to action as well as objects. For example, to JUMP 'means' the act of leaping in the air. Some writers have claimed that semanticity is exclusively human. Animals may only be able to communicate about a total situation. A hen who utters 'danger' cries when a fox is nearby is possibly conveying the message 'Beware! beware! there is terrible danger about!' rather than using the sound to 'mean' FOX. But, as is shown by the call of the vervet monkey who might or might not mean 'snake' when he *chutters* it is difficult to be certain. We must remain agnostic about whether this feature is present in animal communication.

Cultural transmission or *tradition* indicates that human beings hand their languages down from one generation to another. The role played by teaching in animal communication is unclear and varies from animal to animal – and even within species. Among birds it is claimed that the song-thrush's song is largely innate, but can be slightly modified by learning, whereas the skylark's song is almost wholly learned. Birds such as the chaffinch are particularly interesting: the basic pattern of the song seems to be innate, but all the finer detail and much of the pitch and rhythm have to be acquired by learning (Thorpe, 1961, 1963). However, although the distinction between man and animals is not clear-cut as regards this feature, it seems that a far greater proportion of communication is genetically inbuilt in animals than in man. If a child is brought up in isolation, away from human beings, he does not acquire language. In contrast, birds reared in isolation sing songs that are sometimes recognizable (though almost always abnormal).

The fifth property, *duality* or *double-articulation*, means that language is organized into two 'layers': the basic sound units of speech, such as P, I, G, are normally meaningless by themselves. They only become meaningful when combined into sequences such as P – I – G PIG. This property is sometimes claimed to be unique to humans. But this is not so. Duality is also present in bird song, where each individual note is itself meaningless – it is the combination of notes which convey meaningful messages. So once again we have not found a critical difference between animals and humans in the use of this feature.

A more important characteristic of language is *displacement*, the ability to refer to things far removed in time and place. Humans frequently say things such as 'My Aunt Matilda, who lives in Australia, cracked her knee-cap last week.' It may be impossible for an animal to convey a similar item of information. However, as in the case of other design features, it is sometimes difficult to decide whether displacement is present in an animal's communication system. A bird frequently continues to give alarm cries long after the disappearance of a cat which was stalking it. Is this displacement or not? The answer is unclear. Definite examples of displacement are hard to find. But it is undoubtedly found in bee communication (von Frisch, 1950, 1954, 1967). When a worker bee finds a source of nectar she returns to the hive to perform a complex dance which informs the other bees of its location. She does a 'round dance', which involves turning round in circles if the nectar is close to the hive, and a 'waggle dance' in which she wiggles her tail from side to side if it is far away. The other bees work out the distance by noting the tempo of her waggles, and discover what kind of flower to look for by smelling its scent on her body. After the dance, they unerringly fly to the right place, even if it is several miles away, with a hill intervening.

This is an unusual ability – but even this degree of displacement is considerably less than that found in human speech. The bee cannot inform other bees about anything further removed [in time] than the nectar patch she has just visited. She cannot say 'The day before yesterday we visited a lovely clump of flowers, let's go and see if they are still there' – she can only say, 'Come to the nectar I have just visited.' Nor can she communicate about anything further away in place. She could not say 'I wonder whether there's good nectar in Siberia.' So displacement in bee communication is strictly limited to the number of miles a bee can easily fly, and the time it takes to do this. At last, it seems we may have found a feature which seems to be of importance in human language, and only partially present in non-human communication.

The seventh feature is *structure-dependence*. Humans do not just apply simple recognition or counting techniques when they speak to one another. They automatically recognize the patterned nature of language, and manipulate 'structured chunks'. For example, they understand that a group of words can sometimes be the structural equivalent of one:

SHE THE OLD LADY WHO WAS WEARING A WHITE BONNET	GAVE THE DONKEY A CARROT

and they can rearrange these chunks according to strict rules:

A CARROT	WAS GIVEN TO THE DONKEY	BY THE OLD LADY WHO WAS WEARING A WHITE BONNET

As far as we know, animals do not use structure-dependent operations. We do not know enough about the communication of all animals to be sure, but no definite example has yet been found.

Finally, there is one feature that seems to be of overwhelming importance, and unique to humans – the ability to produce and understand an indefinite number of novel utterances. This property of language has several different names. Chomsky calls it *creativity*, others call it *openness* or *productivity*. A human can talk about anything he likes – even a platypus falling backwards downstairs – without causing any linguistic problems to himself or the hearer. He can say *what* he wants *when* he wants. If it thunders, he does not automatically utter a set phrase, such as 'It's thundering, run for cover'. He can say 'isn't the lightning pretty?' or 'Better get the dog in' or 'Thunder is two dragons colliding in tin tubs, according to a Chinese legend.'

In contrast, most animals have a fixed number of signals which convey a set number of messages, sent in clearly definable circumstances. A North American cicada can give four signals only. It emits a 'disturbance squawk'; when it is seized, picked up or eaten. A 'congregation call' seems to mean 'let's all get together and sing in chorus!' A preliminary courtship (an invitation?) is uttered when a female is several inches away. An advanced courtship call (a buzz of triumph?) occurs when the female is almost within grasp (Alexander and Moore, quoted in McNeill, 1966). Even the impressive vervet monkey has only thirty-six distinct vocal sounds in its repertoire. And as this total includes sneezing and vomiting, the actual number used for communication is several fewer. Within this range, choice is limited, since circumstances generally dictate which call to use. An infant separated from its mother gives the lost *rrah* cry. A female who wishes to deter an amorous male gives the 'anti-copulatory squeal-scream' (Struhsaker, 1967).

But perhaps it is unfair to concentrate on cicadas and monkeys. Compared with these, bees and dolphins have extremely sophisticated communication systems. Yet researchers have reluctantly concluded that even bees and dolphins seem unable to say anything new. The bees were investigated by the famous 'bee-man', Karl von Frisch (1954). He noted that worker bees normally give information about the *horizontal* distance and direction of a source of nectar. If bee communication is in any sense 'open', then a worker bee should be able to inform the other bees about *vertical* distance and direction if necessary. He tested this idea by placing a hive of bees at the foot of a radio beacon, and a supply of sugar water at the top. But the bees who were shown the sugar water were unable to tell the other bees where to find it. They duly performed a 'round dance', indicating that a source of nectar was in the vicinity of the hive – and then for several hours their comrades flew in all directions *except* upwards, looking for the honey source. Eventually, they gave up the search. As von Frisch noted, 'The bees have no word for "up" in their language. There are no flowers in the clouds' (von Frisch, 1954, 139) Failure to communicate this extra item of information means that bee communication cannot be regarded as 'open-ended' in the same way that human language is open-ended.

The dolphin experiments carried out by Dr Jarvis Bastian were considerably more exciting – though in the long run equally disappointing. Bastian tried to teach a male dolphin Buzz and a female dolphin Doris to communicate across an opaque barrier.

First of all, while they were still together, Bastian taught the dolphins to press paddles when they saw a light. If the light was kept steady, they had to press the right paddle first. If it flashed, the left-hand one. when they did this correctly they were rewarded with fish.

As soon as they had learned this manœuvre, he separated them. They could now hear one another, but they could not see one another. The paddles and light were set up in the same way, except that the light which indicated which paddle to press first was seen only by Doris. But in order to get fish both dolphins had to press the levers in the correct order. Doris had to *tell* Buzz which this was, as only she could see the light. Amazingly, the dolphins 'demonstrated essentially perfect success over thousands of trials at this task' (Evans and Bastian, 1969, 432). It seemed that dolphins could *talk* ! Doris was conveying novel information

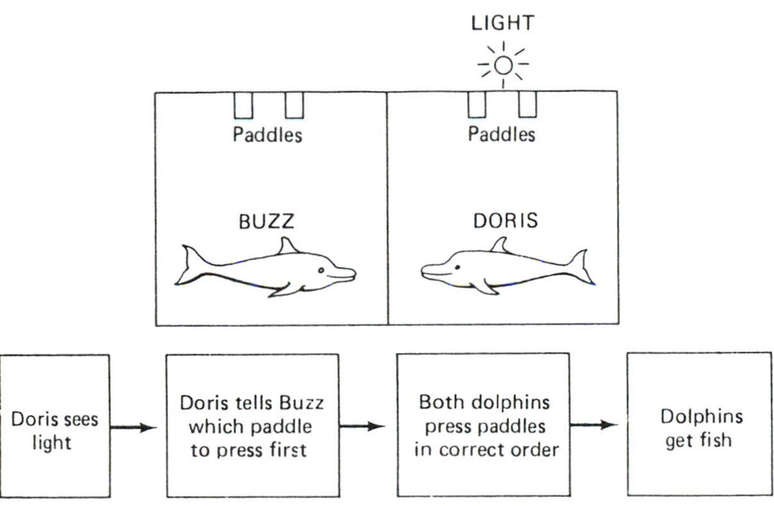

through an opaque barrier! But it later became clear that the achievement was considerably less clever. Even while the dolphins were together Doris had become accustomed to making certain sounds when the light was flashing and different sounds when it was continuous. When the dolphins were separated she continued the habit. And Buzz had, of course, already learnt which sound of Doris's to associate with which light. Doris was therefore not 'talking creatively'.

So not even dolphins have a 'creative' communication system, it seems – though it is always possible that more is known about 'dolphinese' than has been made public. The high intelligence of dolphins has obvious implications for naval warfare, and so has attracted the attention of military authorities, with the result that much research is shrouded in official secrecy. But on the whole it seems unlikely that there exist hidden tanks of 'talking dolphins' (as was suggested in a recent film). Most researchers would agree with the comment of the psychologist John Morton: 'On the question as to whether dolphins have a language, I would like to comment parenthetically from the evidence I have seen, if they do have a language they are going to extraordinary lengths to conceal the fact from us' (Morton, 1971, 83).

It seems, then, that animals cannot send truly novel messages, and that Ogden Nash encapsulates a modicum of truth in his comment:

> The song of canaries
> Never varies.

And so does Alice in her complaint about kittens:

> It is a very inconvenient habit of kittens that, whatever you say to them, they always purr. If they would only purr for 'yes' and mew for 'no', or any rule of that sort, so that one could keep up a conversation! But how *can* you talk with a person if they *always* say the same thing?
>
> Lewis Carroll

It is now possible to answer the question, can animals talk? If, in order to qualify as 'talkers' they have to utilize all the design characteristics of human language 'naturally', the answer is clearly 'no'. Some animals possess some of the features. Bird song has duality, and bee dancing has some degree of displacement. But, as far as we know, no animal communication system has duality *and* displacement. No animal system can be proved to have semanticity or to use structure-dependent operations. Above all, no animal can communicate creatively with another animal.

But although animals do not 'naturally' talk, this does not mean that they are *incapable* of talking. Perhaps they have just never had the chance to learn language.

4

Deceit and misrepresentation

W.T. Scott

From Scott, W.T. 1990: *The possibility of communication*. Berlin: Mouton De Gruyter, 126–31.

In an article of 1972 Donald MacKay addresses the fact that encoders and decoders may adopt very different views of each other's relationship to the message. Thus, I may decide you are lying, whilst you continue to assume you are being believed, because I form the impression that you are trying too hard to convince me and become suspicious, a suspicion I successfully conceal, meantime.

Adapting MacKay's diagram (MacKay, 1972, 24) in which the crucial distinction between informative/significant signalling and communicative/meaningful signalling is incorporated, we can summarize at least some aspects of these relationships as in Figure 4.1.

The horizontal lines (1–1; 2–2) in Figure 4.1 represent, firstly, successful informative behaviour, as when I blink quite naturally and am taken to be blinking: and, secondly, successful communicative behaviour as when I wink (openly or 'broadly') and am taken to be winking. The angled lines represent, firstly, unsuccessful informative behaviour as when I blink and am taken to be winking (1–2): and unsuccessful communicative behaviour as when I wink and am taken to be blinking (2–1).

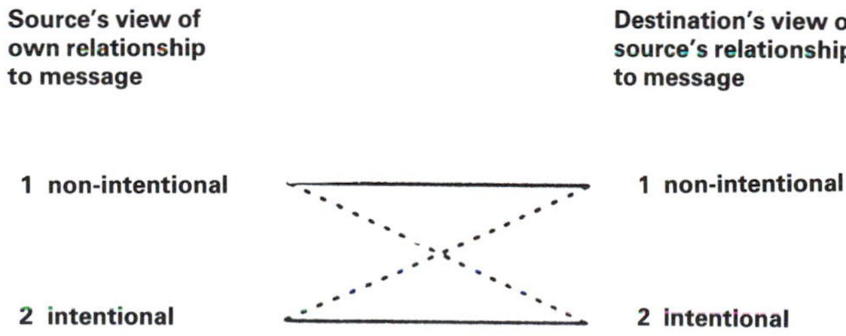

Figure 4.1 Source-message relationships assumed or imputed by participants

These cases are relatively simple and familiar. When we attempt to incorporate the existence, or not, detected or not, of some *covert*, secondary intention the four possibilities multiply to at least sixteen as follows:

Source's view of own relationship to message		Destination's view of source's relationship to message	
General Intentionality	Covert particular Intention(s)	General Intentionality	Covert particular Intention(s)
1 No	No	1 No	No
2 No	Yes	2 No	Yes
3 Yes	Yes	3 Yes	Yes
4 Yes	No	4 Yes	No

1–1 Here we have, as already noted, an example of straightforward informative behaviour, such as we find in natural indexical signs. These are not under the control of the source; are not designed to mislead (as, for example, permanent camouflage markings in plumage might be said to be); and are so deemed at destination. Genuine medical symptoms helplessly manifested and interpreted as such belong here. So do slips of the tongue or pen, at least as isolated lapses within a performance manifesting general intentionality. Recurring lapses or other items of behaviour, especially if these form some systematic pattern and/or correlation with other phenomena, might cause source and/or destination to re-classify the items into one or other of the categories to be discussed below.

2–2 Some informative signs/messages while not generally under the control of the source are, in a sense, designed to mislead. The case of camouflage markings, or indeed any of the forms of deception so abundant in the natural world (baits and lures, for example) illustrates the sense of 'designed to mislead' in mind here, that is, designed by the evolutionary process. Given the 2–2 relationship, however, we are presently considering unsuccessful examples. Camouflage that fails to work, perhaps because of some change in the environment for which the adaptations are specific, is unsuccessful, from the source's viewpoint at least. If predation is thereby made easier, then the opposite is true at the decoding end.

1–2 In this case, we deal with straightforward items of, say, physical motion, such as an unwitting gesture, interpreted as unintentional yet indexical of something non-evident but not concealed or repressed. Doctors make mistakes and, at least in popular mythology, psychiatrists make more than most.

2–1 This case represents the converse of 1–2. To return to medical examples, what are in fact tell-tale symptoms may be noticed but not taken into account. Successful camouflage would also exemplify this relationship, i.e., when the decoder is fooled. In the natural world, the example of the species of ground-nesting bird which lures predators away from its nest by feigning a damaged wing belongs here, on the assumption that the individual bird is behaving in automatic response to danger, rather than deliberately. Given a worldly-wise predator, of course, who has followed the easy pickings once too often, only to see them fly off when a safe distance has been achieved, then the signalling would belong in the 2–2 category.

1–3 In this case we deal with circumstances where behaviour not under the source's control, and not in any way devious or misleading, is interpreted as both deliberate and communicative of some concealed or oblique intention. A source, for example may have a naturally rapid blink rate but be taken to be 'fluttering eyelashes' by a mistaken observer.

Alternatively, we can imagine visiting what we take to be someone's own home, which is so furnished and decorated in a particular period style that we decide it is a humorous allusion or at least an 'inverted commas' reconstruction of the sort where affectionate nostalgia mixes with mockery. Were we to discover that the person is merely flat-sitting for someone who has lived with these items since they were 'authentic', and for whom they retain at least some innocence (if only by familiarity) then our interpretation would surely alter.

1–4 A good illustration of this situation is the poem by Stevie Smith entitled 'Not Waving but Drowning', the first stanza of which is as follows:

> Nobody heard him, the dead man,
> But still he lay moaning:
> I was much further out than you thought
> And not waving but drowning.

4–1 This converse of the previous category captures the fact that we often misclassify a plain and straightforward sign/message as if it were neither under the control of the source nor possessed of any oblique or secondary significance, e.g. not drowning, but waving interpreted as drowning. A more everyday example would be an attempt to speak to someone in a whisper, which is noticed but misclassified as exhalation.

4–2 Here we have a situation where plain and straightforward behaviour intentionally carried on by the source is interpreted as involuntary, e.g. compulsive, and indicative of something else entirely. Trombonists, for example, are no doubt inured to jokes about displaced sexual expression, and so on. A more sinister example might be the practice of treating political dissidents or other deviants as being neither in control of their general behaviour, nor of particular actions, e.g. expression of certain attitudes, opinions and beliefs which may be seen as symptomatic rather than authentic.

2–4 Conversely, of course, there are many situations where mentally-ill persons, some at least of whose general behaviour is out of their control and whose specific actions are symptomatic (if considered in the 'correct' way), are treated normally. Many talented and/or eccentric individuals hover or oscillate between category 2 and category 4, not only to observers but to themselves.

4–3 It is not uncommon for a perfectly innocent statement to have 'other' levels of meaning read into it by decoders. For example, they may see sarcasm where none was intended, or pick up a pun that was there by accident, as is often the way with words. A particularly irritating or disturbing ploy in relationships is remorselessly to read in meanings that all concerned know were never intended, for example by taking literally all the turns of phrases, idioms and figures that abound in ordinary talk. Another example, sometimes used by groups (typically male) scapegoating unsuspecting victims is to laugh over-appreciatively at mildly humorous remarks, thus encouraging the individual to bolder efforts. When the deceit is eventually revealed, often by the use of heavy sarcasm (see 3–3), the effect of this ritual degradation can be painful, to say the least.

3–4 Conversely, successful lies (from the source's viewpoint) and unsuccessful sarcasm (from both viewpoints) are not incorporated. The latter can be particularly annoying for the source, who may fail by too light a touch, to arouse the decoder's suspicion that what has been said might be the opposite of what is meant. Sometimes a source, duty- or convention-bound to be polite, can obtain private satisfaction by indulging in very light sarcasm or irony, the butt of which remains oblivious. If others present, for example a group of subordinates faced with an unpopular superior, catch on, much therapeutic fun may be had, especially by risking heavier forms of sarcasm. Once again, the possibility of oscillation or fuzziness between categories can be exploited by competent performers.

3–3 Detected, hence failed, lies belong in this category, while very heavy sarcasm involves overlap with category 4 in complex ways. In essence, heavy sarcasm achieves its humiliating effect because the source flauntingly disclaims the possibility of protesting innocence (i.e. category 4 status) for the utterance. Thus, at one level, there is a clearly-signalled absence of any intention to mislead the decoder of the fact that, at another, the utterance *is* misleading. A double relation to the utterance is therefore revealed. Such ostentatious removal of an escape route displays considerable contempt and/or power: correspondingly, it is experienced as humiliating and/or provocative by the recipient (who can always, of course, 'act daft' by taking the literal meanings, thus creating a problem for the source).

3–2 Here we encounter a complex relationship that raises very basic questions of degrees and scope of responsibility for actions at different times. One example might be that of an apprehended criminal who knew what he or she was doing in committing the crime and is subsequently attempting to conceal it. A successful plea that (sincerely) admitted the concealment yet (insincerely) denied full or any responsibility for the act at the time, would in itself be categorized in 4, and have the effect of placing the original act and its aftermath in 2.

2–3 An example here might be of the criminal who, unable to control his or her actions in general terms (or at least, vis à vis the circumstances of the crime, such as shoplifting) yet able to attempt a cover up, is wrongly held to be responsible for everything. Another example might be the perjurer who lies under duress, and has the lies detected, but not the duress.

3–1 Here we consider attempts at subtlety and obliquity which are taken at face value or ignored: nudges; winks; warning coughs; 'casual' nods and tilts of the head; meaningful glances; stares (at or away from a speaker); sighs; specially-toned interjections such as *mmm* or *uh-huh*; frowns; raised eyebrows etc; teeth-sucking; lip-pointing; and many others. These, individually and in combination, are usually controlled and oblique, in that they hint at some state of mind in the source which the alert recipient needs to work out, or ask about directly. Not all destinations are recipients, not all are equally alert, and no recipient is consistently and/or maximally alert, hence this category of failure.

4–4 Lastly, we arrive at what may by now seem an impossible situation, the core of the standard contract where people mean simply what they expect they will be taken to mean: plain talk and plain listening, each of which enables the other. This situation seems highly problematic for numerous reasons. At the root of all semiotic activity, as argued already, lies the fact that something stands for something else in some way and in certain respects only. In Peircian terms, what we mean is never a 'first', an unmediated monadic experience clearly apprehended in and of itself. It is always fabricated and allocated, mediated and mediating, even in 'internalese'. For this reason, in turn, what is meant is always someone's responsibility. The cynic who insists, say, that it is the culture, as agent for the forces of ideology, which is responsible, underplays the fact that speakers are usually aware what a word or construction has already been made to mean (synchronically at least) in the language, conventionally and on particular occasions of use. To the extent that they are not aware, they are not responsible for that meaning, therefore do not mean authentically that. Moreover, they cannot be said either willingly to endorse a particular meaning or unwittingly to reify whatever values, beliefs, attitudes etc. are said to be embedded in a given sign or construction if they have no awareness of what possibilities exist, hence of what commitments they display.

5

Verbal and non-verbal communication

Michael Argyle

From Argyle, M. 1972: *The psychology of interpersonal behaviour*. Harmondsworth: Penguin Books (Pelican Original), 37–47.

Different kinds of social act

1 *Bodily contact* is of interest since it is the most primitive kind of social act, and is found in all animals. In addition to aggressive and sexual contacts there are various methods of influence, as when others are pushed, pulled or led. There are symbolic contacts, such as patting on the back, and the various ways of shaking hands. Outside the family, bodily contact is mainly restricted to the hands. Jourard (1966) has surveyed who has been touched by whom and where, and his results for American students are shown in Figure 5.1. It can be seen that there are great differences in who is touched by whom, and on which parts of their anatomy.

There are great cross-cultural differences in bodily contact, and this form of social behaviour is less common in Britain than almost anywhere else. It usually conveys intimacy, and occurs at the beginning and end of encounters. There has been some interest in 'encounter groups' in the USA and Britain during recent years. The greater use of bodily contact here is found to be exciting and disturbing – but it must be remembered that those concerned have been brought up in cultures in which there are strong restraints against bodily contact and will have internalized these restraints.

2 *Physical proximity* is important mainly in relation to intimacy and dominance. It is one of the cues for intimacy, both sexual and between friends of the same sex. The normal degree of proximity varies between cultures and every species of animal has its characteristic social distance. The significance of physical proximity varies with the physical surroundings – proximity to the point of bodily contact in a lift has no affiliative significance, and it is noteworthy that eye-contact and conversation are avoided here. If A sits near B, it makes a difference whether there are other places where A could have sat, whether he is directly facing B or at an angle, and whether there is any physical barrier. Closest distances are adopted for more intimate conversations: at the closer distances different sensory modes are used – touch and smell come into operation, and vision becomes less important (Hall, 1963). It is found that people sit or stand closer to people that they like. There are also large cross-cultural differences – Arabs

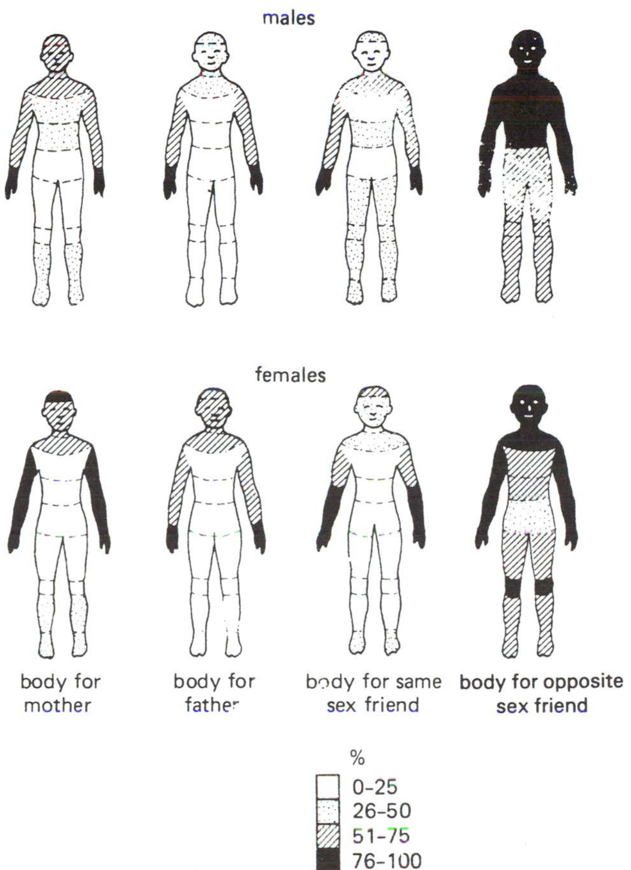

Figure 5.1 Male and female 'bodies-for-others', as experienced through the amount of touching received from others (Jourard, 1966)

and Latin Americans stand very close, Swedes and Scots are the most distant (Lott *et al.*, 1969). *Changes* of proximity are of course used to signal the wish to begin or end an encounter, accompanied by other appropriate messages.

3 *Orientation* signals interpersonal attitudes. If person A is sitting at a table, as shown in Figure 5.2, B can sit in several different places. If he is told that the situation is cooperative he will probably sit at B1; if he is told he is to compete, negotiate, sell something or interview A, he will sit at B2; if he is told to have a discussion or conversation he usually chooses B3 (Sommer, 1965). This shows (a) that one can become more sensitive to the cues emitted, often unintentionally, by others, and (b) that one can control non-verbal as well as verbal signals.

If one person is higher up than another – by being on a rostrum, standing, or perhaps simply by being taller, it puts him in a somewhat dominating position – probably because parents are taller than children. On the other hand, there is the curious cultural convention that more important people should sit while others have to stand. Hutte and Cohen in Holland made silent films of managers entering

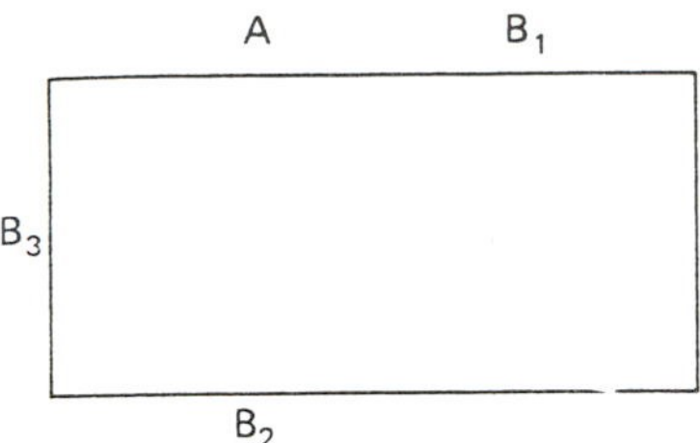

Figure 5.2 Orientation in different relationships

offices of other managers. It was quite clear to subjects who were shown these films which manager was the senior (in a and c), and how friendly they were (in b and d) (Burns, 1964; see also Figure 5.3).

4 *Bodily posture* is another signal which is largely involuntary, but which can communicate important social signals. There are distinctive 'superior' (or dominant) and 'inferior' (or submissive) postures. The desire or intention to dominate can be signalled by standing erect, with the head tilted back, and with hands on hips, for example. There are also friendly, and hostile postures. (See Figure 5.4).

By his general bodily posture a person may signal his emotional state, e.g. tense versus relaxed. He can show his attitude to the others present – as when a person sits in a different way from the others, or puts his feet on the table.

People also have general styles of expressive behaviour, as shown in the way they walk, stand, sit and so on. This may reflect past or present roles – as in

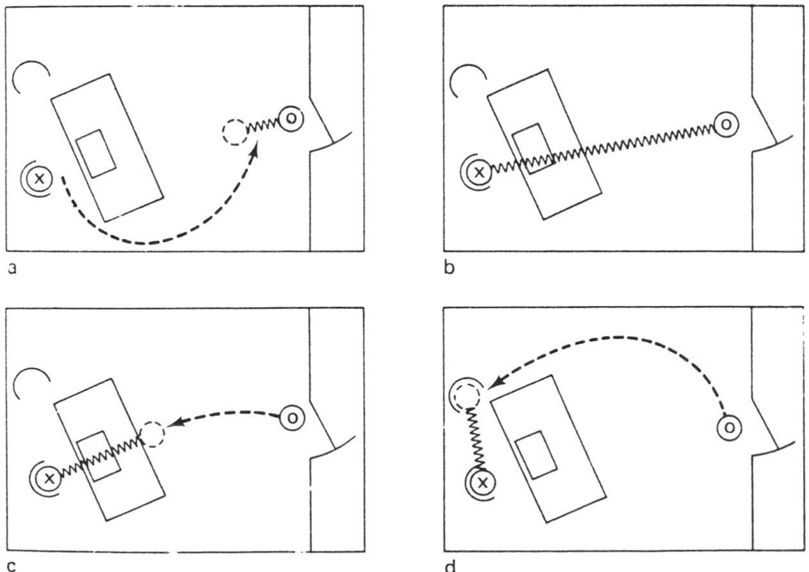

Figure 5.3 Key movements between men in an office indicating their relationship (Burns, 1964)

Figure 5.4 The meaning of bodily postures (Sarbin and Hardyck, 1953)

the case of a person who is or has been a soldier; it also reflects a person's self-image, self-confidence, and emotional state. It is very dependent on cultural fashions:

> In a street market I watched a working-class mum and her daughter. The mother waddled as if her feet were playing her up. Outside a Knightsbridge hotel I watched an upper-class mum and her daughter come out from a wedding reception and walk towards Hyde Park Corner, the mother on very thin legs slightly bowed as though she had wet herself. She controlled her body as if it might snap if moved too impulsively. Both daughters walked identically (Melly, 1965).

5 *Gestures* are movements of hands, feet or other parts of the body. Some are intended to communicate definite messages; others are involuntary social cues which may or may not be correctly interpreted by others.

Communicating emotional states. When a person is emotionally aroused he produces diffuse, apparently pointless, bodily movements. A nervous lecturer may work as hard as a manual labourer. More specific emotions produce particular gestures – fist-clenching (aggression), face-touching (anxiety), scratching (self-blame), forehead-wiping (tiredness) etc. (Ekman and Friesen, 1969).

Completing the meaning of utterances. It has been found that while a person speaks he moves his hands, body and head continuously, that these movements are closely coordinated with speech, and that they form part of the total communication. He may (1) display the structure of the utterance by enumerating elements or showing how they are grouped, (2) point to people or objects, (3) provide emphasis, and (4) give illustrations of shapes, sizes or movements (Scheflen, 1965; Kendon, 1972).

Replacing speech. When speech is impossible for one reason or another, gesture languages develop.

6 *Head-nods* are a rather special kind of gesture, and have two distinctive roles. They act as 'reinforcers', i.e. they reward and encourage what has gone before, and can be used to make another talk more, for example. Head-nods also play an important role in controlling the synchronizing of speech – in Britain a nod gives the other permission to carry on talking, whereas a rapid succession of nods indicates that the nodder wants to speak himself.

7 *Facial expression* can be reduced to changes in eyes, brows, mouth, and so on. The face is an area which is used by animals to communicate emotions and attitudes to others; for humans it does not work so well since we control our facial expression, and may smile sweetly while seething within. Emotions can be recognized to some extent from facial expression alone, as is shown by studies using still photographs of actors. Emotions can be recognized in terms of broad categories – for example, the pleasant and unpleasant ones – but those which are similar are harder to tell apart. The circle below (Figure 5.5) shows which are seen as most similar to one another – those furthest apart are the easiest to distinguish. In addition to these states, it is possible to recognize degrees of emotional tension – by perspiration on the forehead and expansion of the pupils of the eyes.

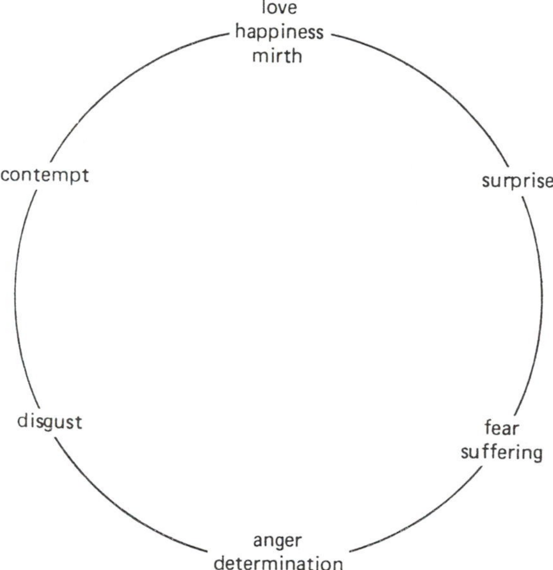

Figure 5.5 The dimensions of facial expression (Schlosberg, 1952)

Facial expression works rather better as a way of providing feedback on what another is saying. The eyebrows provide a continuous running commentary, going from:

fully raised	– disbelief
half raised	– surprise
normal	– no comment
half lowered	– puzzled
fully lowered	– angry.

The area round the mouth adds to the running commentary by varying between being turned up (pleasure) and turned down (displeasure).

8 *Eye movements* have an effect quite out of proportion to the physical effort exerted. When A looks at B, in the region of the eyes, B knows that A is attending primarily to him, and that interaction can proceed. If A gazes for a long time at B, this can have a variety of meanings, depending on A's facial expression and on the situation – it can be an amorous, friendly, aggressive or curious gaze – in each case revealing something of A's feelings towards B. Glances can be long or short, furtive or open, and can combine together to form complex strategies of eye-play. Eye movements play an important part in sustaining the flow of interaction: while A is speaking he looks up to get feedback on how B is responding, and he ends a long utterance with a gaze which tells B that it is his turn to speak. When there is eye-contact between two people this is experienced as a heightening of interpersonal emotions, usually in the sense of greater intimacy.

9 *Appearance.* Many aspects of personal appearance are under voluntary control, and a great deal of effort is put into controlling them – clothes, hair and skin; other aspects can be modified to some extent by clothes and plastic surgery. The main purpose of this manipulation of appearance is self-presentation – signalling how the presenter sees himself and would like to be treated.

10 *Non-linguistic aspects of speech.* The same words may be said in quite different ways, conveying different emotional expressions, and even different meanings, as when 'yes' is used as a polite way of saying 'no'. Davitz (1964) found that when actors read out an emotionally neutral passage to express different emotional states, these were recognizable by judges about 60 to 70 per cent of the time. The emotions in question were: admiration, affection, amusement, anger, boredom, cheerfulness, despair, disgust, dislike, fear, impatience, joy, satisfaction and surprise. The author has used this method as a means of sensitivity training. Some people are much better at making such judgements than others. Several aspects of voice quality are involved – loudness, pitch, speed, voice quality (such as breathiness, or breaking into incipient tears), and smoothness. These aspects of speech are correlated, though not perfectly, with emotional states. For example, an anxious person tends to talk faster than normal and at a higher pitch. A depressed person talks slowly, and at a lower pitch; an aggressive person talks loudly.

The pattern of pauses, stress and pitch is really part of the verbal utterance itself. Pauses provide punctuation (instead of saying 'full stop' as when dictating); stress and pitch show whether a question is being asked and provide emphasis, thus showing which of several possible meanings is intended (Crystal, 1969).

There are non-linguistic aspects of the conversation as a whole – the patterns of speech and silence – how much of the time each person talks, how fast, how soon after the other stops, and so on. Chapple (1956) has shown that people have characteristic ways of reacting to interruption and silence on the part of another. In his 'standard interview' the subject is first interviewed in a relaxed manner; later follows a period in which the interviewer fails to respond to twelve successive utterances by the subject, and another period during which the interviewer interrupts twelve successive speeches by the subject. Some people yield at once if interrupted, while others try to talk the interrupter down. Some people cannot tolerate silence, and will speak again if the interviewer is silent.

Another non-verbal aspect of speech is the rate of speech errors. These are of two main kinds – 'ah's and 'er's, and 'non-ah' errors like changes of sentence, repetitions, stutters, etc. 'Non-ah' errors are caused by anxiety; 'ah's and 'er's are not, and seem to be used to create time to think and decide what to say next (Cook, 1969).

11 *Speech* is the most complex, subtle and characteristically human means of communication. Most animal noises simply communicate emotional states. Human speech is different in that it is learned, can convey information about external events, and has a grammatical structure. But it still consists of a set of learned social techniques which are used to influence others. There are great differences in the skill of individuals at using language, mainly associated with intelligence, education and training, and social class. A large part of most social skills lies in putting together utterances which are tactful, persuasive, or whatever is required.

Speech is used to ask questions. These are of interest as they lead to further interaction, and to information about others. Some forms of encounter, such as the interview, consist entirely of questions and the answers to them. Questions vary in the extent to which they are open or closed – an open-ended question requires a lengthy explanation rather than a choice between alternatives; the best way to get someone to talk is to ask this kind of question.

Speech can be used to convey information to others, in answer to questions, as part of the work of committees or work-teams, in lectures, and elsewhere. The speaker may be reporting facts, giving his opinions, or arguing on the basis of these. Such communications are often imperfectly received, because the speaker has not made himself sufficiently clear, or because the hearer attaches different meaning to words or phrases. Ideally, both should speak exactly the same language, i.e. where every sentence carries the identical penumbra of meanings and implications.

Thirdly, speech is used more directly to influence the behaviour of others, by means of instructions or orders, persuasion and propaganda, as well as by aggressive remarks – which may be used when all else fails. Aggressive speech occurs in a variety of forms, of which the milder ones are gentle ridicule and teasing, and the more severe are direct insults.

A lot of speech is not primarily intended to communicate anything very serious, or to solve any problems. Informal speech, as it is called, is more concerned with establishing, sustaining and enjoying social relationships – chat, idle gossip, and joke-telling between friends or family members, or during coffee breaks at work. It has been found that about half of the very expensive conversation over the trans-Atlantic telephone is of this kind. Formal speech, as in a well-delivered lecture, is more concise, conveys information clearly, and is more like written language in structure.

Speech can also be concerned with the process of social interaction itself. This may be acceptable in clinical or training situations, but can be very disturbing in other encounters, when for example someone says 'There's an awkward silence, isn't there.' Words may be used to provide rewards and punishments; in fact this happens continually in the course of interaction, but is largely unintended.

Utterances vary in a number of ways; they can be intimate or impersonal, easy, abstract or technical, interesting or boring to the hearer. In specialized forms of social encounter particular kinds of utterance may be important, like 'interpretation' during psychoanalysis, 'follow-up questions' in an interview, etc. The joke is a special kind of verbal utterance, which has the effect of relieving tension and creating euphoria in social situations. Speeches may have latent meanings, as when the speaker reveals something additional to the main message. This may be unintended, as with the speaker who was asked if he had ever been to Nigeria and replied 'that's a place I've not been to' (Brown, 1965). Or it may be intentional as when a schoolboy says 'Please, Sir, the board's shining' (when it isn't).

6

The analysis of representational images

Bill Nichols

From Nichols, Bill 1981: *Ideology and the image*. Bloomington: Indiana University Press, 57–64.

Note: There are two references to colour in the discussion of the examples, though the illustrations in the original text were, as here, reproduced in monochrome. The extract begins just after a discussion of the ways in which photographs 'place' viewers in certain viewing relationships to what is in the image by various devices of framing and composition.

The meaning of images

The photographic image, however, does more than place the viewer; and these other functions, no less fraught with ideological implications, also need examination. A still image, for example, is a remarkably mute object, testifying perhaps only to a 'having been there' of the image's reference at that single instant in time of its capture. Meaning, though rich, may be profoundly imprecise, ambiguous, even deceiving. A large component of the work undertaken in the construction

Figure 6.1

and reading of images becomes directed toward a distillation of that ambiguity of meaning into a more defined, and limited, concentrate. Possible meanings are scattered to the periphery of a solid charge of determined, or overdetermined, meaning pinned down by those nodal points of intersection between signifiers and signifieds, the shadowy trace of a complex *moiré* pattern. The strategies available to the image maker to anchor or secure meaning are numerous and have been frequently catalogued in introductory film tests and in the work of theoretical writers like Rudolf Arnheim or Bela Balazs. Rather than repeat them here, it should be more profitable to trace their application in a number of concrete instances.

We readily recognize Figure 6.1 as a member of the genus *advertising image*. No one cue provides the warranty for identification, but some of the cues intersecting at this point of meaning are (1) the carefully balanced and focused composition, (2) the camera's unprovocative proximity – the privilege accorded it of approaching a man who has receded behind his function as signifier of warmth and pleasure, (3) the finely rendered texture with its attendant spill of gentle light, (4) the pre-eminence of these two additional signifiers of warmth and pleasure – the man's drink and the top of a liquor bottle whose status as commercial merchandise is naturalized by the broad tropical leaves ringing it like a bouquet of floral petals, and (5) the corporate imprimatur (*'punch au planteur. . .'*) authorizing the image but incorporated within it, arising from it almost as though (as though!) these very words of authorization had been put into the man's mouth in order to be passed on to us, unmediated by a business world, from one friend to another . . . .

We can go further and tease out those points at which potentially floating signifiers are pinned to the specific signifieds of warmth and pleasure, the dominant association invited between image and product and assured by the image as product, the image as a site readied for and prepared by production, the work of codes. One level of work involves the paradigmatic or metaphoric level: the choice of a specific figure or feature from a repertoire of possibilities, similar to the choice of a specific item from a list of appetizers during a meal. Each choice carries with it a certain meaning – implications or associations that accrue to it on the basis of its difference from other possible choices. Again, this snaring of meaning is overdetermined but would include: (1) the clean (not dirty), frayed-edged (relaxed, not formal) straw hat; (2) the rich brown skin tones (sun-warmed, not aetiolated or charred); (3) the white (not stained) regular (not unpleasant) teeth; (4) the slim cigar (neither cigarette – placid and non-aromatic – nor stogy – coarse, acrid); and (5) the flecks of stippled light radiating, improbably, from the man's shaded eyes, rendering them lively (not dull).

Other choices of features could also be identified paradigmatically (in terms of the significance of the actual choice from among possible choices), but it is useful to dwell further on one of the compositional nodal points of the image: the man's mouth. Here the level of work is at a syntagmatic or metonymic level: the actual arrangements of figures of features chosen from a larger repertoire. Our concern is with the spatial (and in film, temporal) relationships between those figures that are present rather than the relationship of figures present to figures absent. The man's mouth is one such figure and its syntagmatic relationships are what might be more traditionally discussed in terms of style.

The mouth occupies a privileged position – centred, one third of the way up from the bottom edge. It is further centred by the shadows de-emphasizing the eyes, nose, and chin, by its location midway between beverage and bottle, by the loosely curled fingers aimed toward it, the cigar extending from it. This carefully orchestrated centre of attention, although consigned to a deeper plane than either the drink or bottle, boasts possession of the one object bridging this spatial chasm – the cigar. The cigar, lit but not smoking, provides, like a straw, a bridge from (inflamed) oral cavity to (cool) refreshing drink. Our friend, gazing at us, is drawn via this hypertrophied and tubular tongue into a foreground zone of liquid pleasure. This zone, midway between the two pyramid apexes of vanishing point and camera/viewer, stands ready to sheer or buckle in two directions like a sheet of paper tossed into a fire; the curl of his fingers waits to draw the refreshing drink inwards toward him; the curve of the leaves waits to escort the bottle outwards toward us. Between friends a source of warmth and pleasure is to be exchanged. (Need we add, '. . . for a price'?)

Images in combination

The potential ambiguity of the single photographic image can be pinned down by the work of codes internal to that image with or against each other. But, commonly, an image's environment enters into the production of meaning. In film this environment includes the succession of one image, and shot, by another along a diachronic axis and the accompanying sound effects, music, speech, or

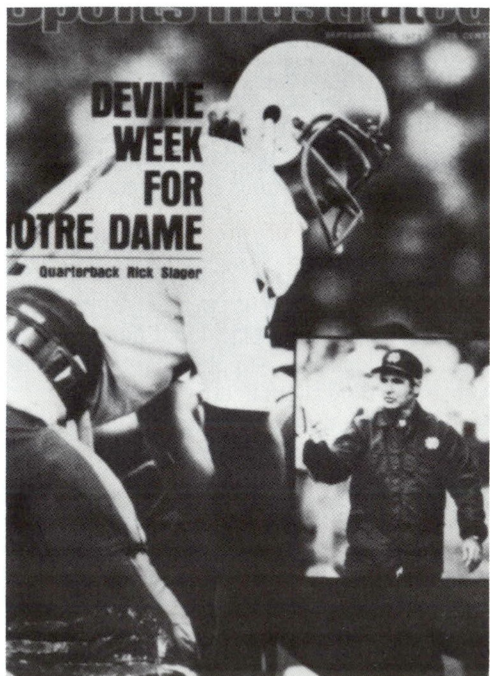

Figure 6.2

written words organized in tandem with the image track. With still images combinations of the same possibilities arise (comic books and slide shows, for example), although without the impression of movement peculiar to motion pictures and television. In most instances image-image and images-written-word combinations have the greatest importance. These combinations can be regarded as relationships of context/text or syntagmatic aspects of a text, depending on our point of view and the specifics of the situation. A magazine cover may provide a contextual frame for those images displayed inside the magazine, for example, whereas a combination of images used to produce the cover itself may be more easily seen as a syntagmatic aspect of one text. The emphasis here will once again be on textual analysis and hence syntagmatics.

This image (Figure 6.2) announces its class (advertising image), genus (magazine cover), and species (*Sports Illustrated* cover) quite readily, if we know the codes of magazine merchandising and football. Its combination of images and words is clear testimony to the funnelling of meaning, the reduction of ambiguity, made possible by the interplay of codes. Even though meaning is pinned down, it may not be paraded forthrightly, especially since the relationship of codes such as word to image need not and cannot be one of pure redundancy, guaranteeing already apparent meaning. Much of the meaning continues to originate from the matrix of analogue codes that we are hard put to recognize as codes and whose messages are often received at less conscious levels despite being precisely formulated. With this proviso in mind, we can begin an analysis of this cover by noting the complementary relationship between the inset of the coach and the quarter-back in the centre of attention. This relationship involves (1) the approximation of an eyeline match (the imaginary meeting of their gazes); (2) the reciprocal pinning down of the quarterback's expression and the coach's gesture from a larger repertoire of possibilities to 'What should I do' and 'Let me help you'; (3) the mutual colour bond of the quarterback's badge of bloodstains and the red border of the inset, which sublimates the brutality of combat from the coach to his frame, to the boundary between brawn and brain; (4) the contrast of scale (looming brawn, diminished brain) rendered ironic by the placement of lettering of the quarterback's jersey (where his own name should be emblazoned, but where the coach's reigns both benevolently – 'divine', guiding – and oppressively – on his back). This unspoken bond invokes much of the lure football holds for the armchair quarterback – the formulation of strategy, the crossing of the boundary between brain/brawn – and its very invocation upon the magazine's cover carries with it a promise of revelation: within the issue's interior, mysteries of strategy and relationship will be unveiled.

The play on words (Devine/divine), far from diffusing meaning, buttons it still more closely. The coach, Dan Devine, has just come to Notre Dame from the professional ranks: two forms of supremacy (of spiritual being and professional sport) coagulate in the red-framed, free-floating portrait of the coach alongside his team. This meaning, like many others employed in advertising topical commodities like a sports magazine, requires a pre-existing familiarity with current events, with the kind of events the magazine would have covered in a previous issue, for example. Punning here signifies not only the relationship of Devine to Notre Dame and his previous job but also the *au courant* status of the issue itself and its placement in an ongoing series of updates.

Images and Ideology

The function of words in relation to images has often been singled out as a factor of singular importance in holding meaning in check. The great precision of a digital code like written language allows a dense mass of meaning to be packed into a relatively small surface area to which the eye is almost inevitably attracted and from which meanings are discharged like a shower of needle points to pin

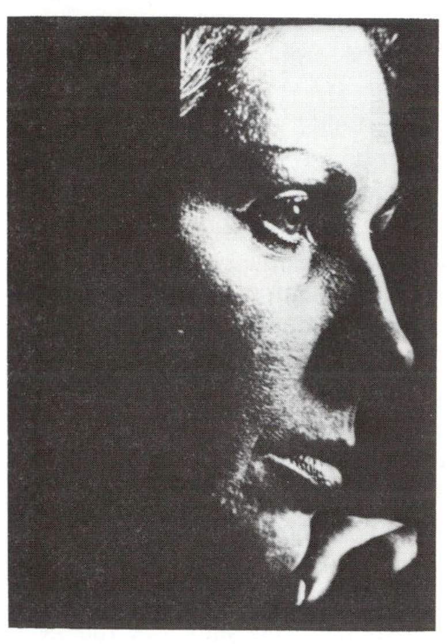

(a)

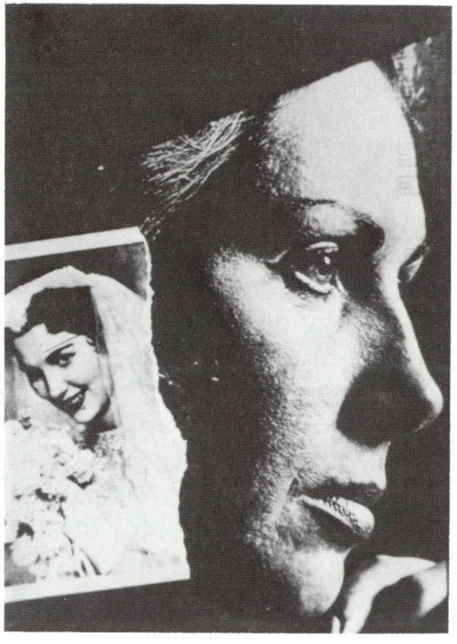

(b)

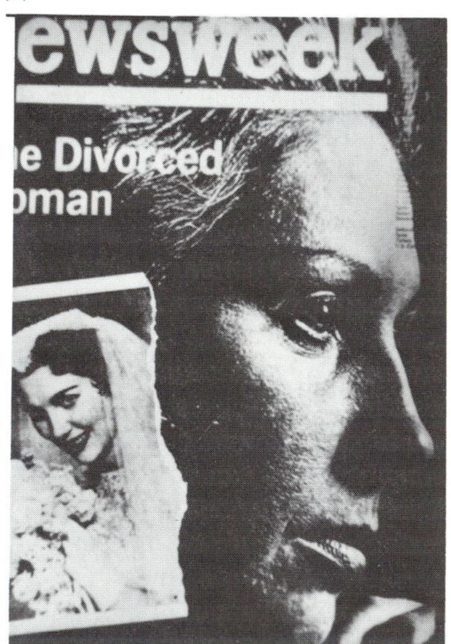

(c)

Figure 6.3 (a), (b), (c) The increments of meaning added to the highly ambiguous image of a woman's face by combination with another image and by yoking both images to a single caption pin down a range of possible interpretation to a far more limited set. This pinning also proposes conceptual constraints around the topic of divorce that are clearly ideological, i.e., sexist

down the ambiguity of images (Figure 6.3). In analysing a widely disseminated photograph of Jane Fonda in North Vietnam (Figure 6.4). Godard and Gorin have this to say about the caption orginally printed in *L'Express* (this image forms the basis of their film *Letter to Jane,* 1972).

> The text doesn't mention the Vietnamese people in the photograph. For example, the text doesn't tell us that the Vietnamese who cannot be seen in the background is one of the least known and least moderate of the Vietnamese people. This photograph, like any photograph, is physically mute. It talks through the mouth of the text written beneath it. This text does not emphasize, does not repeat, because a photograph speaks and says things in its own way. The fact that the militant is in the foreground, and Vietnam is in the background. The text says that Jane Fonda is questioning the people of Hanoi. But the magazine does not publish the questions asked, nor the answers given by the representatives of the North Vietnamese people in this photograph. In fact, the text should not describe the photograph as Jane Fonda questioning but as Jane Fonda listening. This much is obvious and perhaps the moment only lasted 1/250th of a second but that is the 1/250th that has been recorded and sent throughout the western world.[1]

Words can indeed lie, and they can lie about images as well as anything else, though the very ambiguity of an image seems to soften these possible lies to helpful notes of emphasis. The play between word and image remains a site for disintegration as well as integration, of non-cooperation as well as incorporation.

Figure 6.4

Section II

The socio-cultural relations of language

In this section we hope to be able to make the reader aware of the limitations of a 'commonsensical' view of language as a sort of neutral medium for the transmission or storing of ideas or thoughts, much as a delivery-van or a warehouse transmits or stores goods. The early influence of certain technological or behaviourist models of communication on Communication Studies, with their vocabulary of 'senders', 'receivers', 'codes', and so on, encouraged a rather limiting view of human language as a purely formal system which was not different in kind from, say, the codes with which a telecommunications engineer operates, and which could, therefore, be studied in isolation as a self-enclosed system.

The extract from Jean Aitchison's book which we included in the previous section compared and contrasted language with the communication systems of animals. We hope that this has enabled readers to appreciate that human language is vastly more complex and multi-faceted than such systems. Language does not just enable us to *communicate* certain things (which of course it most definitely does); without it we would have far, far less to communicate in the first place – we would not *be* human at all. Language enters into our personal identities, our social relations, our cultural heritage, and our history, and it is this permeation of language into everything that is human that we wish to explore in more detail in the present section.

In their discussion of a popular British radio programme, for example, **Graham Brand** and **Paddy Scannell** show how DJ talk can be used actually to construct a public identity, to 'make a person' for those who listen to the show. Such listeners do not just receive information from the DJ, they help interactively to create the public persona that he assumes. (In the next section, we include an earlier American study by Donald Horton and R. Richard Wohl which analyses a similar phenomenon in psychological rather than linguistic terms.)

The very size, complexity, and comprehensiveness of language has inevitably led researchers to limit the scope of their enquiry into it. Language can be studied in one of its national-cultural variants (English, Japanese, Swahili) or in a variety such as Black American English vernacular; it can be studied in written rather than spoken form; or the focus of concern can be narrowed down even further by concentrating upon one aspect such as grammar or phonology. In recent years other ways of limiting the scope of enquiry have been influential. Language study has traditionally fallen into one of two dominant approaches: the *synchronic* (language seen as a complete system at a given point in time), and the *diachronic* or historical (language as it evolves over time). Early in the present century the Swiss linguistician Ferdinand de Saussure's use of this division allowed him to make a further important distinction: between *langue* (the overall and abstract system that is language) and *parole* (the individual utterances generated by means of this system). A distinction made by the influential American linguistician Noam Chomsky, between *competence* and *performance*, has something in common with this division of Saussure's.

Such delimitations and distinctions have proved enormously productive in the present century, and without them the study of language would have remained piece-meal and positivistic (that is, concerned only with the accumulation of directly observable facts). But in recent years many researchers have become more concerned to remind us of the comprehensive presence of language in social and cultural life and, accordingly, to argue against attempts to separate language from human consciousness, activity, and history. Earlier this century the pioneer theorist of communication Charles Peirce suggested a tripartite distinction between *syntactics* (or syntax) which would study the formal relation of signs to one another, *semantics* which would study the relations of signs to the phenomena or ideas to which the signs are applicable, and *pragmatics* which would study the relation of signs to interpreters.

It is fair to say that within Communication Studies in general, and with regard to the place of language study in Communication Studies in particular, pragmatics has assumed a position of increasing importance in the last decade, and this shift of emphasis is reflected in the present section. Stephen C. Levinson, in his influential book *Pragmatics*, suggests reasons for this development.

> [This interest in pragmatics] developed in part as a reaction or antidote to Chomsky's treatment of language as an abstract device, or mental ability, dissociable from the uses, users and functions of language.
>
> . . .
>
> Another powerful and general motivation for the interest in pragmatics is the growing realization that there is a very substantial gap between current linguistic theories of language and accounts of linguistic communication . . . . For it is becoming increasingly clear that a semantic theory alone can give us only a proportion, and perhaps only a small if essential proportion, of a general account of language understanding.[1]

We can add that a concern with the pragmatic dimension of language, as a result of its interest in different contexts of use, leads naturally to that inter- or cross-disciplinary approach which a field of study such as Communication Studies favours.

Geoffrey Leech has argued that whereas semantics is *rule-governed* (and we call any attempt to formalize these rules a *grammar*), pragmatics is, in contrast, *principle-controlled*.[2] Attempts to study that which is rule-governed typically assume a formal approach, whereas attempts to study that which is principle-governed typically place much more emphasis on the analysis of the actions, events or whatever which are seen to express these principles. During the past decade or so Communication Studies has steadily moved away from more formal and formalist approaches to the study of language, leaving them to theoretical linguisticians, and has instead concentrated upon looking at language in use, in social, cultural and historical contexts. During the time covered by the four editions of this Reader, Communication Studies has witnessed a steady movement away from the study of language as a formal system, away from attempts to isolate *rules*, and towards a more contextual view of the actual use of language, towards in other words, pragmatics and rhetoric.

Gunther Kress's approach is certainly representative of a desire to move away from the study of abstract linguistic systems and to get closer to a study of language as it is actually used in the primary form of interactive utterances, while recognizing the significant effect that the development of writing systems and the emergence of literacy bring with them – both to society and culture at large as well as to the individual.

Many of the extracts in this section are concerned, albeit in different ways, with the linguistic aspects of power: power over nature, power in society, the power which comes from the ability to distance oneself from concrete experiences, to examine oneself dispassionately and objectively, and to be able to change oneself and, in co-operation with others, one's society. **Deborah Cameron** points out that literacy is related to power; those with it do not just have a degree of power over nature, but over the illiterate (including women, for example, in many societies). The extract from Jean Aitchison's book in the previous section showed that the possession of language gave human beings a power over nature that animals lack, and thus it gave, too, power over these animals: even where they were physically much stronger. Aitchison's discussion of *displacement* is particularly important in this context. But while it is true that *any* human speech involves powers of displacement unknown to animals (*all* human beings can talk about things not immediately present, such as 'the things we will do tomorrow'), the extracts from **Basil Bernstein** and **Trevor Pateman** in the present section indicate that some human beings possess this ability to a much greater extent than others. The subjects about whom both Bernstein and Pateman write, although literate, cannot, according to Pateman, handle abstractions and counterfactuals in the way that more educated and linguistically developed people can. Bernstein's work is, to a considerable extent, an attempt to find out why, if so, this should be the case. What are the mechanisms which allow an individual to develop more complex and sophisticated powers of reasoning and abstraction? As he phrases it, 'under what conditions does a given speech form free itself sufficiently from its embodiment in the social structure so that the system of meanings it realizes points to alternative realities?'

It should be clear that, with a different emphasis, this question lies also behind Cameron's concern with the linguistic aspects of women's oppression. And the argument in the extract included from Pateman's book can be seen as a inquiry into the implications of Bernstein's assertion that 'Historically, and now, only a tiny percentage of the population has been socialized into knowledge at the level of the metalanguages of control and innovation, whereas the mass of the population has been socialized into knowledge at the level of context-tied operations.'

More advanced linguistic abilities give social and cultural power, but social and cultural power provide in their turn access to the opportunity to acquire more advanced linguistic abilities. (Traditionally, one of the first things a newly-rich person would buy would be a more expensive education for his or her children.) As Bernstein puts it: 'It is not only capital, in the strict economic sense, which is subject to appropriation, manipulation and exploitation, but also *cultural capital* in the form of symbolic systems through which man can extend and change the boundaries of his experience.' After reading Cameron's piece we may attach more significance to Bernstein's use of the male gender here than he may have done.

The authors of a recent book entitled *Discourse and social psychology: beyond attitudes and behaviour,* writing about 'social texts' rather than about written ones, make what is nonetheless an extremely apposite comment in their introductory pages:

> [S]ocial texts do not merely *reflect* or *mirror* objects, events and categories pre-existing in the social and natural world. Rather, they actively *construct* a version of these things. They do not just describe things; they *do* things. And being active, they have social and political implications. We have seen how description is tied to evaluation and how different versions of events can be constructed to justify or blame these events.[3]

The study of how language, including linguistic texts, 'does things' has a very ancient pedigree, especially in that tradition known as *rhetoric.* We believe it is no accident that the study of rhetoric has attracted much more interest of late. **Deborah Tannen**'s fascinating analysis of a political speech by The Rev. Jesse Jackson shows how rhetorical analysis can move away from a somewhat sterile concern with the compiling of lists of rhetorical techniques, to a concern with the way in which culture, language and power interpenetrate.

Further reading

All of the pieces in this section come from books which contain much additional material of value to the Communication Studies student. In addition to these, the following can all be recommended as accessible to the relatively uninformed reader and relevant to the issues raised in this section.

Bolinger, D. 1980: *Language – the loaded weapon.* London: Longman.
A clear and vigorously-argued book on the 'use and abuse' of language today. Drawing upon contemporary American illustrations, Bolinger discusses the relationships between language and various kinds of social power. His chapters focus, for example, on types of verbal stigma, on sexism, on advertising and on the language of bureaucracies. A close attention to linguistic features is impressively maintained throughout the general, social argument.

Cameron, Deborah (ed.) 1990: *The feminist critique of language.* London and New York: Routledge.
A useful collection of articles and extracts, including material both by linguisticians and also by others who have written on language from a feminist perspective.

Coates, Jennifer and Cameron, Deborah (eds.) 1988: *Women in their speech communities: new perspectives on language and sex.* London and New York: Longman.
A collection containing more specialised contributions using approaches from sociolinguistics and discourse analysis. Many of the contributions focus in on particular speech situations and communities, with a number devoting attention to the position of minorities. Two of the contributions deal directly with the issue of power.

Ellis, Andrew and Beattie, Geoffrey 1988: *The psychology of language and communication.* London and Hillsdale: Lawrence Erlbaum. First pub. 1986.
A clear and accessible textbook covering the variety of human communication from a psychological perspective. Includes a useful summary of the criticisms made of Basil Bernstein's research methods in his early work.

Fairclough, Norman 1989: *Language and power.* London and New York: Longman.
A book which describes the ways in which language is and can be used to maintain and change power relations in modern society, and which also suggests how awareness of such processes can help us to resist them. Central to the book's argument is a view of language as 'the primary domain of ideology'. It includes a case study of the political discourse of Thatcherism.

Montgomery, Martin 1986: *An introduction to language and society.* London and New York: Methuen.
A book which is both introductory but also up-to-date with the latest theoretical developments in this area, moving from language acquisition through linguistic diversity, language and social interaction, and language and representation. The book to start with if you wish to follow up the ideas in this section.

Ong, Walter J. 1982: *Orality and literacy: the technologizing of the word.* London and New York: Methuen.
An engrossing book which investigates the changes in society and the individual which are attendant upon the invention of writing and the attainment of literacy.

Stubbs, Michael 1980: *Language and literacy: the socio-linguistics of reading and writing.* London: Routledge.
A book which aims to provide a sociolinguistic theory of literacy. It contains a persuasive account of the relations between reading and writing which those interested in the issues raised in this section by Gunther Kress should find engaging.

7

Social class, language and socialization

Basil Bernstein

From Bernstein, B. 1971: *Class, codes and control.* Vol I. London: Routledge. This extract taken from the Paladin edition, 193–205.

It may be helpful to make explicit the theoretical origins of the thesis I have been developing over the past decade. Although, initially, the thesis appeared to be concerned with the problem of educability, this problem was embedded in and was stimulated by the wider question of the relationship between symbolic orders and social structure. The basic theoretical question, which dictated the approach to the initially narrow but important empirical problem, was concerned with the fundamental structure and changes in the structure of cultural transmission. Indeed, any detailed examination of what superficially may seem to be a string of somewhat repetitive papers, I think would show three things:

1. The gradual emergence of the dominance of the major theoretical problem from the local, empirical problem of the social antecedents of the educability of different groups of children.
2. Attempts to develop both the generality of the thesis and to develop increasing specificity at the contextual level.
3. Entailed in (2) were attempts to clarify both the logical and empirical status of the basic organizing concept, code. Unfortunately, until recently these attempts were more readily seen in the *planning* and *analysis* of the empirical research than available as formal statements.

Looking back, however, I think I would have created less misunderstanding if I had written about socio-linguistic codes rather than linguistic codes. Through using only the latter concept it gave the impression that I was reifying syntax and at the cost of semantics; or worse, suggesting that there was a one-to-one relation between meaning and a given syntax. Also, by defining the codes in a context-free fashion, I robbed myself of properly understanding, at a theoretical level, their significance. *I should point out that nearly all the empirical planning was directed to trying to find out the code realizations in different contexts.*

The concept of sociolinguistic code points to the social structuring of meanings and to their diverse but *related* contextual linguistic realizations. A careful reading of the papers always shows the emphasis given to the form of the social relationship, that is to the structuring of relevant meanings. Indeed, role is defined as a complex coding activity controlling the creation and organization of specific meanings and the conditions for their transmission and reception. The general sociolinguistic thesis attempts to explore how symbolic systems are both realizations and regulators of the structure of social relationships. The particular symbolic system is that of speech *not* language.

It is pertinent, at this point, to make explicit earlier work in the social sciences which formed the implicit starting point of the thesis. It will then be seen, I hope, that the thesis is an integration of different streams of thought. The major starting points are Durkheim and Marx, and a small number of other thinkers have been drawn into the basic matrix. I shall very briefly, and so selectively, outline this matrix and some of the problems to which it gave rise.

Durkheim's work is a truly magnificent insight into the relationships between symbolic orders, social relationships and the structuring of experience. In a sense, if Marx turned Hegel on his head, then Durkheim attempted to turn Kant on his head. For in *Primitive Classification* and in *The Elementary Forms of the Religious Life*, Durkheim attempted to derive the basic categories of thought from the structuring of the social relation. It is beside the point as to his success. He raised the whole question of the relation between the classifications and frames of the symbolic order *and* the structuring of experience. In his study of different forms of social integration he pointed to the implicit, condensed, symbolic structure of mechanical solidarity and the more explicit and differentiated symbolic structures of organic solidarity. Cassirer, the early cultural anthropologists, and, in particular, Sapir (I was not aware of von Humboldt until much later), sensitized me to the cultural properties of speech. Whorf, particularly where he refers to the fashions of speaking, frames of consistency, alerted me to the selective effect of the culture (acting through its patterning of social relationships) upon the *patterning* of grammar *together* with the pattern's semantic and thus cognitive significance. Whorf more than anyone, I think, opened up, at least for me, the question of the deep structure of linguistically regulated communication.

In all the above work I found two difficulties. If we grant the fundamental linkage of symbolic systems, social structure and the shaping of experience it is still unclear *how* such shaping takes place. The *processes* underlying the social structuring of experience are not explicit. The second difficulty is in dealing with the question of change of symbolic systems. Mead is of central importance in the solution of the first difficulty, the HOW. Mead outlined in general terms the relationships between role, reflexiveness and speech and in so doing provided the basis of the solution to the HOW. It is still the case that the Meadian solution does not allow us to deal with the problem of change. For the concept, which enables role to be related to a higher order concept, 'the generalized other', is, itself, not subject to systematic inquiry. Even if 'the generalized other' is placed within a Durkheimian framework, we are still left with the problem of change. Indeed, in Mead change is introduced only at the cost of the re-emergence of a traditional Western dichotomy in the concepts of the 'I' and the 'me'. The 'I' is both the indeterminate response to the 'me' and yet, at the same time, shapes it. The Meadian 'I' points to the voluntarism in the affairs of men, to the fundamental creativity of man, made possible by speech; a little before Chomsky.

Thus Meadian thought helps to solve the puzzle of the HOW but it does not help with the question of change in the structuring of experience; although both Mead implicitly and Durkheim explicitly pointed to the conditions which bring about pathological structuring of experience.

One major theory of the development of and change in symbolic structures is, of course, that of Marx. Although Marx is less concerned with the internal structure and the process of transmission of symbolic systems, he does give us a key to their institutionalization and change. The key is given in terms of the social significance of society's productive system and the power relationships to which the productive system gives rise. Further, access to, control over, orientation of and *change* in critical symbolic systems, according to the theory, is governed by power relationships as these are embodied in the class structure. It is not only capital, in the strict economic sense, which is subject to appropriation, manipulation and exploitation, but also *cultural* capital in the form of the symbolic systems through which man can extend and change the boundaries of his experience.

I am not putting forward a matrix of thought necessary for the study of the basic structure and change in the structure of cultural transmission, *only* the specific matrix which underlies my own approach. Essentially and briefly I have used Durkheim and Marx at the macro-level and Mead at the micro-level to realize a sociolinguistic thesis which could meet with a range of work in anthropology, linguistics, sociology and psychology.

I want first of all to make clear what I am not concerned with. Chomsky, in *Aspects of the Theory of Syntax*, neatly severs the study of the rule system of language from the study of the social rules which determine their contextual use. He does this by making a distinction between competence and performance. Competence refers to the child's tacit understanding of the rule system, performance relates to the essentially social use to which the rule system is put. Competence refers to man abstracted from contextual constraints. Performance refers to man in the grip of the contextual constraints which determine his speech acts. Competence refers to the Ideal, performance refers to the Fall. In this sense Chomsky's notion of competence is Platonic. Competence has its source in the very biology of man. There is no difference between men in terms of their access to the linguistic rule system. Here Chomsky, like many other linguists before him, announces the communality of man; all men have equal access to the creative act which is language. On the other hand, performance is under the control of the social – performances are culturally specific acts, they refer to the choices which are made in specific speech encounters. Thus, according to Hymes, Chomsky indicates the tragedy of man, the potentiality of competence and the degeneration of performance.

Clearly, much is to be gained in rigour and explanatory power through the severing of the relationship between the formal properties of the grammar and the meanings which are realized in its use. But if we are to study speech, *la parole*, we are inevitably involved in a study of a rather different rule system; we are involved in a study of rules, formal and informal, which regulate the options we take up in various contexts in which we find ourselves. This second rule system is the cultural system. This raises immediately the question of the relationship between the linguistic rule system and the cultural system. Clearly, specific linguistic rule systems are part of the cultural system, but it has been argued that the linguistic rule system in various ways shapes the cultural system. This very briefly is the view of those who hold a narrow form of the linguistic relativity hypothesis. I do not intend to get involved in that particular quagmire. Instead, I shall take the view that the code which the linguist invents to explain the formal properties of the grammar is capable of generating any number of speech codes, and there is no reason for believing that any one language code is better than another in this respect. On this argument, language is a set of rules to which all speech codes must comply, but which speech codes are realized is a function of the culture acting through social relationships in specific contexts. Different speech forms or codes symbolize the form of the social relationship, regulate the nature of the speech encounters, and create for the speakers different orders of relevance and relation. The experience of the speakers is then transformed by what is made significant or relevant by the speech form. This is a sociological argument because the speech form is taken as a consequence of the form of the social relation or, put more generally, is a quality of social structure. Let me qualify this immediately. Because the speech form is initially a function of a given social arrangement, it does not mean that the speech form does not in turn modify or even change that social structure which initially evolved the speech form. This formulation, indeed, invites the question: under what conditions does a given speech form free itself sufficiently from its embodiment in the social structure so that the system of meanings it realizes points to alternative realities, alternative arrangements in the affairs of men? Here we become concerned immediately with the antecedents and consequences of the boundary-maintaining principles of a culture or subculture. I am here suggesting a relationship between forms of boundary maintenance at the cultural level and forms of speech.

I am required to consider the relationship between language and socialization. It should be clear from these opening remarks that I am not concerned with language, but with speech, and concerned more specifically with the contextual constraints upon speech. Now what about socialization? I shall take the term to refer to the process whereby a child acquires a specific cultural identity *and* to his responses to such an identity. Socialization refers to the process whereby the biological is transformed into a specific cultural being. It follows from this that the process of socialization is a complex process of control, whereby a particular moral, cognitive and affective awareness is evoked in the child and given a specific form and content. Socialization sensitizes the child to the various orderings of society as these are made substantive in the various roles he is expected to play. In a sense, then, socialization is a process for making people safe. The process acts selectively on the possibilities of man by creating through time a sense of the inevitability of a given social arrangement, and through limiting the areas of permitted change. The basic agencies of socialization in contemporary societies are the family, the peer group, school and work. It is through these agencies, and in particular through their relationship to each other, that the various orderings of society are made manifest.

Now it is quite clear that given this view of socialization it is necessary to limit the discussion. I shall limit our discussion to socialization within the family, but it should be obvious that the focusing and filtering of the child's experience within the family in a large measure is a microcosm of the macroscopic orderings of society. Our question now becomes: what are the sociological factors which affect linguistic performances within the family critical to the process of socialization?

Without a shadow of doubt the most formative influence upon the procedures of socialization, from a sociological viewpoint, is social class. The class structure influences work and educational roles and brings families into a special relationship with each other and deeply penetrates the structure of life experiences within the family. The class system has deeply marked the distribution of knowledge within society. It has given differential access to the sense that the world is permeable. It has sealed off communities from each other and has ranked these communities on a scale of invidious worth. We have three components:

knowledge, possibility and invidious insulation. It would be a little naïve to believe that differences in knowledge, differences in the sense of the possible, combined with invidious insulation, rooted in differential *material* well-being, would not affect the forms of control and innovation in the socializing procedures of different social classes. I shall go on to argue that the deep structure of communication itself is affected, but not in any final or irrevocable way.

As an approach to my argument, let me glance at the social distribution of knowledge. We can see that the class system has affected the distribution of knowledge. Historically, and now, only a tiny percentage of the population has been socialized into knowledge at the level of the meta-languages of control and innovation, whereas the mass of the population has been socialized into knowledge at the level of context-tied operations.

A tiny percentage of the population has been given access to the principles of intellectual change, whereas the rest have been denied such access. This suggests that we might be able to distinguish between two orders of meaning. One we could call universalistic, the other particularistic. Universalistic meanings are those in which principles and operations are made linguistically explicit, whereas particularistic orders of meaning are meanings in which principles and operation are relatively linguistically implicit. If orders of meaning are universalistic, then the meanings are less tied to a given context. The meta-languages of public forms of thought as these apply to objects and persons realize meanings of a universalistic type. Where meanings have this characteristic then individuals have access to the grounds of their experience and can change the grounds. Where orders of meaning are particularistic, where principles are linguistically implicit, then such meanings are less context-independent and *more* context-bound, that is, tied to a local relationship and to a local social structure. Where the meaning system is particularistic, much of the meaning is embedded in the context and may be restricted to those who share a similar contextual history. Where meanings are universalistic, they are in principle available to all because the principles and operations have been made explicit, and so public.

I shall argue that forms of socialization orient the child towards speech codes which control access to relatively context-tied or relatively context-independent meanings. Thus I shall argue that elaborated codes orient their users towards universalistic meanings, whereas restricted codes orient, sensitize, their users to particularistic meanings: that the linguistic realization of the two orders are different, and so are the social relationships which realize them. Elaborated codes are less tied to a given or local structure and thus contain the potentiality of change in principles. In the case of elaborated codes the speech may be freed from its evoking social structure and it can take on an autonomy. A university is a place organized around talk. Restricted codes are more tied to a local social structure and have a reduced potential of change in principles. Where codes are elaborated, the socialized has more access to the grounds of his own socialization, and so can enter into a reflexive relationship to the social order he has taken over. Where codes are restricted, the socialized has less access to the grounds of his socialization, and thus reflexiveness may be limited in range. *One of the effects of the class system is to limit access to elaborated codes.*

I shall go on to suggest that restricted codes have their basis in condensed symbols, whereas elaborated codes have their basis in articulated symbols; that restricted codes draw upon metaphor, whereas elaborated codes draw upon rationality; that these codes constrain the contextual use of language in critical socializing contexts and in this way regulate the orders of relevance and relation which the socialized takes over. From this point of view, change in habitual speech codes involves changes in the means by which object and person relationships are realized.

I want first to start with the notions of elaborated and restricted speech variants. A variant can be considered as the contextual constraints upon grammatical-lexical choices.

Sapir, Malinowski, Firth, Vygotsky and Luria have all pointed out from different points of view that the closer the identifications of speakers the greater the range of shared interests, the more probable that the speech will take a specific form. The range of syntactic alternatives is likely to be reduced and the lexis to be drawn from a narrow range. Thus, the form of these social relations is acting selectively on the meanings to be verbally realized. In these relationships the intent of the other person can be taken for granted as the speech is played out against a backdrop of common assumptions, common history, common interests. As a result, there is less need to raise meanings to the level of explicitness or elaboration. There is a reduced need to make explicit through syntactic choices the logical structure of the communication. Further, if the speaker wishes to individualize his communication, he is likely to do this by varying the expressive associates of the speech. Under these conditions, the speech is likely to have a strong metaphoric element. In these situations the speaker may be more concerned with how something is said, when it is said; silence takes on a variety of meanings. Often in these encounters the speech cannot be understood apart from the context, and the context cannot be read by those who do not share the history of the relationships. Thus the form of the social relationship acts selectively in the meanings to be verbalized, which in turn affect the syntactic and lexical choices. The unspoken assumptions underlying the relationship are not available to those who are outside the relationship. For these are limited, and restricted to the speakers. The symbolic form of the communication is condensed, yet the specific cultural history of the relationship is alive in its form. We can say that the roles of the speakers are communalized roles. Thus, we can make a relationship between restricted social relationships based upon communalized roles and the verbal realization of their meaning. In the language of the earlier part of this paper, restricted social relationships based upon communalized roles evoke particularistic, that is, context-tied, meanings, realized through a restricted speech variant.

Imagine a husband and wife have just come out of the cinema, and are talking about the film: 'What do you think?' 'It had a lot to say.' 'Yes, I thought so too – let's go to the Millers, there may be something going there.' They arrive at the Millers, who ask about the film. An hour is spent in the complex, moral, political, aesthetic subtleties of the film and its place in the contemporary scene. Here we have an elaborated variant; the meanings now have to be made public to others who have not seen the film. The speech shows careful editing, at both the grammatical and lexical levels. It is no longer context-tied. The meanings are explicit, elaborated and individualized. While expressive channels are clearly relevant, the burden of meaning inheres predominantly in the verbal channel. The experience of the listeners cannot be taken for granted. Thus each member of the group is on his own as he offers his interpretation. Elaborated variants of this kind involve the speakers in particular role relationships, and *if you cannot manage the role, you can't produce the appropriate speech.* For as the speaker proceeds to individualize his meanings, he is differentiated from others like a figure from its ground.

The roles receive less support from each other. There is a measure of isolation. *Difference* lies at the basis of the social relationship, and is made verbally active, whereas in the other context it is *consensus*. The insides of the speaker have become psychologically active through the verbal aspect of the communication. Various defensive strategies may be used to decrease potential vulnerability of self and to increase the vulnerability of others. The verbal aspect of the communication becomes a vehicle for the transmission of individuated symbols. The 'I' stands over the 'we'. Meanings which are discrete to the speaker must be offered so that they are intelligible to the listener. Communalized roles have given way to individualized roles, condensed symbols to articulated symbols. Elaborated speech variants of this type realize universalistic meanings in the sense that they are less context-tied. Thus individualized roles are realized through elaborated speech variants which involve complex editing at the grammatical and lexical levels and which point to universalistic meanings.

Let me give another example. Consider the two following stories which Peter Hawkins, Assistant Research Officer in the Sociological Research Unit, University of London Institute of Education, constructed as a result of his analysis of the speech of middle-class and working-class five-year-old children. The children were given a series of four pictures which told a story and they were invited to tell the story. The first picture showed some boys playing football; in the second the ball goes through the window of a house; the third shows a woman looking out of the window and a man making an ominous gesture, and in the fourth the children are moving away.

Here are the two stories.

1 Three boys are playing football and one boy kicks the ball and it goes through the window the ball breaks the window and the boys are looking at it and a man comes out and shouts at them because they've broken the window so they run away and then that lady looks out of her window and she tells the boys off.

2 They're playing football and he kicks it and it goes through there it breaks the window and they're looking at it and he comes out and shouts at them because they've broken it so they run away and then she looks out and she tells them off.

With the first story the reader does not have to have the four pictures which were used as the basis for the story, whereas in the case of the second story the reader would require the initial pictures in order to make sense of the story. The first story is free of the context which generated it, whereas the second story is much more closely tied to its context. As a result the meanings of the second story are implicit, whereas the meanings of the first story are explicit. It is not that the working-class children do not have in their passive vocabulary the vocabulary used by the middle-class children. Nor is it the case that the children differ in that tacit understanding of the linguistic rule system. Rather, what we have here are differences in the use of language arising out of a specific context. One child makes explicit the meanings which he is realizing through language for the person he is telling the story to, whereas the second child does not to the same extent. The first child takes very little for granted, whereas the second child takes a great deal for granted. Thus for the first child the task was seen as a context in which his meanings were required to be made explicit, whereas the task for the second child was not seen as a task which required such explication of meaning. It would not be difficult to imagine a context where the first child would produce speech rather like the second. What we are dealing with here are differences between the children in the way they realize in language-use apparently the same context. We could say that the speech of the first child generated universalistic meanings in the sense that the meanings are freed from the context and so understandable by all, whereas the speech of the second child generated particularistic meanings, in the sense that the meanings are closely tied to the context and would be fully understood by others only if they had access to the context which originally generated the speech.

It is again important to stress that the second child has access to a more differentiated noun phrase, but there is a restriction on its *use*. Geoffrey Turner, Linguist in the Sociological Research Unit, shows that working-class, five-year-old children in the same contexts examined by Hawkins use fewer linguistic expressions of uncertainty when compared with the middle-class children. This does not mean that working-class children do *not* have access to such expressions, but that the eliciting speech context did not provoke them. Telling a story from pictures, talking about scenes on cards, *formally framed* contexts, do not encourage working-class children to consider the possibilities of alternate meanings and so there is a reduction in the linguistic expressions of uncertainty. Again, working-class children have access to a wide range of syntactic choices which involve the use of logical operators, 'because', 'but', 'either', 'or', 'only'. The constraints exist on the conditions for their *use*. Formally framed contexts used for eliciting context-independent universalistic meanings may evoke in the working-class child, relative to the middle-class child, restricted speech variants, because the working-class child has difficulty in managing the role relationships which such contexts require. This problem is further complicated when such contexts carry meanings very much removed from the child's cultural experience. In the same way we can show that there are constraints upon the middle-class child's use of language. Turner found that when middle-class children were asked to role-play in the picture story series, a higher percentage of these children, when compared with working-class children, initially refused. When the middle-class children were asked, 'What is the man saying?' or linguistically equivalent questions, a relatively higher percentage said 'I don't know.' When this question was followed by the hypothetical question, 'What do you think the man might be saying?' they offered their interpretations. The working-class children role-played without difficulty. It seems then that middle-class children at five need to have a very precise instruction to *hypothesize in that particular* context. This may be because they are more concerned here with getting their answers right or correct. When the children were invited to tell a story about some doll-like figures (a little boy, a little girl, a sailor and a dog) the working-class children's stories were freer, longer and more imaginative than the stories of the middle-class children. The latter children's stories were tighter, constrained within a strong narrative frame. It was as if these children were dominated by what they took to be the *form* of a narrative and the content was secondary. This is an example of the concern of the middle-class child with the structure of the contextual frame. It may be worthwhile to amplify this further. A number of studies have shown that when working-class black children are asked to associate to a series of words, their responses show considerable diversity, both from the meaning and form-class of the stimulus word. Our analysis suggests this may be because the children for the following reasons are less constrained. The form-class of the stimulus word may have reduced associated significance and this would less constrain the selection of potential words *or* phrases. With such a weakening of the grammatical frame there is a greater range of alternatives as possible candidates for selection. Further, the closely controlled, middle-class, linguistic socialization of the young child may point the child towards both the grammatical significance of the stimulus word and towards a tight logical ordering of semantic space. Middle-class children may well have access to deep interpretative rules which regulate their linguistic responses in certain formalized contexts. The consequences may limit their imagination through the tightness of the frame which these interpretative rules create. It may even be that with *five*-year-old children, the middle-class child will innovate *more* with the arrangements of objects (i.e. bricks) than in his linguistic usage. His linguistic usage is under close supervision by adults. He has more *autonomy* in his play.

[The final eight pages of this essay, which develop the argument with further examples, are here omitted.]

8

Impossible discourse

Trevor Pateman

From Pateman, T. 1975: *Language, truth and politics*. Newton Poppleford, Devon: Trevor Pateman and Jean Stroud, 70–84.

Note: Language, truth and politics was published in a new, revised and enlarged edition in 1980. The extract that follows includes some amendments from this new edition.

Language and logic

I don't understand physics because I don't know the language of physics. This is partly a question of *vocabulary*, partly one of *concepts*, partly one of the mental *organization* of vocabulary – and the last two aspects are interrelated, as I shall try to show. These are some of the obstacles to my understanding physics. Do some people face comparable obstacles to understanding radical or revolutionary politics?

Is it, first of all, that some people lack the *vocabulary* with which they could understand and within which they could think certain thoughts? Though not of central importance, the absence of vocabulary is, I think, of more importance than the logical possibility of paraphrase might make it seem. It is plausible for Orwell in *1984* to attribute considerable significance to the removal of words from the lexicon, for though paraphrase remains logically possible, to actively engage in paraphrase requires a greater commitment to thought than does the simple use of a ready-made word-concept. Words are things to think with, and without them one is obliged to produce the means of thought as well as thought itself.

However, even if a vocabulary is known, the concepts belonging to each word may not be fully or accurately known. 'Trotskyism', 'Anarchism', 'Soviet', 'commune', etc. are known to many people (though how many?), but perhaps in the majority of cases they will be known as the *names* of desirable or undesirable practices. They name objects, institutions and practices and they direct or discharge considerable emotional energy, but their conceptual content in use is small; they are used as *proper names*, which do no more than designate or refer. The words (no more than a proper name) cannot be used to think with about the practices to which they are used to refer. (This is perhaps what people are getting at when they object to *labelling*, that is, the use of an emotionally loaded proper name.) In addition, such words may be used inaccurately to refer, though this is not entailed by their being used as proper names.

The ways in which descriptive words are emptied of their conceptual content, and thereby become purely referential expressions, has been amply commented upon, for example, by Marcuse. I think that Anglo-American philosophers have compounded rather than counteracted this process in their analysis of political terms, since they have taken as the starting point for their analyses the actual occurrence of such words on the surface of discourse, where they are used as mere naming and 'boo' 'hooray' expressions. The results of such analysis are bound to be as disappointing as the discourse being analysed. (On this point, see my 1973a). The results of such analysis for political philosophy have been disastrous, but no more so than such a use of words has been for the possibility of political thinking among the population at large.

If not by suppression, or emptying of content, then by other means can potentially critical concepts be rendered practically useless for critical thought. Marcuse has commented on such means in *One Dimensional Man* (1964). There he writes of the role of combining contradictory thoughts in a single expression. This occurs, for example, when a policy described in terms which would justify its designation as 'reactionary' is then named as 'revolutionary'. This could be a simple case in which 'revolution' has been emptied of meaning and is used simply to express an attitude or name an object, without giving rise to formally contradictory predication. On the other hand, there do appear to be more subtle cases (and it seems that Marcuse has these in mind) where one can speak of genuine contradiction, since the conceptual content of each of two terms is simultaneously predicated and denied. That is to say, the effectiveness of the statement made by A in persuading B to adopt an attitude towards x involves simultaneously evoking in B his understanding of the conceptual content of (say) 'revolution' whilst describing x in terms which indicate that it does not possess the properties which would justify the predication 'revolution'.

I think I can make this clearer with a concrete example from the sphere of trade names. In one sense, the trade name *Belair* for a brand of cigarette is an arbitrary proper name. If I ask for a packet of *Belair*, I use the name as a proper name and do not think of the name as having a conceptual content. On the other hand, *Belair* contributes to the task of selling the cigarette to the degree that the conceptual meaning (here, the literal meaning) and also the connotations of meaning (such as 'Frenchness') are known to the buyer.

But, with regard to the literal meaning, it might be questioned whether this would work as a selling force if the buyer became explicitly aware of the meaning. For if someone were to become explicitly aware of the literal meaning of *bel air*, would this not equal awareness of a characteristic so 'obviously' the opposite of the real characteristics of the product as to lead the potential buyer to ridicule the product? (In Orwell's *1984, Victory* is the brand name for the products with which a defeated population is drugged.)

Against this interpretation, consider what could happen if the potential buyer does not accept as 'obviously' true that 'Smoking can Damage your Health', that is to say, does not accept as obviously true the content of HM Government's Health Warning printed on the side of the packet. In such circumstances, conscious awareness of the meaning of *Belair* would produce a situation which could be characterized as follows: the packet of cigarettes carries, printed on it, two statements which both purport to be true of the contents of the packet. One says that what you inhale when you smoke *can* damage your health; the other says that what you inhale when you smoke is *bel air*, and 'good air' *cannot*, by definition, damage your health. (I am expanding and interpreting the two statements to put them in formally contradictory form, but I don't think that my expansion is far from the truth.) Now, all students of philosophy know that if two statements are formally contradictory, then they cancel each other out.

No meaning is 'produced'. Unless the buyer privileges either the Government's statement, or the meaning *Belair,* the formal effect of giving the name *Belair* to the cigarette is to cancel the Government's message. Of course, the meaning of the proposition implied by the name *Belair* is also cancelled. *Belair* remains as a name, and a set of connotations, that is all, though the cigarette manufacturer, like the Government, can try to get the buyer to privilege its statement against that of its opponent, in which case no cancelling of meaning occurs. But, apart from this, the effect of giving the cigarette the name *Belair* is to repulse and reduce the prior critical discourse of the Government.

Marcuse also refers to *telescoping* and *abridgement* of discourse as means by which rational thought with critical concepts is rendered difficult. He writes of this process of telescoping and abridgement that it 'cuts development of meaning by creating fixed images [which "militates against the development and expression of concepts" p. 95] which impose themselves with an overwhelming and petrified concreteness' (1964, 91; compare Barthes, 1972). Perhaps the best example of this process is the photo-journalism in which one is presented with the 'picture which sums it all up'. Of course, the picture is captioned to make sure that there is no misreading of it. But, very literally, meaning is reduced to an *image.* In political thinking, I think that this sort of photo-journalism encourages the reduction of structurally very complex situations to the level of exemplification of very general, a-historical and non-operational concepts, such as 'trouble', 'violence', 'fear', 'hunger', 'bewilderment', etc., all of which have their Faces. Such photo-journalism never improved anyone's understanding of the realities or complexities of political life. Its images fix understanding at the level of surface appearance.

Such practices as those discussed above can become important social phenomena because language, though the socially produced means of thought, is not socially controlled. Increasingly, control over the development of language and its use is held by State institutions, including mass media, and monopolistic private enterprise, as in journalism and advertising. Orwell's *1984* developed the possible consequences of the State's domination over language. The semiologists, who have studied the same kind of linguistic developments as those which interest Marcuse, have sometimes failed to appreciate the possibility and existence of class or other minority control over language, whilst recognizing that minority groups are responsible for the creation of sign systems and fixed combinations of signs in such fields as furniture and clothing. Even Barthes, on whose work Marcuse draws extensively, can write:

> In the linguistic model, nothing enters the language without having been tried in speech, but conversely no speech is possible . . . if it is not drawn from the 'treasure' of the language. . . . But in most *other* semiological systems, the language is elaborated not by the 'speaking mass' but by a deciding group. In this sense, it can be held that in most semiological languages, the sign is really and truly 'arbitrary' since it is founded in artificial fashion by unilateral decision (1967, 31; my italic).[1]

But isn't the situation of 'most other semiological systems' also increasingly true of natural language? Is the language of politics really elaborated by the 'speaking mass'?

Beyond the question of vocabulary, and the effects on it of the way in which it is used, there is the question of how a given vocabulary is organized in the individual's mind, and how this in turn affects the possibilities of thought. What

I mean by this can be illustrated by an example from Vygotsky's psychology. In his *Thought and Language*, Vygotsky points out that 'A child learns the word *flower*, and shortly afterwards the word *rose*; for a long time the concept "flower", though more widely applicable than "rose", cannot be said to be more general for the child. It does not include and subordinate "rose" – the two are interchangeable and juxtaposed. When "flower" becomes generalized, the relationship of "flower" and "rose", as well as of "flower" and other subordinate concepts, also changes in the child's mind. A system is taking shape' (1962, 92–3). Vygotsky has already indicated what he takes to be the significance of this development: 'To us it seems obvious that a concept can become subject to consciousness and deliberate control only when it is part of a system. If consciousness means generalization, generalization in turn means the formation of a superordinate concept that includes the given concept as a particular case' (1962, 92).[2] Two further points need to be made before the significance for political thinking of such phenomena can be indicated. First, that relations of superordination and subordination develop as a result of socialization and not as a result of some inner maturation process, proceeding independently of the particular social environment.

Vygotsky himself stresses the significance of formal instruction in school subjects, arguing that through school instruction concepts are learnt from the start in relations of superordination and subordination, and that this catalyses a similar development of the organization of concepts which the child has learnt 'spontaneously': 'It is our contention that the rudiments of systematization first enter the child's mind by way of his contact with scientific concepts and are then transferred to everyday concepts, changing their psychological structure from the top down' (p. 93). This also implies that there is nothing inevitable about the development of conceptual organization.

The second point to be made is that not all the words of a natural language are organizable into the 'trees' which can always be constructed for scientific words, and which – perhaps – makes them scientific. Lyons, who calls subordinate words 'hyponyms' (thus, 'scarlet', 'crimson', 'vermilion' are co-hyponyms of 'red' (1968, 454–5), writes that

> The main point to be made about the relation of hyponymy as it is found in natural languages is that it does not operate as comprehensively or as systematically there as it does in the various systems of scientific taxonomy . . . The vocabularies of natural languages tend to have many gaps, asymmetries and indeterminacies in them' (1968, 456).

– something which is explored at length in Wittgenstein's later writings (1958).

I think that these psychological and linguistic theses are relevant to the question of the possibility of different sorts of political thinking. I think that even if relevant political words are learnt, they need not be organized hierarchically or systematically, even in an adult's mind. In consequence, they can remain wholly or partly a-conceptual. If this is the case, it necessarily affects the possibility of understanding discourse which employs them as concepts. Lyons seems to make a similar point in a 'neutral' context, though it depends on how one reads the 'as for instance':

> It may be impossible to determine and perhaps also to know the meaning of one word without also knowing the meaning of others to which it is 'related' – as for instance, *cow* is to *animal* (1968, 409).

Let me now try to illustrate this line of argument with an example from the realm of political discourse.

Suppose that the concept of *anarchy* or anarchism, which would be involved in any proposition about the possibility or features of an anarchist society, is (partly) defined as being a *society without a government*. A verbal or conceptual tree[3] built up from the elements 'anarchy', 'society', 'government' and expanded to include the co-hyponyms of 'government', looks like this:

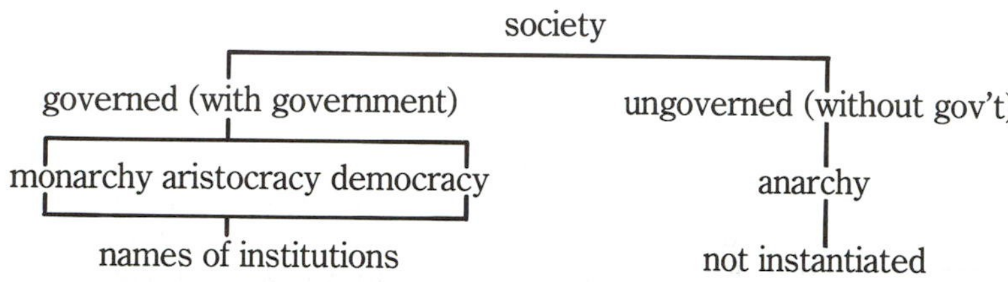

Within the hierarchy presented by the diagram, the higher up the tree you go, the greater the degree of abstraction, though all of the concepts above the level of the names of institutions are abstract ones; they are concepts rather than complexes in Vygotsky's sense of 'complex', which is that a complex word is one for the application of which there is no finite set of necessary or sufficient defining characteristics; in Wittgenstein's terminology, the members of a complex possess no more than a *family resemblance* (1958, paragraphs 66, 67).

Now, I think that though the word 'society' is generally learnt, it is frequently not organized in a person's mind into the kind of tree given in the diagram; it can be and is learnt and retained a-conceptually. The understanding of 'government' which such a person would then have would not be one which related 'government' to 'society' but one where the concepts of types of government or the names of instantiations of governments were used to give it meaning. This would entail that 'government' was understood not as a concept, but as the family name of a collection of particulars. It would be understood as the name of a complex.[4]

Consider now what is required for 'anarchy' to be adequately understood. My theory is that it cannot be adequately understood except when organized within a tree which extends up to and including the level of 'society'. For, first, the concept of 'anarchy' cannot be understood by reference to its own instantiations, since there are none. And, second, the definition of 'anarchy' includes reference to both 'government' and 'society' and it cannot be understood on its own level as a simple *absence* of monarchy, aristocracy, etc. Monarchy, aristocracy, etc., are linguistic co-hyponyms, but I do not see how the concept of 'anarchy' can be produced by opposing these concepts to the word 'anarchy' unless there is already an implied reference to 'society'. In other words, the meaning of 'anarchy' cannot be generated or understood within an opposition to its co-hyponyms. It requires placing in a system which includes not only oppositional features, but superordination as well. If this is correct, my understanding (which may be inaccurate) of structuralist theories of meaning leads me to the conclusion that their account of the generation and understanding of meaning requires revision. For, as I understand such theories, their oppositions are made at only *one* level – indeed, the idea of *opposition* implies that of being on the same level (see Saussure, 1966).

But this point is an aside to my main object in this section, which is to suggest a theory of the following sort:

I am suggesting that adults can possess, and some do possess, a vocabulary in a particular area (I am using the case of political vocabulary) without having that vocabulary conceptually organized, with consequences similar to those which psychologists report for the non-conceptual organization of children's vocabularies. Not only would the existence of such non-organization explain failure to understand the meaning of terms, though those terms might be known, and thus explain failure to understand or generate the relevant sorts of discourse, but it would also explain such phenomena as insensitivity to contradiction which has frequently been remarked in adults for specific areas of discourse. Thus, in their case studies, Laing and Esterson remark the insensitivity of schizogenic parents to their own self-contradictions. Studies of the 'authoritarian personality' report the same thing (see Reich 1970, 1971) as do McKenzie and Silver in their study of working-class Conservatism (1969, especially 120, 121). In this last case in particular there are other possible explanations of apparent self-contradiction. Thus, limitations of vocabulary may lead a person to use contradiction as a means of conveying meaning for the explicit conveyance of which he lacks the necessary vocabulary. There is no self-contradiction in the bad sense when I say of a taste difficult to define 'It's sweet and yet it isn't.'

Classical cognitive-defect theories of schizophrenia also comment on insensitivity to contradiction in schizophrenics, and I think these studies are particularly valuable. (See Hanfmann and Kasanin, 1942, and Kasanin, 1944.) For whilst, as Eluned Price told me, these studies have largely been discredited as studies of phenomena specific to *schizophrenic* cognition (see, for example, Wason and Johnson Laird, 1972, chapter 18),[5] this discrediting takes the form of showing that what is allegedly related to schizophrenia is, in fact, related to level of formal education received. It was possible to think that features of schizophrenia were being described simply because most hospitalized schizophrenics have a low level of formal education. Theories about the 'schizophrenia' of culture thus turn out to have a firmer basis than simple analogy. For the features of 'schizophrenia' which originally prompted such analogies turn out to be non-specific to that state; they are features to be found in the thinking of a large proportion of the population.[6]

There may be objections to my procedure of using material from child psychology and psychopathology to understand sane, adult people, who are neither children nor schizophrenics. But there is no reason why they should not share characteristics with children and schizophrenics, and if the latter are better studied than the former (most psychological studies of 'normal adults' in fact use undergraduates as subjects), this is an added reason for using such studies as a jumping-off point. There may be theoretical objections to such a procedure, but this is different from a simple dislike of it, which can only be founded on a contempt for children and 'schizophrenics'. That is simply the chauvinism of the man in the street, defined as sane, adult, white, male and middle class. So I am not too worried by this dislike. I may well be wrong in thinking that some adults, as a result of their socialization, have important characteristics in common with children and schizophrenics. I am bound to think that the theory is at least worth testing.[7]

To return, then, to Vygotsky. In the report of an experiment, Vygotsky tries to show how children regard the name of an object as a property of that object, and thus lack a fully formed appreciation of the nature of symbolism: writers like Cassirer (1944) and Goldstein (1963) would say that the child lacks the *abstract attitude.*[8] To demonstrate his point. Vygotsky confronts the child with a situation in which the name of an object is arbitrarily changed; a dog is henceforth to be called a 'cow'. For most adults, I presume that such an arrangement would produce no substantial difficulties, and about the first thing a Philosophy student learns to recognize explicitly is the arbitrariness of the word-sign. I doubt that an average adult would respond to Vygotsky's questions as does the typical child, who, having been told that a dog is henceforth to be called a 'cow', responds to questions in the following way:

[Experimenter]	–	Does a cow have horns?
[Subject]	–	Yes
[Experimenter]	–	But don't you remember that the cow is really a dog. Come now, does a dog have horns?
[Subject]	–	Sure, if it's a cow, if it's called a cow, it has horns. That kind of dog has got to have little horns (1962, 129).

Even if this particular experiment was badly designed and conducted, I don't think it follows that the value of such an experiment is destroyed by the claim (made by objectors to the above argument) that the child is more intellectually sophisticated than the experimenter.[9]

I presume that most adults asked whether a dog, arbitrarily renamed 'cow', would have horns, would reply that it would not (Is that your reply?), but I also presume that it does not follow that they would get the answer right whatever the *kind* of word change involved. I think that the ability to perform this kind of intellectual operation could be content-specific. And where the situation is a non-experimental one, from which the motive either to please or displease an experimenter is absent, I think that sane, adult people do sometimes perform operations like those of the child in Vygotsky's experiment. I suggest that in some areas, some adults do treat a word as the property of an object. Aaron Sloman remarked, in comments on draft material for this chapter, that people can be conceptually agile in some domains and strait-jacketed in others. I assume this is so. I shall illustrate the kind of phenomena which the assumption covers, and I shall draw my examples from the realm of political discourse and understanding.

Consider the following case.[10] I come across a person who argues in such a manner that it is clear that he believes that if something is conventionally called a system of Justice, then it must have the property of being Just, again as conventionally defined. This entails such consequences as that, confronted with empirical evidence that an existing institution of Justice contravened in its practice the *conventional* norms of Justice (that is, it is unjust in its own terms), such a person could not admit that such evidence *might be true.* The possibility of empirical evidence being relevant is ruled out *a priori,* and this entails that for this person it is tautologically true that if something is called a system of Justice, then it must be Just. This is not quite the same kind of case as that where the practice of existing institutions, whatever it may be, *defines* what is right, for I am not denying that this person has a concept of Justice defined independently of practice.[11] I am claiming that this person cannot admit that the word 'Justice' could be erroneously applied. This seems to me analysable as a case in which the arbitrariness of the word-sign 'Justice' is not recognized, that is, a case where the word-sign is treated as a property of the object. Hence, any discussion which contrasts Reality with an independently defined Ideal is impossible, which means that any rational discussion is impossible. Though this example, based on a single conversation, may seem extreme (and does leave many questions unanswered), I think there is a widespread tendency to use political concepts in the above fashion. Marcuse sensed it and produced an analysis in terms of one-dimensional thought.

Take another example, a little easier to describe and substantiate from experience. To me it seems that some people are unable to understand, or argue in terms of propositions involving explicit counter-factual conditionals, though again such inability need not be across-the-board; it can be content-specific. What I mean is that some people respond to arguments of the sort which begin 'Suppose you had . . .' with a straightforward 'But I didn't . . .' It has been put to me that this *genre* of response does not show people's cognitive limitations, but rather their cognitive sophistication. They refuse to be drawn into the kind of hypothetical argument in which the most skilled in dialectical debate necessarily wins. Now, whilst I agree that some people use the response 'But I didn't . . .' as a refusal, I don't accept that this is always the case. My reasons are two-fold. First that one elicits the response 'But I didn't . . .' even in circumstances where the respondent recognizes that it is in his interest to understand the hypothetical argument (as in direction giving: 'Imagine you're standing at the corner of Charing Cross Road and Oxford Street, with Centre Point on your right . . .' Second, that I find it inconsistent that people should be so intellectually sophisticated as my opponents' argument makes out, yet so frequently be taken for a ride by all sorts of con-men, from politicians upwards.

I have produced only two examples, but I am sure others can produce further instances of linguistic or logico-linguistic ways in which adults with little formal education and no experience of other relevant learning situations, can be inhibited or prevented from thinking certain thoughts and understanding certain sorts of discourse or argument. Such incapacities are content-specific and can be given a social explanation. The incapacities may not always be apparent, not least if the person consciously or unconsciously seeks to hide them. Thus, for example, when he does not understand an idea, he may say that it is 'Nonsense'. Roland Barthes has commented on this particular mode of response as a characteristic of petty-bourgeois thinking (Barthes, 1972, especially the essay *Blind and Dumb Criticism*, 34–5). Enoch Powell seems to employ this strategy of labelling as 'Nonsense' that which he does not understand (Wood, 1970, especially *The Enemy Within*, 104–12) though here it is difficult to distinguish the use of 'Nonsense' as an emotive synonym of 'False' from the use of 'Nonsense' to disguise lack of understanding.[12]

The reader could try to add his own examples and analyses to my list. Why is it, for example, that some people reply to Why? questions about causation with restatements of facts?

Theory

Part of my ignorance of physics consists in not knowing the substantive theories of physics. It is out of the question that I should re-invent such theories myself – at least, any but the most elementary ones. Quite aside from the material resources needed to set up and conduct experiments, one person does not have the time or intelligence to re-invent the product of centuries, if not millennia, of collective work. There is no practical alternative to reading other people's books, going to other people's lectures, and using apparatus under other people's supervision.

There are many theories of politics, both scientific and normative. Most people have never had any formal instruction in any of them, let alone the ones, like Marxism, which can help them understand and act upon their own political situation in the world. The *omission* of political education in *theory* from the curriculum of schools is more important, I think, than any substantive instruction they administer to their pupils, such as 'British Constitution', the ideology of which seems to consist in presenting the abstract formal description of institutions as a model of their actual functioning. Again, such political theories as the media present seem to me either to be of too high a level of abstractness, or to be trapped in the concrete example. It is the middle range of the theoretical concepts which is *missing*, and this quite aside from any overt *bias*. The abstractness of media concepts consists in the use of reified 'forces' such as 'progress', 'reaction', 'order', 'disorder', etc., as explanatory concepts; the concreteness of the media consists in the description of social reality at the level of isolated *events*. Of course, it is at the most abstract and most concrete levels that there is least disagreement (at least overtly so) and the resort to abstract and concrete may be the means whereby they discharge their obligation to be overtly unbiased. The middle range of concepts (which includes sociological concepts like 'class', 'status', etc.) is the area of greatest overt controversy and it is therefore *excluded*. (The exclusion of substantive *contents*, perhaps the most important way in which bias operates, is a different issue.)

Professional politicians are possibly even worse as sources of theoretical understanding of political reality. Their position depends on limiting understanding of politics and their own role within it. The fact that the speeches of very few politicians bear reprinting is some indication of the absence of theoretical (even informational) content from what they say. The reading public has nothing to learn from them, only about them: the Collected Speeches of Edward Heath or Harold Wilson could be used only to damn.

To the degree that people feel the need for explicit political theories, I think they are left very much to their own devices. The vocabulary of politics is determined externally, and certain ideologies of voting are propagated, but for the rest, politics, far from being the field *par excellence* of deliberately propagated ideology, is left to a great degree to be occupied by *spontaneous ideologies*.

These spontaneous ideologies reflect closely the immediate circumstances of daily life and do not transcend them; they take appearance for reality and concrete incident for theory. They reify the existing order of things. Thus, racism and working-class conservation. Reification involves at least three processes absent from radical and revolutionary theories. First, there is no theoretical recognition of the historical character of the present social order. Even if there is an awareness of the historical emergence of the present system, that system is seen as the culminating goal of the historical process beyond which no substantial change is possible. Second, there is no conscious awareness of the dependence of social change on human action, whether collective or individual. This is true of religious conceptions in which 'God has his plan', of technocratic conceptions in which technology has its own immanent plan, and in deference views in which 'You've got to have the people with money' – where a contingent feature of society, that wealth is the dominant source of power, is reified into a necessary feature of social existence. Third, the needs, wants or interests which might stimulate action to change the social order are repressed and sublimated. Wilhelm Reich theorizes in detail about the mechanisms by which this is accomplished (1970, 1971).

The argument of this chapter has been that many people are linguistically inhibited from thinking theoretically about politics, and if this is correct it helps explain the weakness of spontaneous ideologies. For essential to theory (which is what such ideologies aspire to) is the employment of elaborate causal and functional concepts and the corresponding vocabulary. In my teaching, I noticed that when asked for explanations of social phenomena, apprentices often redescribed the phenomena or restated the question minus the interrogative (e.g., Why is there a housing shortage? There aren't enough houses. Compare Halliday, 1969, 35). There was, at least, a lack of readiness to volunteer causal explanations, though no lack of readiness to offer answers. In terms of theory, I have read and been influenced by the work of Basil Bernstein, but I do not think that my arguments depend for their validity on acceptance of his theory of restricted and elaborated linguistic *codes*. For Bernstein, it is a defining feature of a restricted code, to which the lower working class (at least) is allegedly confined that it lacks 'elaborate causal conceptions' (1971, 47). I agree that elaborate causal conceptions are lacking, but doubt that the lack is necessarily or empirically systematic enough to merit the use of the concept of 'code'.

If there is such a thing as a restricted code, lacking in elaborate causal concepts, then I think its existence would have to be explained by reference to the fact that some people think they have no *use* for elaborate causal conceptions. They may have reconciled themselves, for example, to their real exclusion from politics (see section following). Such privatization would explain the use of a restricted code, as Bernstein himself indicates (1971, 147), though I would expect a dialectical relationship between code and circumstances: that is to say, that whilst the circumstances produce the code, the code strengthens the grip of circumstances.

Practice

> I don't think explanations are needed of people's conceptual limitations. What is needed is an explanation of how such limitations are ever broken out of. Aaron Sloman, in remarks on draft material for this book. (Compare Wertheimer, 1961).

If I were to make the attempt to learn physics, I should need a motive (and a pretty strong one) if I were to make any progress in the subject. I might be driven by ambition, pure or idle curiosity, or necessity. It is unlikely that I shall ever be strongly enough 'motivated' to learn physics.

Could the relation of people to the understanding of politics be similar? A motive is needed equally for someone to set out to understand politics, but it would seem, at first glance, that such a motive is always and universally present. For the life of every person is inescapably a political life, and chances of being happy and free depend in direct and important ways on the form of society within which a person lives. Failure to 'recognize' such facts is precisely what constitutes the psychological side of real alienation and political exclusion (see my 1973b).

Because of real and psychological alienation, the ever-present motive for understanding politics does not 'surface' in the form of an interest in politics, whilst alienation does surface in the flight[13] from politics into 'private' modes of existence, doomed to defeat because based on belief in a non-existent possibility of privacy. Rather than challenge a real state of affairs, or the psychological states which it generates, people try to escape them both. Perhaps they can conceive of no other way out. Certainly, their attempted solutions only compound their powerlessness and sense of powerlessness; but why might they not be able to conceive of any other way out?

On the one hand, I am sure that some people make a rational calculation and conclude that change is practically impossible. On the other, I am sure – as indicated in previous sections – that there are people who can't do or don't envisage radically different situations, let alone believe that it is possible to bring them about. The non-existence of fully acceptable alternatives elsewhere in the world is a potent factor in confirming both these perspectives, which is why it is far from irrelevant for left-wing organizations to devote a great deal of intellectual effort to analysing the structure of the [former] Soviet Union, China, Cuba, etc.

Again, it is not as if the work or the family situation provides alternative experience which might be applied critically to the political realm from the point of view of changing it. If anything, work and family are less democratic situations than is politics, though, interestingly, the family situation at least produces reactions in children structurally very similar to the reactions of adults to the political situation in which they find themselves. For example, in their studies of the families of schizophrenics (many of the characteristics of which must be regarded as typical rather than deviant, at least of practice if not of ideological norms), Laing and Esterson (1970) show how children respond to denial or prohibition of their practice, or invalidation or negation of their own self-definitions, by *withdrawal* (which becomes catatonia in the clinical stage). When there is no way out through the door, or there seems no way out, or where they lack the means to fight back (including the cognitive means), children may withdraw as the only alternative to complete submission. It resembles political apathy or privatization in being a fugitive practice in response to a denial of the *need* for self-determination.[14]

Last but not least, changing the world demands a great deal of time, effort, strength and courage. Serious efforts to change society frequently encounter severe repression. Who'd be a revolutionary?

The powerful, over-determined character of the obstacles to political action and knowledge does illustrate the meaning and realism of the Marxist perspective which emphasizes not an abstract political-educational effort as that which generates a political movement, but, rather, the force of economic circumstances, and the needs and interests which economic developments simultaneously generate and frustrate. But if this is so, then the greatest of the obstacles to political action would then appear to be those practices which encourage people to accept frustration, or make them fearful of fighting for their needs, or lead them to repress their self-knowledge of their own frustrations. Earlier I discussed and criticized the established assumption that men *will* act to satisfy their needs and further their interests. The most difficult problem facing radicals and revolutionaries is that where men do not or will not act to satisfy their needs, where they accept frustration or repress their knowledge of it.[15]

This is why I think the writings of Wilhelm Reich deserve such careful study, and also why I think that any political movement today must itself be a source of satisfaction to its members and not purely a sacrifice. There should be nothing 'religious' about a political movement.[16] Activities like theatre, film and music should be neither decorative features of a political movement or narrowly instrumental. For in them people may come to know their own desires and become willing to act collectively to satisfy them.[17]

Conclusions

I have compared politics to physics, but I do not wish to encourage the scientization of politics in the sense that that phrase is understood by writers like Habermas. That is, I do not wish to regard politics as a sphere in which exclusively instrumental problems arise; nor do I wish that political control should be vested in 'experts'. On the other hand, I do believe that theory is needed to understand what is going on in political society, and I come reluctantly to the conclusion that people will always remain unequal in the level of their theoretical sophistication, and that, in consequence, there will be theoretical leaders. But they need not be leaders in everything else as well; and if they claim to be, and enforce their claim, the effects are often enough disastrous. Leaders can also be subject to greater control and enjoy less permanence than they now do.[18] Yet it still remains true, I think, that people cannot spontaneously generate the range and depth of political knowledge required for them to function effectively as political agents. In this sense, I think my heuristic analogy between physics and politics is not so misleading, and I have to come down, against some of the anarchists, on the side of Lenin.

My ignorance of physics is historically explicable and of no importance. Comparable ignorance about politics is common, explicable and important. I think that the theories I have sketched in this chapter are relevant to political activists in their daily work, and that the effort to confirm or refute them would be worthwhile. This book would more than have achieved its purpose if some of its remarks were found useful by those engaged in the nitty-gritty work of achieving radical social change.

9

The structures of speech and writing

Gunther Kress

From Kress, Gunther 1982: *Learning to write*. London: Routledge, 22–33.

The immediate presence of the audience in the case of spoken language is reflected in one of its most significant characteristics. Superimposed on the clausal structure of speech (which will be discussed below) is a structure determined by the speaker's wish to present information in the form most accessible to the hearer. Consequently the speaker parcels up by intonational means the information which he or she wishes to present to the hearer in what the speaker regards as the relevant chunks or units of information. This structure of *information units* derives from the interactional nature of speech, and encodes some of its major features, namely the exchange of information from one participant to the other, the mutual construction of meaning, and the mutual development of a topic. It also takes account of the transient nature of speech. A hearer cannot usually pause or check back over the message to make sure that he or she has understood it. The structure of speech makes allowance for this factor by structuring information, providing sound cues which highlight the relevant informational structuring.

This structure is expressed through intonation. The speaker uses intonation to bracket together segments of his utterance which he regards as constituting one relevant parcel or unit of information. The function of intonation in providing clearly marked structures in speech has been the subject of research by a number of linguists, foremost among them Michael Halliday, whose innovative work has in turn been taken up and furthered in important directions by David Brazil. Halliday calls the basic unit of the spoken language the *information unit*. It is expressed intonationally in the *tone group*, a phonological entity. It encodes what the speaker wishes to present as one segment or unit of information. It is marked off in speech by a single unified intonation contour, and above all, by containing one major pitch-movement. The unit has no previously determined length. It may correspond to a clause but need not do so at all. For instance, in the following text, clausal structure and information structure coincide partially. (The speaker is a six-year-old child.)

> //let's have a look in your mouth to make sure//no//like that//no//none// . . . put the bottom lip down// . . . no//it's not your teeth//

(The double oblique strokes are used to mark the boundaries of information units.) The clauses in this brief text are *let's have a look in your mouth, to make sure, like that, put the bottom lip down, it's not your teeth.* That is, there are five clauses, but eight information units. A clausal analysis inevitably strikes problems with items like *no none,* which either have to be treated as clauses by some theoretical contortion, or else fall uncomfortably outside the description. In terms of informational structure there is no problem of description. It is clear from this example that the information units need not coincide with the clauses or the clause-boundaries; for instance, the first information unit contains two clauses.

The information units are motivated directly by the interaction between the speaker and other participant. Some response on the addressee's part causes both the specific content and the structure in which it is presented. The speaker can monitor the hearer's reaction from moment to moment, and adjust the informational structure accordingly.

The information units are, as mentioned, segments of speech which have (are spoken with) one unified intonational 'tune'. Each unit has an internal two-term structure, of 'known' and 'unknown' information. That is, within each unit the speaker represents some part of the information contained in the unit as being unknown to the hearer, the other part by implication as being already known to the hearer. The unknown part is marked by intonational prominence, in that the greatest and most pronounced pitch-movement occurs at the beginning of the unknown segment. So in the example just given

> //let's have a look in your mouth to make sure//

sure receives intonational prominence with a noticeable fall in pitch, thus marking it as informationally unknown. The whole of this text has a known/unknown structure as shown here (with the unknown segment underlined).

> //let's have a look in your mouth to make sure//no //like that//no//none// . . . put the bottom lip down// . . . no//it's not your teeth//

It can be seen from this that the unknown segment tends to occur at the end of the information unit. An exception is the last information unit, where the second from last segment is prominent. Where the intonation prominence falls other than on the last word, the prominence is noted by the hearer as unusual or marked, and is interpreted as being special in some sense. In this example a contrast is established between *your* teeth and *other* teeth (in this case, the speaker's teeth). From this text it is also clear that information units need not include anything other than unknown information. The known, by definition, can be left unsaid.

The two-part structure implies that there is always some common ground, some known information, between speaker and hearer. The unknown must be said, the known may be said or it may be left unsaid and implicit. It is, of course, the speaker's judgement which determines that something is shared knowledge, and what that shared knowledge is. It is also not clear from the speaker's decision to treat some information as known just how well known this may be to the hearer. David Brazil makes a similar point in *Discourse Intonation*:

> Another way of expressing the same distinction is to say that some parts of what a speaker says merely *make reference* to features which he takes to be already present in the interpenetrating worlds of speaker and hearer. Others have the status of *information* in that they are presented as if likely to change the world of the hearer. (p.6)

Brazil is here introducing a twofold distinction into the element which Halliday regards as unknown. In doing so he draws attention to the fact that it is the speaker's decision to present content either as though it belonged to the shared knowledge of speaker and hearer, or was drawn from outside that area.

In this example the ***known*** is (among other, much more taken for granted information) that the previous conversation has been about losing teeth, and about whose teeth may have been replaced with a coin by the fairies. *Mouth, looking in one's mouth, your*, etc., are all assumed to be known to the hearer as they have been the topic of the preceding conversation and the shared experience of speaker and hearer. What is new, and hence 'unknown' to the hearer, is the suggestion to make *sure*. Of course, the speaker may be wrong in his assumptions about what is and what is not known to the hearer. In that case there will be some perplexity on the hearer's part, a momentary check in the flow of the interaction. Also, the speaker may be deliberately misconstruing and misrepresenting the hearer's knowledge, and by presenting information as known may attempt to coerce the hearer into accepting this insinuated assumption.

In this extract there are a number of information units which consist only of unknown information. That is not unusual. An information unit must consists of an unknown item of information, and it normally includes a known segment of information as well. To some extent the known segment is redundant, by definition. However, in interactional terms the ***known*** is not redundant, it serves as a link between the two participants: it provides them with common ground and shared knowledge, which serves as a starting point, a bridge almost, for each speaker in turn. (That is, making reference to 'the interpenetrating worlds of speaker and hearer'.) Thus while the known segment is informationally redundant, it is most important from an interactional point of view. Features of this structure of speech have been regarded as faults in speaking. Some parents and teachers have a habit of correcting children who give 'monosyllabic' answers, insisting that children should always speak, and in particular answer, in full sentences. This clearly rest on a misunderstanding of interactive language, as well as on the assumption that sentences are the basic units of spoken language so that any utterance which is not a sentence is defective. That assumption is incorrect, and the view could only persist because of the absence of research and hence understanding of the structure of spoken language.

Many of the interactional aspects reflected in the structure of spoken language are clearly absent from the context in which written language occurs, so that there is no need for linguistic forms to express them. However, written language does have a need for emphasizing particular items; equally, it must have the means for indicating the writer's topical and thematic organization of his material. The written language does not, of course, have intonation available for these purposes (other than parasitically by getting the reader to superimpose intonation – on italicized, underlined items, or words and phrases in inverted commas, or by the use of exclamation marks, etc.) Instead, written syntax uses sequential ordering of items or more thoroughgoing restructurings to express these meanings. Emphasis may be expressed by placing an item out of the normal syntactic order: *John she couldn't stand*, or *couldn't stand John, could Mary* (from *Mary couldn't stand John*). The writer may restructure the sentence more thoroughly: *It was John whom Mary couldn't stand.* These reorderings and restructurings still employ intonation indirectly, in the sense that these forms invite or even determine a particular reading which involves a more or less silent intonational implementation and actualization of the reading. However, even less drastically reordered or restructured sentences allow the writer to organize his material in a manner which presents his content in a particular fashion. All sentences display a two-part structure. One part, the theme, announces what the writer has chosen to make the topic of the sentence; the other part, the rheme, consists of material pertaining to that theme, furthering and developing it in some sense. In the sentence above, *All sentences display a two-part structure, all sentences* is the theme and *display a two-part structure* is the rheme. That is, the first syntactic constituent of a sentence is always theme, the rest rheme. Consequently theme is not another label for subject; frequently the two will coincide, but often they will not. In the preceding sentence *consequently* is the theme, and *theme is not another label for subject* is rheme.

Hence the textual structuring of speech and that of writing proceed from two distinctly different starting points. The structure of speech starts from the question: 'What can I assume as common and shared knowledge for my addressee and myself?' This question, and its answer, are at the basis of the structure of speech. Writing starts with the question: 'What is most important, topically, to me, in this sentence which I am about to write?' This question, and its answer, are at the basis of the structure of writing. The cohesive and continuous development of a topic is thus paramount in writing, while the construction of a world of shared meaning is paramount in speaking.

The immediacy of the interaction and the speaker's awareness of the hearer lead to another major structural difference between speech and writing, a clausal structure which is both evidence of immediate thinking, in that the speaker does not have time to assemble complex structures, and evidence of the needs of the hearer as the recipient of information. Speech is structured by sequences or 'chains' of clauses, which are, generally speaking, co-ordinated, or else adjoined without any co-ordinating particles (rather than subordinated or embedded). An example of a co-ordinated structure is *let's have a look in your mouth and make sure,* where the two clauses have equal status, and are joined by a conjunction. An example of an adjoined structure is *let's have a look in your mouth(,) make sure,* where the two clauses have equal status, and are adjacent but without a conjunction. A subordinating construction would be *when we look in your mouth we'll make sure,* where there is a main clause, *we'll make sure,* and a subordinate when-clause. In the text from which the example is taken the structure is an embedding one, *let's have a look in your mouth to make sure,* where *to make sure* is syntactically integrated into the main clause *let's have a look in your mouth* and forms a syntactic constituent of it. 'Adjoined' is, however, open to a misinterpretation, namely that it means 'simply adjoined, adjoined without indication of the structural and semantic relationship of the clauses.' Such an interpretation is incorrect. The absence of a co-ordinating particle is possible because of the use of intonation to provide the structural connections. The pitch movement and the pitch height at the information unit/clause boundaries give a precise indication of how the following clause connects both structurally and semantically with the preceding clause. The direction of the intonation contour in the new clause makes this structural connection, and further develops it. The pitch movements involved in this are minute but absolutely clear indications of the structure. Taking a further extract from the text already discussed as an example (numbering the information units)

> //No[1]//we got them[2] separate//see[3]//the big[4] one// we got him from my// . . . uh . . . // we got him from my[6] brother's friend//my oldest[7] brother//

1 would have a falling intonation to a medium low position with a slight rising 'hook'; 2 would begin at that level, rise slightly, fall on *separate* to a medium low position; 3 would be level, at that height; 4 would start at a medium high position, fall on *big,* continue to fall on the first part of *one,* with a slight rising hook at the end; 5 would start at that level, continue to rise slightly to medium level on *my; uh* would be on that level; 6 would continue at that level, rise steadily to the end of *brother's,* at a medium level, and fall on *friend,* to a medium low level; 7 would start at the same height as *friend* and fall on *oldest,* to a full low position (without final rising hook) on *brother.* This over-arching link is not broken until an information unit ends in a high rise without any downturn, or in a full fall without any upturning hook, and where the following information unit is not an exact intonational copy of its preceding unit. For instance, the fact that 7 starts at the same height as the highest point of 6 is interpreted by the hearer as indicating that 7 is a gloss on the item which had the same height in 6, that is, that it is offering further information and elaboration on aspects of one element of 6, namely that with the intonation (and information) focus.

Consequently, far from the linking structure being missing or weakly articulated, the structure and the meaning of clausal connections in speech are highly articulated and capable of the finest nuance. In writing, this structure is mirrored to some extent in the system of punctuation, which is not, however, capable of expressing the same detail and precision. In *Discourse Intonation,* Brazil discusses a related feature, namely the semantic connectedness of tone-groups (that is, information units) in discourse. He points out that tone-groups are spoken in one of three *keys,* that is, relative pitch-levels: mid, high, or low.

> Any occurrence of a high-key tone group can be thought of as being phonetically bound to a succeeding tone-group; any low-key tone group as bound to the preceding one. The former carries the implication, 'There is more to follow'; the latter 'This is said in a situation created by something that went immediately before'. In discourse we can say that one sets up expectations, the other has prerequisites. (p. 10)

This suggests that in addition to the linking discussed above, there exists another, superimposed level of intonationally expressed linking, relating the content of tone-groups in the way indicated by Brazil.

Topic development in speech, within this clausal structure, is by sequence, restatement, elaboration and intonational articulation. The evidence of 'thinking on your feet' is everywhere evident in speech. This is in contrast to writing, where there are, typically, no traces of immediate thinking. Writing is the domain of circumspection, of (self-)censorship, reworking, editing. The development of the topic in writing is by another order: not by sequence but by hierarchy. That which is more important is given structural prominence, the less important is structurally subordinated. Consequently writing is the domain of a more complex syntax, typified by the sentence, by subordination and embedding of various types, by syntactic and conceptual integration. Speech is typified by the syntax of sequence, of the clausal chain, of addition and accretion. To the extent that there is circumspection and self-censorship in speech, it emerges in the form of hesitation phenomena. The greater the need for circumspection, the more prominent hesitation phenomena are likely to be; the less need for care, the less likely hesitation phenomena are to occur. Bernstein and others have noted that hesitation phenomena are more common with middle-class than with working-class speakers.

Perhaps the most obvious and most characteristic unit of the written language is the sentence. It features prominently in linguistic theory (it also features quite prominently in folk linguistics), and we tend to assume that we know what a sentence is. In fact, and this is also well known, there is no agreement in linguistic theory on the definition of sentence. I wish to say that the sentence (whatever its definition) is not a unit of informal spoken language, but that it is the basic textual unit of the written language. Because of the influence of writing on speaking and because of its higher status, the assumption is that speech is also organized by the sentence as its basic linguistic unit. I think that is not correct. There is much evidence, for instance from the description of Australian Aboriginal languages, which seems to suggest that, in some of these languages at least, there are chains of clauses and paragraphs but there are no sentences. One of the problems in attempting to write these languages down, to make them literate languages and to give them a 'literature', is precisely this absence of the sentence, so that in transcribing narratives and myths the transcriber wants to impose a sentence structure on the narrative; this falsifies the narrative, because in its spoken form it is not structured in sentences. The interesting point is that after a while the original tellers of the myth also want to organize the written versions of the narratives in sentence form. There seems to be something about writing which demands a different syntactic and textual organization from that of speech.

To show the difference between these two forms of the language, here is an example of the same adult (myself) speaking and writing. The spoken text is from the transcript of a lecture given without notes: the written text was prepared from the transcript, for publication. I deliberately kept reasonably close to the spoken style of the lecture; a fully fledged translation into writing would have been very different.

Text 1

Spoken text

Now of course an exhortation to be open in the way we look at things is easier said than done *because* we have all finished I suppose much of our learning *with* most, I think all of us have finished probably all of our significant learning *and* learning has of course positive aspects it has the positive aspect of enabling us to live in the culture that we are born into but learning also has um I feel quite negative aspects um the positive ones as I say are clear enough they enable us to function. In our world the negative ones have also been pointed to um frequently enough, I'll just er perhaps talk about them very briefly in relation to language um the negative aspects of learning I think are concerned with a kind of reduction that goes on with a kind limiting that goes on when we learn cultural things. We come to learn things *and* once we have learned them they seem to be the only way to do things um the way we say things seem to be the natural way to say things *and* so forth.

Text 2

Written text

> Now of course, an exhortation to be open in the way we look at things is easier said than done *because* we have all finished most of our significant learning. Learning has positive aspects, enabling us to live in the culture that we are born into; *but* learning also has quite negative aspects. These are concerned with a kind of reduction, a kind of limiting that goes on when we learn cultural things. We come to learn things *and* once we have learned them they seem to be the only way to do a thing; the way we say things seems to be the natural way to say things *and* so forth.

I have asked a typist to transcribe the tape, which she kindly did. The 'punctuation' of the spoken text is therefore as the typist heard it. Clearly, the main organizing unit of the spoken text is not the sentence. Though it may be unusual to find the structure of speech in the formal register of 'lecture', the structure itself is not at all untypical of speech. It consists of clauses of equal or near equal syntactic status 'chained' together in sequence. This chaining may take the form of co-ordinated clauses, or of main and subordinated clauses, linked by conjunctions such as *and, but, or, if, so, because, though.* That is, the structure of speech is characterized by 'chains' of syntactically relatively complete and independent clauses. For instance, Text 1, lines 3–5: *I think all of us have finished probably all of our significant learning and learning has of course positive aspects it has the positive aspect . . .* – here three syntactically complete clauses are in the chain. Where a clause is significantly changed syntactically in order to make it part of another clause we have examples of either subordinating or embedding constructions; these are less typical of speech. For instance, in Text 1, lines 4–5; *it has the positive aspect of enabling us to live in the culture, enabling us* is a clause which is heavily modified to fit syntactically into its syntactic unit: *to live in the culture* is, similarly, a clause which has been significantly modified to fit into its matrix clause. In Text 1 the chaining syntax predominates, though it is clearly a mixture of chaining and embedding syntax. Examples of the latter are *an exhortation to be open, the way we look at things, enabling us to live, the culture that we are born into.* Informal speech tends to have even fewer instances of embedding than this text.

The written text by contrast is not only much shorter (eight lines compared to fourteen), it has more sentences (four compared to two) and the number of co-ordinated and adjoined clauses is significantly lower. Proportionately, full subordination and embedding is more prominent in the written text (twenty-one clauses, five co-ordinated or adjoined clauses, a ratio of 3.2:1) than in the spoken (thirty-four clauses, thirteen co-ordinated or adjoined clauses, a ratio of 1.6:1). It is worth bearing in mind that this spoken text was a relatively formal one, so that it contains a higher proportion of subordinated and embedded clauses than informal speech. Also, as the written text derives from the spoken one, it preserves some of the latter's informality. A formal written text would show a higher degree of subordination and embedding.

To summarize: speech, typically consists of chains of co-ordinated, weakly subordinated and adjoined clauses; writing by contrast is marked by full subordination and embedding. There are other characteristics too. The spoken text is longer, to make allowances for the different mode of reception, it shows repetition, allowing the hearer time to assimilate information. It also contains many features of an interpersonal kind, referring to the interrelation of speaker and audience.

The discussion so far suggests that for our society linguistic competence includes a number of things over and above those which are normally assumed to be included in competence. The traditional Chomskyan version of competence assumes that a speaker of a language has internalized the rules of sentence formation and transformation of that language and uses them as a basis for the production of an infinite set of sentences. There is also a sociolinguist's definition of competence, for instance that of Dell Hymes (1972), who talks about the rules we use in correctly applying these sentences in given situations. I would like to extend this notion of competence, and speak about differentiated competencies for speaking and writing. These would include at least two additional types of rule. One would indicate how the structure-forming rules are to be applied in either speaking or writing. Further, if *sentence* is not a linguistic category of speech, then the competence for writing would include rules of sentence-formation, and the competence for speaking would not. In either case the competences for speech and writing would include knowledge about strategies of conjunction – predominantly chaining for speech and embedding for writing. However, beyond these, the competence for writing would need to include a second type of rule, which indicates the knowledge of those syntactic structures which occur only in the written mode of language; and similarly for speaking. For instance, if a speaker said *Mary was supposed to do the trifle* with intonational prominence on *Mary*, then the most likely form of this in writing is *It was Mary who was supposed to do the trifle*. This structure belongs to writing; when it occurs in speech it is there as an importation, much like a loan-word from another language. Lastly, there would need to be a differentiated textual competence, as textual structures and processes differ significantly from one mode to the other, in areas such as cohesion, topic development, staging. In speaking, the text is usually constructed in interaction with another participant, whereas typically in writing that is not the case.

In the discussion so far I have proceeded as though all speakers and writers were identical. That is not the case. I mentioned above that access to writing is unevenly distributed in Western technological societies. Some members of these societies have no competence at all in writing (or reading) – the figure for countries such as Britain or Australia is between 10 and 15 per cent. Full competence in writing may be restricted to as few as 10 per cent of the population. Obviously there are strong links between the structures of those societies and this uneven distribution of written competence. Speakers too are differentiated according to regional and social dialects. For certain social groups – the professional classes, for instance – the structure of the spoken form of their dialects is very strongly influenced by the structures of writing. As a result, the difference between the syntax of speech and that of writing is far less for such groups than it is for groups whose dialects are little if at all influenced by the structure of writing. This factor has important implications for children learning to write. For some children the syntax of writing will be more familiar than for others, to whom it may be totally unfamiliar. Hence in a group of children some may start with knowledge which others have yet to acquire. This difference in knowledge is unacknowledged, because the fundamental differences between the two forms of language is not a well understood and widely known fact. Teachers are likely to attribute this difference in the performance of children to differences in intelligence. There exists therefore an initial unnoticed hurdle in the learning of writing on which many children stumble and never recover. They will not be fully competent writers and will be regarded as failures in the eyes of our literate society and in terms of our educational system.

10

Beyond alienation: an integrational approach to women and language

Deborah Cameron

From Cameron, Deborah 1985: *Feminism and linguistic theory*. London: Macmillan, 141–6, 149–61.

Note: Omitted from this extract are some introductory comments on what is meant by 'language' and 'meaning' and a section on 'Women and literacy' in which it is pointed out that the majority of illiterates in the world today are women, and that even where women are literate they are often denied access to particular language registers.

Meaning, understanding and alienation

We may now return to the experience described by Simone de Beauvoir: 'I saw greys and half-tones everywhere. Only as soon as I tried to define their muted shades, I had to use words, and I found myself in a world of bony-structured concepts.'[1] This frustration at being unable to make language express the exact nuances of experience is something we have already encountered in women's own testimony: 'Sometimes when I am talking to people I really feel at a loss for words. . . . A vast number of the words I use all the time to describe my experience are not really describing it at all.'[2] Audre Lorde, a poet, notes that we speak of our experience only 'at the risk of having it bruised and misunderstood.'[3] Difficulty in finding words and difficulty in being understood are often spoken of by women as signs of their alienation, the proof that 'this is the oppressors' language' with meanings and limits defined by men. The solution is to create a new language in which women can express their meanings and be understood.

But if we accept the idea that meaning is complex, plural and ultimately perhaps impossible to pin down, the new language solution appears utopian. There will never be a perfect fit between private experience and linguistic expression, and there will never be perfect mutual understanding. Sociologists of language have long been familiar with the idea that participants in any kind of talk take for granted a degree of comprehension which, when you look more closely at the interaction, cannot ever be quite justified. Thus the problems of expressing oneself and being understood are not exclusively women's problems. They are built into all interaction and affect all speakers. Which is not to say that women do not suffer to a greater degree than men: for the causes of their particular problems, however, I think it is necessary to consider the whole social situation of women and men, and not just their relative positions in an abstract symbolic order.

The notion that perfect mutual understanding – telepathy – is not the normal or the ideal outcome of speaking, frightens and confuses many people. It is clear that without the indeterminacy that stops us communicating telepathically we would not be able to adapt our language to the novel situations we need it for; imperfect communication is the price we pay for a creative and flexible symbolic system. But this important insight frequently meets with a great deal of resistance. If we cannot ever really understand each other, are we not trapped in our own private worlds with no hope of making contact? And is this not the ultimate nightmare of alienation?

This fear, and the comforting certainty that perfect understanding *is* possible, goes very deep. For instance, a well-known myth of human prehistory refers to a time when humans did understand each other perfectly, and this understanding conferred enormous power on them. God not unnaturally saw that power as a threat and destroyed it by destroying the unity of human language.

> And the Lord said, Behold, the people is one, and they have all one language; and this they begin to do: and now nothing will be restrained from them, which they have imagined to do. Go to, let us go down, and there confound their language, that they may not understand each other's speech. So the Lord scattered them abroad from thence upon the face of all the earth.[4]

The perfect understanding which supposedly results from 'speaking the same language' is seen here as an essential prerequisite for any kind of collective action. God, in replacing linguistic unity with linguistic diversity, undermined the power of the builders. Feminists have their own version of the Tower of Babel story. They feel that men have undermined women by confounding their language, the language of their bodies, their unconscious, their desire or their experience. In order to act together, an authentic language of women must be forged. If there is no common language, there can be no true collective action.

I do not feel, however, that the view of meaning I have put forward excludes the possibility of collective action, nor does it negate the communication that obviously *does* occur between individuals. Rather, it says that if we are ever to understand the nature of collective action and interpersonal communication, we must first acknowledge its inherent difficulties and limitations. Until we do acknowledge that communication is to some extent an everyday triumph, until we get rid of our fantasies of what it never can be, we can hardly study it at all, but will be content either to avoid the issue or to take it for granted.

Where does this leave the feminist theories of language and oppression we have discussed in this book? I suggest that it leaves us with three propositions corresponding to the three feminist axioms of linguistic determinism, male control and female alienation.

1 Linguistic determinism is a myth. Where there is no determinacy, there can be no determinism. In a system where language and linguistic acts are integrated with non-linguistic acts and social life generally, language can be only one of the multiple determinants of any individual's perceptions and experience. An important determinant it may be, but it cannot be privileged to the extent that both Saussurean and Whorfian[5] theories privilege it.

2 Male control over meaning is an impossibility. No group has it in their power to fix what expressions of a language will mean, because meanings cannot be fixed, and interpretation will be dependent not on the authority of some vast internal dictionary, but on the creative and ultimately idiosyncratic use of past experience and present context.

Learning to communicate and to participate in social life is something which both male and female children do. They do it by actively interacting with their environment and the people in it, and thus they construct – rather than learn – meanings that are highly contextualized, dependent on that environment and those people, subject (as the environment is) to variation and to change.

It would be very surprising if this learning process did not exhibit sex-linked differences. Girls and boys, after all, are very specifically socialized into female and male gender roles; one would expect them to construct meanings which were different not only idiosyncratically, because each individual has a different experience, but more generally, because in patriarchal societies males and females are allowed a different range of experiences. Perhaps, then, we may talk to some extent of male and female meanings. But we cannot speak of women being socialized into male meanings, or of both sexes being socialized into patriarchal meanings (except in the sense that their experience is one of living under patriarchy). Meanings have to be constructed by the individual language user (in this way language is radically unlike, say, folk tales or table manners) and any child who does not learn to construct meanings out of her own interaction with the world cannot be said to possess language at all.

3 Female alienation from language does not exist in the form postulated by the theories we have considered (it should not be denied either in theory or in practice that many women do feel extremely alienated in some modes of language use). Since language is a flexible and renewable resource, and since girls must come to grips with it as their socialization proceeds, there is no reason in principle why language cannot express the experience of women to the same extent that it expresses the experience of men.

In saying this, however, I do not wish to deny that women have real problems in speaking and being heard. Although I reject the usual explanations of them, I believe that the means do exist for men to oppress, silence and marginalize women through language. The sources of silence and oppression are what I want to turn to now.

Linguistic oppression: what it is that men control

One consequence of the integrational approach to language is that the linguist has to take seriously the fact that languages are not used in a social and political vacuum, i.e. she must recognize the institutional aspects of language I have already mentioned in this chapter.

In every society, one finds law, rituals and institutions which regulate language (especially its more public modes) in particular ways. It is not always easy to separate these 'metalinguistic' or 'discursive' practices from language proper, since there is a constant interaction between the two. From an integrational standpoint it is not even worth trying to exclude the metalinguistic: institutional phenomena are part of what the linguist must be concerned with.

If we look closely at the regulatory mechanisms which grow up around languages, it is clear that they are rather closely connected with the power structures of their society. The institutions that regulate language use in our own society, and indeed those of most societies, are deliberately oppressive to women. Men control them, not in the rather mystical sense that they are said to control meaning, by making esoteric semantic rules or possessing the vital signifier, but simply because it is the prerogative of those with economic and political power to set up and regulate important social institutions.

Language, the human faculty and communication channel, may belong to everyone; because of the crucial part it plays in human cognition and development, it cannot be appropriated. But *the* language, the institution, the apparatus of ritual, value judgement and so on, does not belong to everyone equally. It can be controlled by a small elite. As Trevor Pateman remarks,

> Language, though the socially produced means of thought, is not socially controlled. Increasingly control over the development of language and its use is held by state institutions, including mass-media and monopolistic private enterprise, as in journalism and advertising. . . . The semiologists have sometimes failed to appreciate the possibility and existence of class or other minority control over language.[6]

If we acknowledge the importance of institutional control, the crucial question is: how is male control over metalinguistic processes manifested, and what effects does it have on women? I want to consider this now, and true to my integrational aims I shall not be limiting my remarks to linguistic phenomena alone. It is impossible to understand the practices that regulate women's relation to language except with reference to gender roles and regulatory mechanisms in general.

High language and women's silence

Cora Kaplan, in a short but influential essay, makes a point that has since become the received view about women's oppression by linguistic institutions; the point is that women are denied access to the most influential and prestigious registers of language in a particular culture.[7] That is to say, everything defined as 'high' language (for instance, political and literary registers, the register of public speaking and especially ritual – religious, legal or social) is also defined as *male* language. Kaplan observes,

> The prejudice seems persistent and irrational unless we acknowledge that control of high language is a crucial part of the power of dominant groups, and understand the refusal of access to public language is one of the major forms of the oppression of women within a social class as well as in trans-class situations.[8]

If it is in their relation to high language that women are linguistically disadvantaged, it seems that we must ask three questions: *what* are the registers that men control, *how* do they gain and keep control of those registers, and *why* does male control constitute a disadvantage for women? I propose to explore these questions by focusing on particular areas of language use and linguistic control. First there is the area of written language and women's relation to it. An investigation of this shows how a denial of language can constitute a denial of knowledge and of certain kinds of consciousness. Then, there is the problem of bureaucratic/institutional language. Recent work on interethnic communication demonstrates how very tightly controlled norms are used to define subordinate groups as inadequate communicators (and thus to *make* them inadequate). Finally, we must examine the exclusion of women from public and ritual speech, investigating the extent to which femininity has been produced as incompatible with the sphere of rhetoric.

. . .

Institutional language: communicating in urban societies

If literacy is a problem for the underprivileged in less developed countries, bureaucracy is a major linguistic headache for the underprivileged of modern western cities. In two recent books advocating a new approach to communication ('interactional sociolinguistics')[9] the linguist John Gumperz points out how important 'communicative skills' have become with the growth of state and other bureaucracies (health, education, employment and tax services, for instance) in modern industrial society. More and more frequently, individuals in their everyday lives are having to negotiate linguistic interactions with these bureaucracies, and the result is that

> The ability to manage or adapt to diverse communicative situations has become essential and the ability to interact with people with whom one has no personal acquaintance is crucial to acquiring even a small measure of personal and social control. We have to talk in order to establish our rights and entitlements. . . . Communicational resources thus form an integral part of an individual's symbolic and social capital.[10]

The individual needs to be able to interact effectively with institutions and their representatives. Since it is the individual, for the most part, who wants something from the encounter, 'effectively' will mean 'in conformity with the norms of the institution'. Those who cannot express themselves in a way the bureaucracy finds acceptable (or minimally, comprehensible) will be disadvantaged.

Gumperz and his associates have produced a good deal of work on 'crosstalk' or, in plain language, *misunderstanding* between individuals whose norms of interaction are different. Their work has focused mainly on interactions between bureaucrats of various kinds (social workers, clerks, personnel managers) and Asian speakers of English, but the two main points that emerge from it are equally applicable to other ethnic minorities, working-class speakers in certain situations and, of course, women talking to men.

The first point is simply that socially distant individuals (especially those differentiated by ethnicity) do not share rather subtle strategies for structuring and interpreting talk, and this results in misunderstanding which can be frustrating for both parties and seriously disadvantageous for the Asian trying to get a job or a Social Security cheque. The second point, which is rather less explicit in Gumperz's books, is that bureaucracies use their experience of interethnic misunderstanding to generate representations of Asians as defective or inadequate communicators – representations which derive from racist stereotypes and reinforce racism. It is important to note that the right to represent and stereotype is not mutual, and that the power asymmetry here has serious consequences. Undoubtedly the Asians have their own less than complimentary ideas about the *gore* (white people), but these are the ideas of people without power. They do not serve as a base for administrative procedures and decisions, nor do they get expressed routinely in mass media: whereas institutional stereotypes of Asians *do* inform procedures, decisions and media representations. If Asians are defined as inadequate language users, they become *de facto* inadequate (crudely, no one listens any more to what they are actually saying, simply filtering it through the negative stereotype) and as Gumperz points out, in a modern industrial society this can have disastrous consequences.

It is important for feminists to ask whether women have the same ability as men to interact with people 'with whom one has no personal acquaintance' and 'adapt to diverse communicational situations'. If not, why not? Are women routinely misunderstood by men who have power over them, with disadvantageous results? Do men represent women as inadequate communicators, thus reducing their precious 'symbolic and social capital'? The question of misunderstanding between women and men has been addressed by two of Gumperz's associates, Daniel Maltz and Ruth Borker, and they believe that the subcultural differences between Asian and white speakers are paralleled by male/female differences.[11] Using the available literature on children's talk and play patterns they argue that females and males in western culture do in fact form separate subcultures, and that this significantly affects male/female interaction. Women and men do not attempt to do the same things in the same way when they talk, and thus there is likely to be a rather poor fit between what the speaker intends and what an opposite-sex hearer picks up.

It seems likely that there is a good deal in this idea, though Maltz and Borker do depend heavily on the competition versus co-operation stereotype which I have criticized [earlier] and which is itself inspired by a lot of rather dubious literature. My main reservation is that the sociolinguistic analysis of subcultural differences ought to include far more discussion about the political structure superimposed on these differences. In other words, Maltz and Borker say very little about the power of male *definitions* of female speakers, the use of such definitions to exclude women from certain registers and devalue their contribution to others. This omission is what I shall try to make good in the remainder of the chapter.

Discourses and registers

I have already discussed the exclusion of women from written and learned 'registers' (i.e. kinds of language appropriate in content, style and tone to a particular domain of use, say 'scholarship' or 'legal documents' or 'religion') and pointed out that whereas it makes no sense at all to speak of women not possessing 'language' it is quite in order to say that, for historically specific reasons, they may be forbidden to use certain registers at particular times and in particular places. Registers of language historically created by men very often represent women as marginal or inferior, and may well continue to do so even after women have begun to use them (in this book we have already looked at the registers of news reporting and lexicography, and while these practices were undoubtedly masculine originally, they have long been open to women without any noticeable diminution in their sexism). Partly this conservatism reflects the importance of tradition, 'custom and practice' in institutions. The conventions codified in style-books, rule-books, standing orders, editing and sub-editing manuals are quite literally handed down from generation to generation of professionals. They are part of a professional mystique, sanctified by history and enforced very often by the authoritarian training and advancement procedures of hierarchical organizations (the Civil Service and political parties, for instance). Partly, however, it reflects more general ideological matters. This point has been made very forcefully by semiologists like Roland Barthes (whose work on the ideological determinants of literary style in France remains a classic demonstration), Michel Foucault, Michel Pêcheux, Colin MacCabe, Maria Black and Rosalind Coward.

Semiologists refer to what I have been calling 'registers' as *discourses* (MacCabe representatively defines a discourse as a set of statements formulated on 'particular institutional sites of language use').[12] Each discourse needs to be understood in relation to its own conventions (thus TV chat shows and court cases exemplify different linguistic norms) and its functions in society.

Maria Black and Rosalind Coward, in the article on Dale Spender [I discussed earlier] explicitly say that discourse and not language (by which they mean *langue*) is the proper place for feminists to concentrate their efforts.

> Linguistic systems . . . serve as the basis for the production and interpretation of related utterances – discourses – which effect and sustain the different categorizations and positions of women and men. It is on these discourses and not on language in general or linguistic systems, that feminist analyses have to focus.[13]

By concentrating on discursive regularities (for instance, the use of generic masculine pronouns or the linguistic representation of women in the reporting of rape trials) we will discover more about the relation between language use and patriarchal ideology to which not only men but a great many women also subscribe.

What is stressed both in the semiologists' approach and in my own register-based approach is the *materiality* of the practices in question. Thus rather than posit, as Dale Spender does, a historically ubiquitous and unobservable operation by which males regulate meaning through an underlying semantic rule pejorating words for women, linguistic materialists look for the historical moment and circumstances in which a particular practice arose and the specific group who initiated it or whose authority and interests maintain it. We rarely find that a practice is initiated/maintained by all men (an exception might be the practice of pornoglossic intimidation) or that extends into every linguistic register. The negative relation of women not to 'language' or 'meaning' but to various discourses is a variable and piecemeal affair.

Nevertheless, it seems to me that ultimately this piecemeal linguistic disadvantage must be related to the general roles and representations of women in their various cultures. Not every area of language use is regulated by obvious and consciously invoked conventions, and we must now turn to the part played by folklinguistic value judgements and gender-role expectations in silencing women and representing their speech as inadequate. We must consider in particular the Dalston Study Group's assertion that women's day-to-day language is undervalued even by women themselves, and that the disadvantaged 'doubt the strength and potential of their own language'.[14] Is women's language in fact strong and full of potential, or is it repressed/suppressed, impoverished and inauthentic? How could women be persuaded to suppress or undervalue their own speech? To answer these questions we will need to look at the creation and regulation of femininity itself.

Silence: a woman's glory?

Some very obviously male-dominated metalinguistic practices are the customs and traditions of public speaking, which normally require women to be silent in public gatherings and on formal occasions. The key to this cross-culturally widespread phenomenon is, as Jenkins and Kramarae observe, the boundary between the private or familial, and the public or rhetorical. 'We find that women's sphere includes the interpersonal but seldom the rhetorical.'[15] In many societies different linguistic registers, dialects or even languages are used to mark the private/rhetorical boundary. And it is part of women's role generally, not just linguistically, to symbolize the private as opposed to the public. When this split is important in organizing a society (as in most capitalist societies) women are important in defining the boundaries of the private.

An illustration of female marginality in the public and ritual speech of our own culture is provided by the etiquette of the wedding reception. Here we have a number of visible roles distributed between males and females equally: bride and groom, mothers and fathers, bridesmaids and best man, etc. Yet the women are ritually silent. The bride's father proposes a toast to the happy couple; the groom replies, proposing a toast to the bridesmaids which is replied to by the best man. Men speak, women are spoken for: here we have an epitome of women's position as 'seen and not heard'.

Is this the same as children being 'seen and not heard?' Cora Kaplan, in her essay 'Language and Gender', argues that it is.[16] Children are subject to restrictions on their speech in adult company, but whereas boys are eventually admitted to public speaking rights (Kaplan fixes this at puberty, the onset of adulthood and symbolically the beginning of manhood) girls are never allowed to grow up in the same way. Their participation in political, literary, formal, ritual and public discourse is not tolerated in the same way that children's participation is not. This view seems to me to be open to a number of objections, the main one being that restrictions on women speaking are often far stricter than those affecting children, and appear to be linked with explicitly *sexual* rites of passage. I am thinking here of the many taboos on women's speech discussed by Ardener and Smith.[17] It is not uncommon to find women being forbidden to speak for a set period after marriage, or to find men censuring the conduct of married women who allow their voices to be heard outside the private house.[18] Books of etiquette and advice to brides also warn that a married woman must underline her wifely deference with wifely silence. In other words, silence is part of femininity, rather than being an absence of male privilege.

Unaccustomed as I am . . .

Women as public speakers suffer not only from the customs that silence them, but also from negative value-judgements on their ability to speak effectively at all. Whatever style a culture deems appropriate to the public arena, women are said to be less skilled at using; whatever style is considered natural in women is deemed unsuitable for rhetorical use. So, for example, Jespersen thinks indirectness typical of women's style, mentioning 'their instinctive shrinking from coarse and gross expressions and their preference for refined and . . . veiled and indirect expressions'.[19] This lack of 'vigour and vividness' is what makes women unfit to be great orators. Among the Malagasy, however, things are rather different. Here the favoured style for ritual speech or *Kabary* is indirect and allusive. Women's speech is thought to be direct and vigorous, and thus women are once again debarred from public speaking.[20]

It is necessary, as always, to treat the interaction between actual usage and folklinguistic stereotype with the utmost care. The excessively ladylike style 'described' by Jespersen is unlikely ever to have been used consistently by women: it is the usual idealization based on the usual mixture of prejudice and wishful thinking. But folklinguistic beliefs are never without significance, and certainly this kind of belief, expressed in a score of passages masquerading as description in anti-feminist tracts, etiquette books, grammars and even feminist writings, have an effect on how women think they speak and how they think they ought to speak. In formal situations where speech is monitored closely, women may indeed converge towards the norms of the mythology, obeying the traditional feminine commandments (silence, not interrupting, not swearing and not telling jokes).

Value

Folklinguistics inculcates an important set of value judgements on the speech and writing of the two sexes. A whole vocabulary exists denigrating the talk of women who do not conform to male ideas of femininity: nag, bitch, strident. More terms trivialize interaction between women: girls' talk, gossip, chitchat, mothers' meeting. This double standard of judgement is by no means peculiar to linguistic matters. It follows the general rule that 'if in anti-feminist discourse women are often inferior to men, nothing in this same discourse is more ridiculous than a woman who imitates a male activity and is therefore no longer a woman.'[21] This can apply not only to speaking or writing, but also to the way a woman looks, the job she does, the way she behaves sexually, the leisure pursuits she engages in, the intellectual activities she prefers and so on *ad infinitum*. Sex differentiation must be rigidly upheld by whatever means are available, for men can be men only if women are unambiguously women.

This imperative leads to an attitude toward the upbringing of women summed up in 1762 by Jean-Jacques Rousseau:

> In order for [women] to have what they need . . . we must give it to them, we must want to give it to them, we must consider them deserving of it. They are dependent on our feelings, on the price we put on their merits, on the value we set on their attractions and on their virtues . . . Thus women's entire education should be planned in relation to men. To please men, to be useful to them, to win their love and respect, to raise them as children, care for them as adults, counsel and console them, make their lives sweet and pleasant: these are women's duties in all ages and these are what they should be taught from childhood on.[22]

In this notorious passage Rousseau gives us an account of why this sort of femininity must be constructed (to make men's lives 'sweet and pleasant'), how it is constructed (by indoctrination from childhood) and why women conform (because they are entirely dependent on men for the things they need).

Language, like every other aspect of female behaviour, has to be produced and regulated with this male-defined femininity in mind. Parental strictures, classroom practices and so on are designed to make the girl aware of her responsibility, and failure to conform may be punished with ridicule, loss of affection, economic and physical hardship. In short, then, we must treat the restrictions on women's language as part of a more general restricted feminine role. We cannot understand women's relation to language or to any other cultural phenomenon, unless we examine how in different societies those with power have tailored customs and institutions so they fit Rousseau's analysis and obey his prescription.

The limits of control

This model of male dominance locates the linguistic mechanisms of control both in explicit rules and well-known customs restricting women's speech, and in the 'voluntary' constraints women place on themselves to be feminine, mindful of the real disadvantages attendant on failure. Since these mechanisms are not located in immutable mental or unconscious structures, control can only be partial, and even women's silence has its limits.

The more radical feminist theorists have often been unduly pessimistic because they did not acknowledge the limits of control and silence. When Dale Spender claims,

> One simple . . . means of curtailing the dangerous talk of women is to restrict their opportunities for talk. . . . Traditionally, for women there have been no comparable locations to the pub which can encourage woman talk; there have been no opportunities for talk like those provided by football or the union meeting. Because women have been without the space and the place to talk they have been deprived of access to discourse *with each other.*[23]

she is simply wrong. If Spender is thinking here of the 'captive wife' alone with small children in an isolated flat, this is a relatively recent and restricted phenomenon. Even in middle-class Anglophone culture women's talk with each other is an important part of social organization.[24] In other cultures, where segregation is often the norm both occupationally and socially, women's lives revolve around interaction with each other.

Spender is trying to make a case for the subversive nature of women's talk with each other, but by ignoring the age-old oral culture of women (as she must, to argue this case convincingly) she misleads us into accepting what is only a half-truth. Women's talk is not subversive *per se*: it becomes subversive when women begin to attach importance to it and to privilege it over their interactions with men (as in the case of consciousness-raising). Men trivialize the talk of women not because they are afraid of any such talk, but in order to make women themselves downgrade it. If women feel that all interaction with other women is a poor substitute for mixed interaction, and trivial compared with the profundities of men's talk, their conversations will indeed be harmless.

Women's talk: the myth of impoverishment

Recently, feminists have begun to research women's talk. The picture that emerges from their studies is not one of silent or inarticulate women who struggle to express their experiences and feelings. On the contrary, it is of a rich verbal culture.[25] Moreover, that culture has a long history (if obscure: male metalinguistic practices strike again, this time by omission). It may be appropriate to see early women poets breaking through silence and absence, working in a genre where they were insecure and had no rights, but the ordinary women speaker in her peer group cannot be adequately treated in this way.

To sociolinguists this story will have a familiar ring to it. One of the most celebrated achievements of sociolinguistics in the 1960s and early 1970s was to put working-class black American speech on the map through painstaking study of the vernacular black speakers used amongst themselves. Before the sociolinguist Labov and his associates undertook this research, using a methodology specifically designed to win the informants' co-operation, conventional wisdom among commentators on black language was that its speakers developed silent and inarticulate because they grew up in a linguistically deprived culture, were seldom addressed by their parents and not encouraged to speak themselves. The dialect they came to school speaking was labelled 'a basically non-logical form of expressive behaviour' or, in the terms of Bernstein a *restricted code.*[26]

Bernstein's code theory (which was developed with the English class structure, rather than American ethnic differences, in mind) holds that the two types of socialization typical of the middle and working classes respectively, give rise to differing relations to language. The middle-class child controls both a restricted code (roughly emotional, illogical, inexplicit and incorrect, useful for expressing group solidarity and feeling) and an elaborated code (which facilitates higher cognitive operations through its logic and explicitness). The working-class child controls only restricted code, and thus her ability to perform the sort of intellectual tasks expected at school is limited.

The claim that black children were restricted code only speakers, therefore, was an attempt to explain why they failed, or underachieved relative to white children, at school. It led to a compensatory education project in which children were taught the appropriate elaborated code (i.e. white middle-class English). The sociolinguist Labov showed that this project, and the premises underlying it, were fundamentally misguided.[27] For one thing, the linguistic features defining elaborated code turned out to be nothing more than an amalgam of middle-class habits (like use of the passive and the pronoun *one*): it was hard to argue they had any inherent value. The features stigmatized in black English were not failed attempts at standard English, but systematic variations, or more accurately, parts of a related but different dialect. In other words, it was (and is) fundamentally unclear whether there is such a thing as a restricted code which is unable to express complex ideas, logical relationships and so on.

In the second place, Labov demonstrated that black children grow up in an extremely stimulating verbal culture with its own rituals. Individuals who at school were silent or inarticulate were very likely to metamorphose, within their peer group, into skilled verbal performers. To unearth the rich verbal culture of black adolescents, Labov had to go to a great deal of trouble, for they did not willingly display it to white outsiders – which was why successive experiments had failed to elicit anything but silence and inarticulacy. Labov used a young black fieldworker to elicit a wide range of data, and in analysing it he deliberately abandoned his educated middle-class notions of correctness and formality. Labov concluded that black children failed in school mainly because they had no motivation to succeed. They defined themselves in opposition to dominant white values, and to be fully integrated members of their peer group they had to express disdain for formal education.

Other studies of non-standard language users (to use the term *restricted code* would be to beg the question linguistically, as I have already pointed out) stress that there is a linguistic problem, but not in the language itself so much as in the stigma people attach to it. In other words, the theory of codes could be boiled down to an essentially political truism: those who do not speak the language of the dominant elite will find it difficult to get on.

I have dealt with Bernstein's code theory in detail because I think there are parallels in it with the case of women, and that feminists could learn a number of lessons from the controversy it provoked.

Women are not in quite the same position as working-class and black speakers. Their language is less obviously different from men's than working-class from middle-class or black from white varieties, and the differences are much more often below the level of speakers' consciousness. Nor has anyone yet suggested that women's speech variety is responsible for massive educational under-achievement. In many respects, however, women's language has been treated as if it were a type of restricted code. And this evaluation has come both from feminists (who speak of the silence and inarticulacy of women and their culture, and of the inauthenticity with which they have been forced to express themselves) and by old fashioned gallants and chauvinists. Jespersen, for instance, presents the features which typify female speech as products of an impoverished cognitive apparatus whose shortcomings are surprisingly similar to those detailed by Bernstein in his descriptions of restricted-code speakers. Even the linguistic hallmarks of restricted code and women's language are the same: a preference for conjoining over the more complex embedding, unfinished sentences, and a heavy reliance on intonation rather than more 'explicit' syntactic devices.[28] Writers agree, in short, that there are various things women's language is inadequate to express.

Except in so far as it applies to all communication, this strikes me as a false and dangerous belief. Perhaps research now being done on women in small groups, on female folklore and culture, will break it down, both by showing that women have rich and complex verbal resources, and by proving that the folklinguistic consensus on women's speech style is inaccurate. Researchers in this latter area should focus, as I have tried to do in this book, on the connections and similarities between feminist and anti-feminist folklinguistic beliefs, and on the importance of value judgements in producing what disadvantage women do suffer as speakers and writers.

It is also important to make explicitly the connection between *women* as disadvantaged speakers and the disadvantaging of other subordinate groups such as ethnic minorities and the working class. Such groups could certainly learn from the WLM's* refusal to ignore questions of language and politics; on the other hand, the WLM in making those links might be moved to revalue certain theoretical excesses.

As the Dalston Study Group observe, 'Immigrants and working-class people too have a negative point of entry into our culture, something no one has yet explained with reference to the penis/phallus.'[29] If language is an important political and personal resource, feminism cannot afford a theory that tells women only how they are oppressed as speakers: it must convince them also of 'the strength and potential of their own language'.

Although the nature of communication is such that men cannot appropriate

*Women's Liberation Movement

meaning nor completely control women's use of language, they (or a subset of them) control important institutions and practices. The effect of that control is to give men certain rights over women and to hedge women around with restrictions and myths. Its mechanisms range from explicit rules against women speaking in public or on ritual occasions to folklinguistic beliefs and values denigrating women's language and obscuring female verbal culture. These rules, prescriptions and beliefs can be related on the one hand to femininity in general, and on the other, to the linguistic and cultural subordination of other oppressed groups.

Whereas the current feminist belief in determinism, male control and female alienation offers very little prospect of struggle and liberation, we get a more hopeful picture when we concentrate on metalinguistic and discursive processes linked to women's identity and role in particular societies. These processes can be challenged much more easily and effectively than *langue*, meaning, alienation and other such abstractions.

11

Involvement strategies in a speech by the Reverend Jesse Jackson

Deborah Tannen

From Tannen, Deborah 1989: *Talking voices: repetition, dialogue, and imagery in conversational discourse.* Cambridge: Cambridge University Press, 176–94.
Note: Omitted are some introductory comments drawing attention to the use made by Jackson of Martin Luther King's 'I have a dream' speech, and a concluding section. The speech analysed was given by Jackson to the 1988 Democratic National Convention.

Common ground

The theme of Jackson's speech was unity: unity among supporters of different primary contenders (including Jackson) to ensure that the Democratic party win the presidential election. The term 'common' was a recurrent verbal representation of this theme. Jackson first introduced it in a parallel construction that, like other phrases and images he used, was a repetition and variation of parts of his 1984 convention address. In 1984, Jackson used a parallel construction that employed a paradigmatic substitution within the same syntactic frame:

(5) We must leave **the racial battle ground**
and come to **the economic common ground**
and **moral higher ground.**

The parallelism, with its repetition and reframing of the word 'ground', transforms something negative ('racial battle ground') into something positive ('economic common ground') and then into something exalted ('moral higher ground'). Jackson used the same triple parallelism as the basis for a slightly elaborated and altered figure in 1988:

(6) Tonight there is a sense of celebration.
Because we are moved.
Fundamentally moved,
from racial battle grounds by law,
to economic common ground.
Tomorrow we'll challenge to move,
to higher ground.
Common ground.

In this speech, the medial term 'common ground,' was key, so this is the one the parallelism focused on. Jackson repeated it immediately and then raised it to the level of a formula. Davis (1985) finds, 'The most important characteristics of the African-American sermonic formula are the groups of irrhythmic lines shaped around a core idea.' Thus 'common ground' is the core idea, a repeated phrase that captured Jackson's theme and punctuated the points he made throughout his address. This can be seen in (7). (The irrhythmicity of lines can be seen here as well but will be discussed in a later section.)

(7) **Common ground.**
Easier said than done.
Where do you FIND
common ground,
at the point of challenge.
. . .
We find common ground at the plant gate
that closes on workers without notice,
We find common ground,
at the farm auction,
where a good farmer,
loses his or her land
to bad loans,
or diminishing markets,
Common ground.
At the schoolyard,
where teachers cannot get adequate pay,
and students cannot get a scholarship,
and can't make a loan,
Common ground.
At the hospital admitting room . . .

In all, there were nineteen occurrences of the phrase 'common ground,' in addition to 'common grave,' 'common table,' 'common thread,' 'common good,' 'common direction,' 'common sense' (itself part of a repeated formula), and 'one thing in common.'

Jackson frequently used a repetitive strategy that derives impact from a surprising reversal. For example, there are metatheses of phonemes:

(8) No matter how t**ire**d or how t**rie**d,

of morphemes:

(9) With so many **guided mis**siles,
and so much **misguided** leadership,

and of lexical items, frequently resulting in the figure of speech, chiasmus:

(10) **I was born** in **the slum,**
but **the slum** was not **born in me.**

Repetitions of words and phrases are seen throughout the address and throughout this analysis, dovetailing with other involvement strategies.

Dialogue

At five points Jackson used dialogue to anticipate and animate others' points of view. Four instances of dialogue came toward the end of the speech, gradually shifting focus to Jackson's personal life, as his address culminated in his 'life story,' scenes from his childhood. By successive uses of dialogue, he gradually brought listeners closer, preparing them to hear his life story. (At the same time, the television producers made the performance dialogic for viewers by interspersing shots of the speaker with 'reaction shots' of the audience.)

The first instance of dialogue was the longest and different in function from the other four. Here Jackson spoke in the voice of young drug addicts, 'the children in Watts' to whom he says he listened 'all night long':

(11) They said, 'Jesse Jackson,
as you challenge us to say no to drugs,
you're right.
And to not sell them,
you're right.
And to not use these guns,
you're right,
. . .
We have neither jobs,
nor houses,
nor services,
nor training,
no way out, . . .'

By framing these and other details of their situation in the voice of young drug addicts, Jackson lent authority to his claim that their situation is hopeless, and the government bears responsibility for allowing the availability of guns and drugs:

(12) 'We can go and buy the drugs,
by the boxes,
at the port.
If we can buy the drugs at the port,
don't you believe the federal government
can stop it if they want to?'
They say,
'We don't have Saturday night specials any more.'
They say,
'We buy AK-47s and Uzis,
the latest lethal weapons.
We buy them ACROSS the COUNTER
on LONG BEACH BOULEVARD.'

In this, as in all the other instances of dialogue, Jackson animated others addressing him by name. In this first extended use of constructed dialogue, he animated a voice addressing him by his full name; later he brought the audience closer by animating voices addressing him by first name only.

In the second instance of dialogue, like the others that followed, Jackson animated projected objections to his political positions:

(13) I'm often asked,
→ 'Jesse, why do you take on these
→ tough issues.
→ They're not very political.
→ We can't win THAT way.'

Jackson used this projected objection as the frame in which to answer the objection. In (14), he went on to align himself, through parallel constructions, with others who took stands that were unpopular but 'morally right.' In the first four lines, he used chiasmus to reverse the order of phrases 'be political' and 'be right'. The passage culminated in yet another instance of constructed dialogue. (A number of repeated words and phrases are printed in bold.)

(14) If an issue is **morally right,**
it will eventually be **political,**
It may be **political,**
and never **be right,**
Fannie Lou Hamer **didn't have the most votes**
in Atlantic City,
but her **principles** have **out**-lasted
every delegate who **voted** to lock her **out,**
Rosa Parks
did not have the most votes,
but she was morally right,
Dr. King **didn't have the most votes**
about the Vietnam War,
but he was morally right,
If we're **PRINCIPLED** FIRST,
our **politics** will fall into place.
→ **'Jesse,** why did you take these big bold initiatives.'

The dialogue in the last line of (14) is a paraphrase of the line of dialogue seen in (13), restating the question he is speaking to at this point in his address, and reinforcing the closeness he is constructing with the audience by casting himself in dialogue.

Jackson moved from this section of his speech to a section in which he addressed the audience directly, telling them (repeatedly) to 'Dream,' 'Go forward,' 'Never surrender,' and 'Don't give up.' He then animated dialogue in which the audience addressed him directly:

(15) Why can I challenge you this way.
→ **'Jesse Jackson, you don't understand my situation.**
→ You be on television. [laughter]
→ **You don't understand,**
→ I see you with the big people.
→ **You don't understand my situation.'**

The audience laughed, amused perhaps by Jackson's verbalization of a thought some of them had, perhaps by his animation of vernacular Black English. Their laughter contributed to the effect of the dialogue: Engaging the audience in dialogue with him provided a kind of iconic analogue to inviting them to pull up a chair and listen to his life story which ended and capped his speech.

'Understand,' the key word and concept, was picked up from the animated dialogue in (15) to form a phrase that was repeated over and over as the story unfolded, driving home its point. This begins in the introduction to the story:

(16) I **understand**
You're seeing me on TV
but you don't know the **me**
that makes **me me,**
→ They wonder 'Why
does Jesse **run**',
Because they see **me running** for the White **House,**

they don't see the **house** I'm **running** from,
I have a story,

Here Jackson used dialogue to express the projected thoughts of others ('Why does Jesse run?') while echoing the title of the novel *What makes Sammy run?* He used parallel construction to reinterpret the meaning of the word 'run' and to juxtapose the elegance of the White House with the impoverishment of the house he grew up in. Phonologically, the repeated /i/ sound created end-rhymes in 'TV' and 'me,' and the 'ru /rʌ/ of 'run' was repeated in 'from' (/frʌ m/), creating another end-rhyme.

'I understand' occurred fourteen times as Jackson described his childhood. I present only a short excerpt, from the beginning:

(17) You see,
I was born to a teenage mother,
who was born to a teenage mother.
I understand.
I know abandonment,
and people being mean to you,
and saying you're nothing and nobody,
and can never be anything,
I understand.
Jesse Jackson,
is my THIRD **name.**
I'm adopted.
When I had no **name,**
my grandmother gave me her **name,**
My **name** was Jesse Burns,
'til I was twelve.
So I wouldn't have a blank space,
she gave me a **name.**
To hold me over.
I understand,
when nobody knows your **name.**
I understand when you have **no NAME.**
I understand.

In addition to the repetition of the phrase, 'I understand,' (17) shows an incremental repetition of the word 'name' which finally blends into a variation of the title of a book by the Black writer James Baldwin, *Nobody knows my name.*

Details and images

In describing his childhood, Jackson used details to create images that would let listeners imagine what he must have felt:

(18) I wasn't born
in the hospital.
Mama didn't have insurance.
→ I was born in the bed,
→ at house.
I really do understand.
→ Born in a three room house,
→ bathroom in the back yard,

→ slop jar by the bed,
→ no hot and cold running water,
I understand.
Wallpaper used for decoration?
No.
For a windbreaker.
I understand,

Jackson dramatized the poverty of his childhood by depicting specific details that allows hearers to imagine a scene they could elaborate in their minds with other images and associations.

Using specific details, Jackson also described a scene in which his family celebrated Thanksgiving:

(19) I understand.
At three o'clock on
Thanksgiving day,
we couldn't eat turkey.
Because Mama was preparing somebody else's turkey
at three o'clock.
We had to play football to entertain ourselves.
And then around six o'clock,
she would get off the /Alta Vista/ bus,
and we would bring up the leftovers
and eat our turkey,
leftovers:
the carcass,
the cranberries,
around eight o'clock at night.
I really do understand.

The tolling of the clock from a repeated three o'clock, when the family should have been eating Thanksgiving dinner, to six o'clock, when the mother returned home, to eight o'clock, when they finally ate, provided an iconic analogue to the delaying of the children's Thanksgiving dinner. The specific naming of the hours, naming the game the children played while waiting for their mother, the name of the bus she rode, specifying the leftovers: 'turkey carcass' and 'cranberries,' created the images from which listeners could construct a scene and imagine what they might have felt in that scene.

Jackson's description of his childhood was the last major section of his address. By involving the audience in his personal life, especially his vulnerability as a suffering child, he climaxed the process, begun by dialogue, of bringing the audience gradually closer to him. This climax in figurative movement is analogous to the emotional climax that Jackson created: The audience was moved by the rhythms of his speech which involved them in musical ensemble, and by participating in sensemaking as they constructed in their minds scenes of his childhood, based on the details and images he depicted. It was during this segment that the 'reaction shots' shown on the television screen displayed weeping faces, evidence of the emotional impact of the speech.

Figures of speech

Many of the repetitive strategies I have illustrated are figures of speech – what Levin (1982) calls 'style figures of speech,' arraying words in identifiable syntagmatic patterns. Levin identifies another type of figure as 'thought figures of speech,' figures that Friedrich (1986) and Sapir (1977) call 'tropes.' These are figures that play primarily on meaning. Among them are similes and metaphors. Many of Jackson's similes and metaphors arose in connection with his personal life story, contributing to the climactic impact of that part of his speech.

The 'common ground' theme is itself figurative. This metaphor was developed most elaborately in the section excerpted as (7). Several other metaphors were elaborated at other points in the address. Jackson's use of metaphor played a significant role in the emotional impact of his address: Each of these elaborations sparked a crescendo of audience applause at the time of delivery.

I discuss the following three metaphorical elaborations (1) lions and lambs, (2) boats and ships, and (3) the patchwork quilt. All of these metaphors were woven back into the 'common ground' theme.

Lions and lambs

Jackson echoed and then elaborated the conventional metaphor for peace of the lion lying down with the lamb. The metaphor is a repetition from popular culture and from the Bible, its original source:

(20) 1 The Bible teaches that when lions
2 and lambs
3 lie down together,
4 none will be afraid
5 and there will be peace in the valley.
6 It sounds impossible.
7 Lions eat lambs.
8 Lambs flee from lions,
9 Yet even lions and lambs find common ground. Why?
[pause]
10 Because neither lions
11 nor lambs want the forest to catch on fire.
12 Neither lions nor lambs
13 want acid rain to fall,
14 Neither lions nor lambs can survive nuclear war,
15 If lions and lambs can find common ground,
16 surely we can as well,
17 as civilized people.

Readers will have noted numerous repetitions in the elaboration of this metaphor, including repetition of the words 'lion(s)' and 'lamb(s),' and of several syntactic paradigms in which they are reframed, such as chiasmus:

7 **Lions** eat **lambs.**
8 **Lambs** flee from **lions,**

Jackson used repetition to press the old lion-and-lamb metaphor into service as a frame into which he fit the contemporary issues of acid rain and nuclear war:

10 neither lions, nor lambs want the forest to catch on fire.
12 Neither lions nor lambs want acid rain to fall,
14 Neither lions nor lambs can survive nuclear war,

Finally, he merged the lion-and-lamb metaphor with the one he had previously established and elaborated, common ground.

9 Yet even lions and lambs find common ground.
. . .
15 If lions and lambs can find common ground,
16 surely we can as well,

Thus Jackson used a conventional metaphor as the basis of novel elaboration and ultimate integration into his theme of party unity.

Ships

Fairly early in his address, Jackson praised Michael Dukakis, the man everyone knew would be nominated to run for president. Then he compared Dukakis to himself in a series of parallelisms contrasting the circumstances in which they grew up. Having thus emphasized their differences, he used a metaphor to express the bond between them:

(21) His foreparents came to America
on immigrant ships.
My foreparents came to America
on slave ships.
But whatever the original ships,
we're in the same boat tonight.

The audience cheered when Jackson reframed the literal ships on which his and Dukakis's foreparents came to America in terms of the conventional metaphoric expression, 'We're in the same boat.' The aesthetic pleasure of the reframing contributed to highlighting the theme of unity.

Jackson then elaborated a slightly different boat metaphor, depicting himself and Dukakis as navigating ships:

(21) **Our ships**
could pass in the night,
if we have a false sense of independence,
or **they could collide and crash,**
We would lose our passengers,
But we can seek a higher reality,
and a greater good.
Apart,
we can drift on the broken pieces of Reaganomics,
satisfy our baser instincts,
and exploit the fears of our people.
At our highest,
we can call upon noble instincts,
and **navigate this vessel,**
to safety.
The greater good,
is the common good.

Here, too, the boat metaphor, like the lion and lamb, reinforced the theme of party unity: Bad things happen if boats navigating the same waters do not coordinate their movements; good things happen if they do.[1]

The boat metaphor resurfaced later, in answer to the question Jackson posed in the form of dialogue which was seen in (14): 'Jesse, why did you take these big bold initiatives?' In answering this projected question, Jackson cited 'a poem by an unknown author':

(23) As for Jesse Jackson,
'I'm tired of sailing my little boat,
far inside the harbor bar,
I want to go out where the big ships float.
Out in the deep,
where the great ones are,
And should my frail craft,
prove too slight,
the waves that sweep those /?/ o'er,
I'd rather go down in a stirring fight.
Than drown to death
in the sheltered shore.'
We've got to go out my friends
where the big boats are.

In this second elaboration, the boat became a metaphor for Jackson's life: He would rather risk failure in a dramatic effort than find his end in safety and obscurity. This metaphor provided a transition to the climax of Jackson's address. He moved from it to challenging the audience to 'Dream' and 'Never surrender' (in other words, like him, to move out from a small familiar harbour) and then to his life story.

The patchwork quilt

Another extended metaphor compared America to a patchwork quilt. The metaphor grew out of an image from Jackson's childhood:

(24) Common ground.
America's not
a blanket
woven from one thread,
one color, one cloth.
When I was a child growing up
in Greenville, South Carolina
and Grandmother could not afford,
a blanket,
she didn't complain and we did not freeze.
Instead she took pieces of old cloth.
Patches,
wool, silk, gabardine, /crockersack/,
only patches,
barely good enough to wipe off your shoes with.
But they didn't stay that way very long.
With sturdy hands,
and a strong cord,

she sewed them together.
Into a quilt.
A thing of beauty
and power
and culture.

Jackson transformed his grandmother's quilt into a metaphor for the Democratic party and used it as the basis for a repetitive strategy listing groups to whom the Democrats might appeal. Each group and its demands were underlined and punctuated by reference to the patchwork metaphor. (In the first line below, rather than using 'sew' or another verb appropriate to quilting, he invited party members to 'build' a quilt – a verb that is assonant with 'quilt' and has more forceful connotations.)

(25) Now, Democrats, we must build such a quilt.
Farmers,
you seek fair prices
and **you are right,**
but you cannot stand alone.
Your patch is not big enough.
Workers,
you fight for fair wages,
You are right,
But your patch labor
is not big enough.
Women,
you seek comparable worth and pay equity.
You are right.
But your patch
is not big enough.
Women,
mothers,
who seek head start,
and day care,
and pre-natal care,
on the front side of life,
rather than jail care and welfare
on the back side of life,
You're RIGHT,
but your patch
is not big enough.
Students,
you seek scholarships.
You're right,
but your patch is not big enough.
Blacks and Hispanics, when we fight
for civil rights,
we are right,
but our patch is not big enough.
Gays and lesbians,
when you fight
against discrimination,
and a cure for AIDS,
you are right,
But your patch

is not big enough.
Conservatives and progressives,
when you fight for what you believe,
right-wing,
left-wing,
hawk,
dove,
you are right,
from your point of view,
but your point of view **is not enough.**
But don't despair,
Be as wise as my Grandmama.
Pool the patches,
and the pieces together,
bound by a common thread,
When we form a great quilt
of unity,
and common ground,
we'll have the power
to bring about health care
and housing
and jobs
and education
and hope to our nation.

Here again, the metaphor was elaborated with repetitive strategies and sound play (for example, 'patches and pieces,' 'right-wing, left-wing').[2] And here again Jackson brought the audience closer by shifting from 'Grandmother' to 'Grandmama,' figuratively bringing them into his family.

Other metaphors

In addition to these extended metaphors, there were many fleeting ones:

(26) Whether you're a hawk or a dove,
you're just a bird,
living in the same environment,
the same world,

Phonological and metric repetition set this bird metaphor into a poem-like frame. Sound repetition creates a near rhyme between 'bird' and 'world,' while the beats per line result in a 3-2-3-2 pattern:

Whéther you're a háwk or a dóve,
you're júst a bírd,
líving in the sáme envíronment,
the sáme wórld,

In (27), Jackson describes the drug addicts whose voice he animated in (11) and (12) in terms of a grape/raisin metaphor.

(27) 1 I met
2 the children in Watts,
3 who are unfortunate
4 in their despair.
5 **Their grapes of hope have become raisins of despair.**

The impact of the grape/raisin metaphor was also intensified by rhythm. There is an unexpected break in prosody between subject and object ('I met/ the children in Watts') that makes lines 1–4 rhythmically fragmented and choppy. This contrasts with the unexpected length of line 5, the clause containing the metaphor. Furthermore, the raisins of despair echo the poem, 'A dream deferred,' by the Black poet Langston Hughes, and the play about Black experience which borrowed an image from that poem for its title, *A raisin in the sun.*

Surprising prosody

Jackson's delivery was characterized by what Davis (1985, 50) calls 'irrhythmic semantic sensibility.' The first word of a syntactic sentence was often rhythmically linked to the preceding one and bounded by a pause and sentence-final falling intonation. This prosodic contour characterized the repetition of the word 'dream' in Martin Luther King's 'I have a dream' speech, so that sentence-final intonation and pause followed the word 'Dream,' even though it was syntactically linked to the phrase that followed. In other words, each prosodic unit began with the word 'Of' and ended with the word 'Dream'. This highlighted the word 'dream' as well as the various images the audience was told to dream of. This prosodic contour also characterized the repetition of 'common ground' in (7) and in (28):

(28) **We find common ground**
at the farm auction
where a good farmer
loses his or her land
to bad loans
→ or diminishing markets. **Common ground.**
[pause]
At the schoolyard
where teachers cannot get adequate pay,
and students cannot get a scholarship
→ and can't make a loan, **Common ground.**
[pause]
At the hospital admitting room . . .

Similar prosody marked the repeated use of the word 'leadership':

(29) → **Leadership.**
Must meet the moral challenge of its day.
. . .
→ **Leadership.**
What difference will we make?
→ **Leadership.**
Cannot just go along to get along.
We must do more than change
presidents,
We must change direction,
→ **Leadership,**
must face the moral challenge of our day,
The nuclear war,
build-up,

is irrational,
→ Strong **leadership,**
cannot desire to look tough,
and let that stand in the way of the pursuit of peace,
→ **Leadership.**
Must reverse,
the arms race.

The effect of this prosody was to highlight the repeated word and also to highlight, by isolating, the points that were punctuated by it.

Recursive formulas

Just as a word can be lifted from a phrase or extended figure to become a punctuating formula, so too a word that has been pounded home by repetition can blend back into the flow of discourse and give way to another formula. Thus the 'leadership' theme merged into a brief figure built around the phrase 'real world' and then into a series of parallel constructions in which the word 'support' became the punctuation, each instance interspersed with a supporting backup phrase:

(30) This generation,
must offer **leadership** to the real world.
We're losing ground in Latin America,
the Middle East,
South Africa,
because we're not focusing on the **REAL WORLD,**
that **REAL WORLD.**
We must use BASIC PRINCIPLES.
Support,
international law.
We stand the most to gain from it.
Support
human rights.
We believe in that.
Support
self-determination.
You know it's right.

By being prosodically separated from its grammatical object, the word 'support' became part of a three-part repetition. The last phrase, 'You know it's right,' also echoed a number of repetitions of the formula 'You're right' which were seen in (25).

Immediately before he began his life story, Jackson intensified his voice and also intensified repetition and variation of the phrase 'Don't surrender':

(31) **Do not surrender** to drugs.
The best drug policy is a no first use.
Don't surrender with needles and cynicism,
Let's have no first use
on the one hand,
our clinics on the other.
Never surrender,
young America.
Go forward.

America must **never surrender** to malnutrition.
We can feed the hungry and clothe
the naked,
We must never surrender,
We must go forward,
We must never surrender to illiteracy.
Invest in our children,
Never surrender,
and go forward,
We must never surrender to inequality,
. . .
Don't surrender, my friends.
Those who have AIDS tonight,
you deserve our compassion.
Even with AIDS
you must not surrender in your wheelchairs.
. . .
But even in your wheelchairs,
don't you give up.
. . .
Don't you surrender and don't you give up.
Don't you surrender and don't you give up.

Interspersed with the 'don't surrender' formula were repetitions of 'go forward' and other repetitive strategies. In this section, Jackson's voice became strong and loud. The last repetitions of 'Don't you surrender and don't you give up' punctuated loud applause from the audience (a far more active and interactive way to manage audience response than simply waiting for it to die down). It was immediately after this section that Jackson turned to the telling of his life story that climaxed his performance.

Following the section in which he told his life story, Jackson reiterated the phrases, 'You can make it,' and 'Don't surrender.' In these he incorporated a parallel construction that he also used in his 1984 convention speech, one marked by anadiplosis, beginning an utterance with the same unit that ended the preceding utterance. In both addresses he said, near the end of each speech:[3]

(32) Don't you surrender.
→ Suffering breeds character,
→ Character breeds faith,
→ In the end,
faith will NOT disappoint.
You must not surrender.

The 1988 speech then ended with a short play of a number of repeated phrases culminating with four repetitions of 'Keep hope alive!,' the phrase that was the rallying cry of his campaign.

12

Talk, identity and performance: *The Tony Blackburn Show*

Graham Brand and Paddy Scannell

From Scannell, P. (ed.) 1991: *Broadcast talk*. London: Sage, 201–3; 215–22. The original article consists of nine sections; reprinted here are sections 1, 7 & 8.[1]

Note: The Tony Blackburn Show was a D.J. phone-in programme on BBC Radio London.

I

Erving Goffman has shown the constructed nature of identity, the self as a presentation or performance designed to be appropriate to the circumstances and settings in which it is produced in the presence of others (Goffman, 1969). This self is not the privileged possession of its owner-presenter. It is not an essential, inalienable quality of an individual – it is not the soul – but intrinsically social, sustained in relationships with others. If human beings are sacred objects, they can be desecrated, the territories of the self invaded and, in 'total institutions', stripped away and destroyed (Goffman, 1970). The self is, from moment to moment, perishable, dependent on others who, since their self-projections are vulnerable too, have a common interest in collaborating to sustain the general character of the performance in most mundane social settings.

Goffman's early work was subject to two criticisms: the first concerning the nature of the relationship between the individual and his or her projected self, and the second concerning the episodic nature of the social settings he took for consideration. One strong inference from *The presentation of self in everyday life* (1969) was that the self was a mask behind which lurked an unknowable individual possibly manipulating the performance for undisclosed ends – 'there is no art to read the mind's construction in the face'. Such a 'naughty' view – as Harold Garfinkel calls it – suggests that individuals may be radically disaffiliated from their performance, and Garfinkel wished to show that most individuals in most contexts are committed to their performance, that they play themselves 'to the life', that the self projected is offered as a case of 'the real thing' (Garfinkel, 1984: 116–85).

A non-committal stance in relation to self is more likely to seem plausible when social situations are treated as discrete events or episodes with no before or after – as in the theatre where the actor assumes the role for the duration of the play and quits it when the curtain falls. In such an instance we are inclined, like Hamlet, to ask for the actor. 'What's he to Hecuba or Hecuba to him That he should weep for her?' But in real life we cannot walk away from the part(s) we play. The crucial issue of continuity – of the management and maintenance of self through a lifetime – was examined by Garfinkel in the case of Agnes, an intersexed person who wished to discard her biological maleness and become a natural, ordinary 100 per cent female. Agnes displayed, by a perspective of incongruity, what for 'normals' is profoundly taken for granted – the effortless production of a sexed identity. A major crux for her was the lack of a plausible and consistent feminine biography for use in appropriate circumstances. How to produce 'girl talk' with other girls, for instance, is dependent on the incremental accumulation of know-how from such experiences which accrue to the individual through time:

> The troublesome feature encountered over and over again is the cloudy and little known role that time plays in structuring the *biography and prospects of present situations* over the course of action as a function of the action itself. It is not sufficient to say that Agnes's situations are played out over time, nor is it at all sufficient to regard this time as clock time. There is as well the 'inner time' of recollection, anticipation, expectancy. Every attempt to handle Agnes's 'management devices' while disregarding this time, does well enough as long as the occasions are episodic in their formal structure; and all of Goffman's analyses either take episodes for illustration, or turn the situations that his scheme analyses into episodic ones. (Garfinkel, 1984: 166 [emphasis added])

The same problem arises in relation to Goffman's last published work on forms of talk (Goffman, 1981). Goffman is concerned to dissolve the unitary categories of 'speaker' and 'hearer' and in so doing opens up issues of fundamental importance for the analysis of the construction of self, social interaction and the role of talk. The complexities of the relationship between speaker and utterance and between speaker and addressee begin to emerge in the essays on 'Footing' and 'Radio Talk'. Both are richly suggestive and their ideas, as will be apparent, have been absorbed into the bloodstream of this article. Nevertheless, what is missing still is attention to the problem of long-term continuity in time, the reproduction of identity and the way that this is accomplished through talk.

Routinization is the basis of continuity. As Anthony Giddens put it: 'Routine is integral both to the continuity of the personality of the agent, as he or she moves along the paths of daily activities, and to the institutions of society, which are such only through their continued reproduction' (Giddens, 1984: 60). Routines have a double articulation: they have a structure and content that is produced across a single episode. But this structure and content is formatted so that it can be reproduced again and again, thereby achieving the recursive effect of 'things as usual', familiar, known from past occasions, anticipatable as such now and in future. To study the double articulation of routine requires attention to Goffman's concerns with self-presentation in episodic social settings, and to Garfinkel's with the continuing maintenance, on all occasions, of 'the self-same identity'.

The study of such issues in relation to broadcasting is particularly interesting because routinization is at the very heart of programmes and programming. An individual programme is the briefest of ephemerides that perishes in the moment of its transmission. As such its identity is so transient as to be unnoticeable. Broadcast programmes build identity through repetition and regularity via formatting and scheduling. The art of scheduling was, in Britain at least, not immediately obvious and was something learnt, through trial and error, by the programme planners of the pre-war National Programme. Crucially it involved 'locking' programmes into regular time slots so that they recurred, from week to week on the same day and at the same time. Thus what began to be established was a familiar, regular pattern of daily output, reproduced through the weeks and months of the year that meshed in with the day-to-day routines of the population (Scannell, 1988). Once this basic principle was accepted – and it encountered stiff opposition from certain areas of programme output – the more subtle arts of continuity between programmes began to be discovered, and of positioning and sequencing them in the schedule to attract specific or more general audiences.

The routinization of programming went hand in hand with the routinization of programmes, and here the key discovery was seriality (Scannell and Cardiff, 1991: 377–9). The problem of production in broadcasting is, more exactly, that of reproduction. The magnitude of broadcast output is obscured by over-familiarity, but continuous, unceasing production day in day out can only be sustained, in the long term, by formatting. This involves the creation of a template for the production of a programme whose basic structure and content remains the same but which varies in its particularities from one episode to another. Once, say, the format for news as a programme or for a situation comedy (a form invented for radio and television) has been laid down the programme can last indefinitely. On British radio, as is well known, there are programmes that have run continuously now for well over 40 years; *Desert Island Discs, Woman's Hour* and *The Archers* are famous instances.

How is a format created? Key elements include the use of signature tunes, programme presenter(s), standard sequences for the programme material (lead with the big news story and end with a human interest one), techniques of 'anchorage and relay' for moving through the sequence and maintaining continuity (Brunsdon and Morley, 1978), standardized beginnings and endings. The combined effect of such techniques is to create, through time, a familiar, recognizable programme identity that is perceived as such by audiences. When the BBC's newly established Listener Research Unit began, in the late 1930s, to investigate the impact of serialized productions in regular time slots they found that 91 per cent of respondents to their questionnaire favoured the increasing use of serial formats for drama, entertainment and talks, and 85 per cent assured them that the BBC was not overusing the format (Scannell and Cardiff, 1991: 378).

[. . .]

VII

We began by considering two sociological views of self-identity, the cynical and the serious. A third possibility has emerged, namely the playful self. The notion of playfulness retains the dramaturgical echo of Goffman and the spontaneity (sincerity) of Garfinkel: its differential characteristic is self-reflexiveness, that is, awareness of the performed nature of the displayed self. Such self-reflexiveness is hidden in the cynical performance (which is thereby manipulative or instrumental), repressed in the sincere performance (no this isn't a performance, it's the real me) but manifest in playful performances. A playful identity involves a momentary going out of character. It is less likely to be a career unless that career is, as in the case of DJs, a performance.

Garfinkel has argued eloquently that there is not time out from the burden of responsibility for the management and maintenance of identity. Nevertheless, there are, as Goffman shows us, all kinds of occasional opportunities for stepping outside of self: that is one major way in which we relax or have fun:

> When an individual signals that what he is about to do is make believe and 'only' fun, this definition takes precedence; he may fail to induce the others to follow along in the fun, or even to believe that his motives are innocent, but he obliges them to accept his act as something not to be taken at face value. (Goffman, 1974: 48)

Having fun involves pretending, putting temporary brackets round reality, a momentary suspension of the ordinary daily round. We have tried to bring out, in our presentation of *The Tony Blackburn Show* [earlier in full text] not only how fun is defined, organized and projected as such, but also how it is bracketed out from ordinary reality, how it deliberately refuses to be serious.

It is clear, from the strips of conversation considered [earlier in full text], that the participants in the fun – Blackburn and callers – do not stand in the same relationship to each other. Broadcasting is an institution – a power, an authority – and broadcast talk bears its institutional marks, particularly in the way that it is not so much shared between participants as controlled by the broadcasters. Because the institution is, ultimately, the author of *all* the talk that goes out on air (it authorizes it) it is responsible for the talk in a way that those invited to speak are not. If an invited participant should transgress the norms (by saying 'fuck' for instance) it is the broadcasting institution rather than the transgressor who will be held accountable. Thus control and management of all talk in broadcasting must rest, first and last, with the representatives of the institutions, that is, the broadcasters.

Blackburn's control of the talk – in terms of topic management and closure – though idiosyncratic in its manner, is not in any sense particular to him. Broadcast telephone conversations, while sharing many characteristics of private phone calls, have some that mark them out as public displays produced for a listening audience. Blackburn shows this awareness routinely and it is manifest whenever he switches from talking to callers to talking to the studio or the listeners. When he pulls the plugs on a 'boring' caller this may be (subjectively) intended and heard as impolite. It may also be (objectively) intended and heard as dramatic, as 'livening things up a bit' – not so much for the caller, of course, as for other listeners for whom there may be the added frisson that – if their turn should come – they too might provoke, deliberately or not, the same rough treatment.

It is, from moment to moment, from one day to the next, week in week out, Blackburn's responsibility – and no one else's – to maintain the fun. For listeners and callers the fun is optional: for its presenter it is not and this is why Blackburn patrols its boundaries so carefully, since he alone must manage and maintain the show's expressive idiom. To do so he has devised – formatted, we would say – an identity for the programme and himself that is routinely talked into being by himself and others. The talk is the routine, the routine is the identity. Goffman, in a particularly suggestive passage, discusses how talk routines are produced on radio. It may seem as if what he calls 'fresh talk' is constantly produced in unscripted radio talk:

> But here again it appears that each performer has a limited resource of formulaic remarks out of which to build a line of patter. A DJ's talk may be heard as unscripted, but it tends to be built up out of a relatively small number of set comments, much as it is said epic oral poetry was recomposed during each delivery. (Goffman, 1981: 324)

Goffman has in mind the work of Milman Parry (1971) and Albert Lord (1960) who demonstrated how it was possible for the ancient oral tradition to produce such heroic tales as *Iliad* and *The Odyssey*. The problem they addressed was, simply, how did the old tellers of tales know and remember such lengthy narratives which, when transcribed, were thousands of lines long. Since each retelling in the oral tradition must be a fresh version of the tale, what are the techniques that enable them to be learnt, stored and reproduced afresh in each retelling? By a study of the still living tradition of oral epic in Southern Yugoslavia Parry was able to show, as Goffman puts it, how 'prose narratives, songs and oral poetry can be improvisationally composed during presentation from a blend of formulaic segments, set themes and traditional plots, the whole artificially tailored to suit the temper of the audience and the specificities of the locale' (Goffman, 1981: 228).

The production of the same kind of smoothly continuous talk, day in day out, on every broadcast occasion, over a 3-hour stretch poses similar problems for today's DJs. It is not difficult to show that much of Blackburn's monologue talk is a patchwork of formulaic utterances woven into set routines:

(*music fades*)
T.B. Paris and *I Choose You.*
It's now seven minutes before eleven o'clock.
Your main funking funketeer.
Your Boss, with all the hot sauce.
Your Leader . . . (*pause*) . . . Me.
Right. Now Dave in Greenford says 'Drive safely 'n
love you' to wife er Jill who's on her way to Radlett
at the moment.
Mark in Bermondsey sends all his love to fiancée
Sally Ann.
And also Rachel or – yes it is – Rachel in Barnet
says 'Love you' to husband Peter who's working at
Shenley Hospital hrhmm oh dear must clear my throat.
Right. Now, Em Garry's in Camden. Hello Garry.

(1985 Polytechnic of Central London tape)

This strip of talk is embedded within a larger half-hour formatted section of the programme called London Love in which listeners are invited to 'show you care for the one you love' by phoning in if, for instance, they have just got engaged or are getting married or are back from a honeymoon or want to make up a quarrel. The phone calls are taken in pairs between suitably romantic soul 'twelve inchers' and the methodological problem for the presenter is to get from the music to the calls to the music always with an eye on the studio clock to keep to the overall format of the show and the scheduled number of plays within it. Here three routines are displayed: (a) Continuity talk (1–5), (b) Audience message (6–13) and (c) Telephone chat (14). Continuity routines generally contain three elements in sequence: i) record identification, ii) time check, iii) programme-presenter identification. There is more scope for variety in the third than in the first two elements. A jingle may be used or, as here, a few formulae – 'your funking funketeer', 'your Boss with all the hot sauce', 'your Leader' – from Blackburn's stock of stock phrases.

The switch from one routine to another is succinctly signalled by 'Right. Now . . . (6) which indicates ending (Right.) and beginning (Now . . .). The next routine, audience messages, has its standard format: A in X → 'message' → B in Y. The message may be quoted or reported. There are usually three messages, as there were three components of the preceding routine and three in the programme-presenter identification. Triads are, as Max Atkinson has shown, an extremely useful and common rhetorical device for packing memorable and memorizable utterances (Atkinson, 1984). This routine too is closed down and the next introduced in the same way as the preceding one: 'Right. Now . . . (14)' and into the phone-in routine.

Blackburn brings off these routines with effortless ease, including the self-monitoring utterances ('Oh dear must clear my throat') that repair momentary disruptions of the flow (cf. Goffman, 1981: 290). This is the mark of his professionalism, and if lay speakers were suddenly given the DJ's talk tasks they would doubtless be dumbstruck. But this, Goffman suggests, is more for a want of tag lines than for a want of words (Goffman, 1981: 325). Regular listeners, however, know the tag lines and their appropriate usage of them and show this knowledge in conversation when they go on air in phone calls with the programme presenter.

VIII

In his concluding remarks on broadcast talk, Goffman compares it with 'everyday face-to-face talk' without, however, commenting on or distinguishing between monologue talk (with which he has been, in fact, very largely concerned) and talk as social interaction between two or more participants. The absence of such a distinction suggests that is not significant in Goffman's terms of analysis and indeed he concludes that DJ monologue is basically the same as 'what the speaker is engaged in doing' from 'moment to moment through the course of the discourse in which he finds himself.' If face-to-face talk then is something a (male) individual 'finds himself in' he makes the best of it by selecting that footing 'which provides him with *the least threatening position* in the circumstances, or, differently phrased, *the most defensible alignment* he can muster' (Goffman, 1981: 325, our emphases). Talk appears, in Goffman's terms, as yet another threat to face, as a kind of external imposition, to which the individual must respond self-defensively. In this curiously grim view of talk there is no perception of it as sociable interaction, as something collaboratively produced by two or more participants which, at best, is what it is mutually and enjoyably achieved in Tony Blackburn's radio show:

T.B. Right. Now, Em Garry's in Camden. Hello Garry.
Garry. Hello Tony.
T.B. (*chat up voice*) Hello. I gather you're getting married tomorrow.
Garry. Oh yeah 'n I'm really scared I tell you.
T.B. After all – I'm not surprised – after all you 'nd I

have meant to one another as well.
Garry. I know but (*?*) my Leader what can I do. We tried to get down 'nd see you last night as well.
T.B. Really?
Garry. Yeah we couldn't. We wanted to see your twelve incher but –
T.B. I'm – Garry!
Garry. Ahh.
T.B. I'm amazed you're getting married. All those times that we spent in the sand dunes in Swanage together.
Garry. Ah d'you remember that time in Bahamas?
T.B. Yes.
Garry. On the beach just me 'nd you.
T.B. When you used to whisper and nibble my ear.
Garry. Ahhh.
T.B. Underneath the coconut trees.
Garry. And you you used to show me your twelve incher.
T.B. And you threw it all away and you're getting married tomorrow. Don't you think you should reconsider this?
Garry. I think I should Tone, I think I should mate.

In analysing this strip of talk we wish to bring out how the two participants collaborate to co-produce talk that is 'in frame', as Goffman would say, i.e. within the terms of the discursive world of *The Tony Blackburn Show*. In this respect we attend both to the content (what the talk is about) and the style (how it is talked about). We further show how both speakers, in working to produce appropriate talk, draw upon their knowledge of what Garfinkel calls 'the biography of the present situation' that we have sketched above, and thereby how identities are routinely reproduced and reaffirmed by talk.

Hello Garry
Hello Tony
Hello

An exchange like this is so utterly familiar that its oddity escapes us, for the fact is that neither Tony nor Garry know each other, nor have they ever met or spoken to each other before this moment. How then can they hail each other as familiars? We must assume, as must they, that – if not familiar with each other – they must be familiar with the programme and that this is a common knowledge and thereby a shareably relevant resource for the production of talk, both in content and manner. Thus the embedded implicatures, as working conversational hypotheses initially made by each speaker, can be posed as follows:

T.B. Hello Garry [I have not spoken to you before, have never met you and don't know who you are, but I take it that you have listened to this programme before and to that extent know me, and I let you know that I make these assumptions in calling you Garry]

Garry: Hello Tony [I have not spoken to you before, we have not met and you don't know me but I have listened to this programme before and I confirm your assumptions in calling you Tony, thereby displaying knowledge of the programme]

If T.B. starts with this assumption it enables him to mobilize a routine without further ado, because he can reasonably assume that caller will recognize the routine-to-be-initiated as such. One of the most economic ways of getting into a routine that Blackburn uses is voice change, which simultaneously indicates both a change of footing and the character of the new alignment. We have discussed above, in relation to several data samples, Blackburn's voice as an aspect of his chat-up routine with callers. Blackburn's repeated 'Hello', here said in a lower pitch and with a softer inflection than the first 'hello', hearably implicates intimacy. This change of voice accomplishes a number of things: first it shifts out of the first paired greetings exchange which is a display for the general audience into particular conversation with this displayed caller. The change from DJ voice to intimate voice 'keys' the tone of the talk to be initiated, it sets the frame. Note, at this point, that an intimate tone of voice is being used with a male caller (Garry's voice, like his name, is hearably masculine).

'I gather you're getting married tomorrow' is said in the same intimate tone. Let us deal first with the technical question – how does T.B. know this? – before attempting to account for why he says it here in this tone of voice. Callers to this, as to other radio shows with phone-ins, get through to a switchboard in the station that handles the calls. The operators will ask callers for their names, where they come from, their telephone number and if they have anything special to want to say. These bits of information are written down on paper and handed to Blackburn in the studio who is cued, by the producer, as to who is next in the bank of callers on hold to talk to him. 'I gather' implies that Blackburn's source for the statement-query that follows is not directly Garry – by inference, then, the station – and requests confirmation which Garry immediately produces (18).

But what is the object of this utterance at this point? Consider the predictable conversational lines that might be taken by recipients of the information that the person to whom they are speaking is to be married next day. A next turn might be to ask 'to whom?' – a question not posed until later – and certainly the offer of congratulations should be forthcoming very soon, but these are not offered by Blackburn until much later. Garry's marriage – which is topically relevant *today*, in programme terms, by virtue of being tomorrow – serves as the envelope for the conversation as a whole. It is the first thing referred to after initial greetings exchange and the last thing referred to before final thank-yous and good-byes:

106 *T.B.*	Be happily married Garry.
107 *Garry.*	Thank you very much.
108 *T.B.*	Thanks very much indeed for phoning.
109	Jill's in Woodford. Hello, Jill
	(*Continues conversation with Jill.*)

The introduction of Garry's marriage at the beginning of the conversation serves not so much as a topic to be sustained in its own right, but as a foil for the routine that Blackburn wishes to establish.

What that routine is is not apparent at this point and Garry, after confirming Blackburn's statement-query, produces a response – 'nd I'm really scared I tell you – that keeps up the topic of marriage-as-an-imminent-prospect. Blackburn's next turn (19–20) is, for anyone unfamiliar with the biography of the occasion, downright peculiar or 'weirdo!' as Blackburn would say, but in context it is routine and indeed only makes sense as a routine. It does not at first attend to Garry's response but builds on Blackburn's opening move and begins to reveal how he wants to use Garry's marriage as a conversational resource. 'After all . . . after all you'nd I have meant to each other' is said in a hearably reproachful voice that continues and makes explicit the claims to intimacy implicated in the tone of voice adopted in Blackburn's preceding turn. His interpolated reaction to Garry's response – 'I'm not surprised [you're scared]' – is a rapid change of footing, a 'normal' response in his 'normal', slightly jokey voice, a return to the real world from which the conversation is beginning to depart if Blackburn can establish his routine.

That depends on Garry's support, and that depends on Garry recognizing and keying into the fantasy routine. Garry is not in the least fazed by Blackburn's line, 'I know but . . . what can I do?' (21) acknowledges the line of reproach and plays along with it. The playfuness is underlined by the smoothly interpolated 'My Leader', said in a tone of mock deference, which claims membership of 'the gang' and displays knowledge both of the content of the discourse of *The Tony Blackburn Show* and of its jokey, 'send-up' style. Garry has now shown to Blackburn his understanding of the rules of his conversational game and a general disposition to play it. But it is not yet clear, to Garry, that Blackburn wishes to sustain his line, so Garry continues with a bit of real-world chat – 'We tried to get down 'nd see you last night as well'.

A notable feature of the way *The Tony Blackburn Show* reaches out to its audience, attempts to create a listening community, is the Soul Night Out that Blackburn regularly announces on the show.[2] This is a disco, presented by Blackburn often with a guest soul artist, in a venue somewhere in London, to which fans of soul music and Tony Blackburn are invited. It is this that Garry tried to attend, presumably with his bride-to-be, and which he offers here as a topic (it is one that often crops up in phone-in talk on the programme). In referencing it Garry further displays his membership of the programme's listening community, but his object in introducing it here is not yet clear. Blackburn's response token – 'Really?' (23) – is a pass that allows Garry to continue, and to make explicit what was implicit in 'We *tried* to get down' – 'we couldn't' (24). A reason is produced for wanting to get to the show, namely the desire to see Blackburn's twelve-incher. As heroes in the old sagas have their trusty weapons – Achilles his spear and shield, Beowulf and Arthur their swords – so Blackburn has his tool of heroic proportions which he may offer to show to callers on the programme. Garry's use of the formulaic phrase – like 'my Leader' – shows his familiarity with the programme's word-hoard. More particularly, it switches from real-world talk back to fantasy-world talk, keying in to Blackburn's general line though not yet his particular tune.

Blackburn now, taking up the talk after a slight pause after 'but', tries to replay that tune, having momentarily given way to Garry, 'I'm – Garry! – [. . .] I'm amazed you're getting married. All those times that we spent [. . .] together' (26–29) repeats the pattern of the first effort: 'After all – I'm not surprised – after all you 'nd I meant to each other' (19–20). The interpolated 'Garry!' (26), however, is in a tone of mock reproof (for mentioning Blackburn's unmentionable) that is consonant with the rest of the utterance whereas the interpolated 'I'm not surprised' (19) required a momentary change of footing back to the real world. Garry's production of Blackburn's twelve-incher helps to retrieve the tone of the talk which the introduction of the Soul Night Out seemed temporarily to have abandoned. But why the sand dunes in Swanage (28) of all places? Well, the young Tony Blackburn grew up in that part of the world, his father being a doctor with a practice in Poole (Blackburn, 1985: 4–13).

Garry's response (31) tunes in to Blackburn's line and now the conversation has clicked. Both will collaborate in the game of Let's Pretend to produce an imaginary relationship with an imaginary past, places and memories. The account we have offered of the talk thus far has attempted to show how it gets to this point where both participants have sought and found an agreed conversational framework and a shared attitude towards it. That they *can* get to this point depends, from moment to moment, as we have tried to show, on mutual knowledge and understanding of the programme's content and manner. Such knowledge is incremental. It accumulates in time as it is reproduced through time. The past of the programme is not the dead past. It is a pervasively relevant resource for renewing its identity in the particularities of the present. That identity is not wholly constructed and mediated by Tony Blackburn. Listeners, like Garry from Camden, playfully interact with the show to keep up the fun.

Section III

Perception and interaction

'Seeing is believing', we are told from early childhood, and for many people what we see is unproblematically *true* in a way that what we read or are told is not (an assumption that has important implications for the study of the mass media).

In the first of our extracts **Hastorf, Schneider** and **Polefka** move very interestingly from people's perception of objects to their perception of one another. Perhaps the most stimulating aspect of their piece is its stress upon the *active* nature of our perceptions. We are not like blank sheets of paper, passively receiving impressions, but active agents who, individually and collectively, search out and process our perceptions. We may not *make* our own reality, but we do have to *work* to possess aspects of reality, and we interpret this information in the light of our needs and interests. The authors argue that 'The categories we use are derived from our past history and are dependent on our language and our cultural background.' This is perhaps unwittingly exemplified in their own piece by the labellings of the ambiguous drawings of a female (see p. 120 following). Why 'old hag' and 'young woman' rather than 'old woman' and 'young woman'? Changes effected in our own cultural situation and language use, largely as a result of the development of the Women's Movement in the last two decades, help us to perceive these figures rather differently.

The extract from **Stuart Sigman**'s book takes some of these issues a stage further. We now move from how we perceive, to how we interact with others. Rather than seeing interpersonal communication as something which *individuals* do within a space called *society*, Sigman suggests an alternative viewpoint. Communication is a *social process* involving individuals: 'Communication in this view is seen not as an individual-level phenomenon, but as a societal-level phenomenon.' Moreover, 'Just as individual persons are moments in society, so interpersonal behaviour is a moment in social communication.'

Although **Walter Lippmann**'s piece on 'Public Opinion' predates our first two extracts by many years, it also follows on logically from both of them. Lipmann points out that our perceptions of other human beings are dependent upon learned and standardized regularities in the social world much as our perception of physical objects is dependent upon learned and standardized regularities in the physical world. '[W]e pick out what our culture has already defined for us, and we tend to perceive that which we have picked out in the form stereotyped for us in our culture'. Here Lippmann thinks of a culture in terms that associate it very closely with a nation, but the article by **Albert J. Hastorf** and **Hadley Cantril** makes it clear that there are local as well as national cultures, and that membership of these influences our perception of events too – especially events involving competition and conflict. (This could also be argued the other way round: competition and conflict arise on the basis of differential perceptions of people, events, and processes.)

The previous section of this book should have suggested that once we understand how various factors influence language, a particular example of language use can be analysed so as to identify the local influence of these factors. We can make a similar point in this section. For if various social and cultural factors condition and enable our perceptions and interactions, then evidence of these factors may in turn be discernible in accounts of our perceptions and interactions.

The purposive element in human perception stressed in the first extract in this section is also insisted upon in the opening lines of **Erving Goffman**'s discussion of 'self-presentation'. Interacting human beings are not like billiard-balls on a table, whose movements are determined only by the forces transmitted to them and the strictly mechanical effects of other physical forces. We do not interact as isolated and autonomous beings. Nor do we act or react in a mechanical manner, our mutual behaviour regulated only by pre-established, universal and unchangeable rules. We inter*act* with socially directed and culturally conditioned individuals on the basis of shared as well as private perceptions, disputed as well as common goals. And as Goffman reveals so convincingly, we interact in a manner determined by both private and public intentions, with both conscious and unconscious goals in mind.

Not all interaction is between equals, however. Even where two individuals are in conversation, complex issues of status, power and influence enter into the way each presents him- or herself, and into how the other is perceived. Our extract from the article by **Donald Horton** and **R. Richard Wohl** focuses attention on to one of the most common, and least symmetrical of interactive relationships in our culture: that between viewers and television or radio performers. They confirm how even in the very early days of national television, TV personalities (the term is instructive) learned how to mimic more democratic and symmetrical forms of personal interaction in their public performances. Here what the viewer is encouraged to interpret from the screen is, in part, misleading or false – a simulation of true interaction.

Whether looking at pictures, reading novels, or watching films, human beings are just as much influenced by their culture and history as they are when they respond to optical illusions or interact with one another: probably more so. Moreover, they are just as active – again, probably more so. According to **John Berger**, the painter Holbein is also involved in a form of 'presentation of self' in the picture discussed in the extract from his book *Ways of Seeing* (although here it involves the depiction of *others*). Berger assumes that from this painting we can 'read off' those aspects of social and political history which actually entered into its creation. This of course involves a certain degree of resistance on our part: the figures in Holbein's painting are as much involved in inviting us to participate in a form of para-social interaction as are the television personalities discussed by Horton and Wohl.

Our final extract involves an interpretation which we might expect to be rather different: **Sigmund Freud**'s interpretation of a patient's account of a dream. And yet here again one is struck by the involvement of social and cultural factors. The rôle and position of women in Freud's society, the significance of such things as marriage, virginity, and sexuality in this society: the effect of all of these can be traced, if not in the dream itself, certainly in Freud's analysis. Can we imagine a woman from a very different sort of society having this dream, or an analyst from a very different society interpreting it in the same way?

We would like to conclude by drawing attention to the fact that many of the items in this section involve situations of *conflict*. How we define ourselves and others, how we perceive events, how we *interpret* – these are typically matters involving struggle and competition. Human beings have different interests, and these interests inevitably condition processes of definition (of self and other), perception, and interpretation.

Further reading

Our recommendations favour books which share our own view of the importance of studying perception, interaction and interpretation in the context of sociocultural influences, at the expense of more purely technical studies and approaches dominated by behaviourist theories.

Bordwell, David, and Thompson, Kristin 1990: *Film art: an introduction*. Third edition. New York: McGraw–Hill.
A highly recommended introductory textbook which assumes no previous knowledge of film methods, and works through a full range of topics such as the basic styles of camera work, editing, the use of sound, the organization of narrative, and the history of film-making. Includes some excellent extended analyses, and contains very helpful and comprehensive recommendations on further reading in the area.

Burton, Graeme and Dimbleby, Richard 1988: *Between ourselves: an introduction to interpersonal communication*. London and Baltimore: Edward Arnold.
A straightforward and very accessible introductory textbook covering interpersonal communication, the perception of others, social interaction and social skills, transactional analysis, communication in groups, and communication theory. The authors' starting point is that of the individual self, and thus their approach is different from that outlined in the extract from Stuart Sigman's book reprinted in this section.

Gombrich, E.H. 1972: *The story of art*. Oxford: Phaidon Press.
A superb book that is as far from an aridly technical history of the visual arts as can be imagined. Gombrich explores the interpenetration of historical, cultural and technical matters in a masterly fashion, and shows how artistic representation and interpretation are always bound up both with current knowledge of the world and also with the development of technical expertise.

Grimshaw, Allen D. 1990: *Conflict Talk: sociolinguistic investigations of arguments in conversations*. Cambridge: Cambridge University Press.
Much study of human interaction plays down or ignores the issue of conflict, and thus this collection of essays serves to draw attention to what we believe is an element in most contacts between human beings. Deborah Tannen, whose analysis of a speech by Jesse Jackson is included in the previous section of this

Reader, has an interesting essay here on silence as conflict-management in two literary works.

Gudykunst, William B., and Ting-Toomey, Stella 1988: *Culture and interpersonal communication.* Newbury Park, Cal., and London: Sage.

A useful textbook which covers most of the standard topics dealt with by comparable textbooks on interpersonal communication but which sets these in a context of cultural variety.

Morgan, John, and Welton, Peter 1986: *See what I mean: an introduction to visual communication.* London and Baltimore: Edward Arnold.

An excellent introductory textbook, very accessible and fully illustrated. This book covers many of the basic terms and issues with which it is now essential for Communication Studies students to be familiar.

Rock, Irvin 1983: *The logic of perception.* Cambridge, Mass. and London: MIT Press.

Of the many textbooks on perception currently available this is one of the most recommended, although Rock's emphasis is more technical and less sociocultural than our own.

13

The perceptual process

Albert H. Hastorf, David J. Schneider and Judith Polefka

From Hastorf, A.H., Schneider, D.J. and Polefka, J. 1970: *Person perception*. Reading, Mass., and London: Addison-Wesley, 3–9.

Both philosophers and psychologists have long been intrigued with the nature of the human perceptual process. One explanation for their interest is that man is naturally curious about his contact with the outside world and wonders how his experiences are caused and to what degree they reflect the world accurately. Beyond general curiosity, the reason for the interest stems from an apparent paradox, the basis of which lies in the difference between the nature of our experiences and our knowledge of how those experiences are caused.

Anyone who takes the trouble to think about and to describe his own experiences usually finds himself overwhelmed with both their immediacy and their structure. One's experience of the world is dominated by objects which stand out in space and which have such attributes as shape, colour, and size. The immediacy of such experiences becomes obvious if one closes his eyes, turns his head in a new direction, and then opens his eyes again. A structured world of objects is immediately present in awareness, without delay and without any consciousness of interpretative or inferential activity. The world appears to be given to us in experience. Yet a causal analysis of these events indicates a very different state of affairs.

You have opened your eyes and you experience a blue vase about six inches high situated on a table. The vase appears to be at a certain distance, and its shape and colour are equally clear. Let us remind ourselves of the causal events that are involved. Light waves of a certain wavelength are reflected off the vase. Some of them impinge on the retina of your eye, and if enough retinal cells are irritated, some visual nerves will fire and a series of electrical impulses will be carried through the sensory apparatus, including the subcortical centres, and will finally arrive at the cortex. This description paints a picture of a very indirect contact with the world: light waves to retinal events to sensory nerve events to subcortical events and finally to cortical events, from which visual experiences result. What is especially important is that this causal description reveals a very different picture than does our naïve description of experience. (This causal description led a famous German physiologist to remark that 'we are aware of our nerves, not of objects'.) Thus we have a conflict between our everyday-life experiences of objects together with their properties and an analysis of how these experiences come to exist. How *does* the human being create a coherent perceptual world out of chaotic physical impingements?

Our world of experience has structure

Let us begin with this fact of experience and explore how the structure may be achieved. First of all, we know that our experiences are ultimately dependent on our sensory apparatus, which for visual experiences would include both the retina of the eye and the sensory neurons connecting the retina to the visual areas of the cortex. This apparatus plays, in a manner of speaking, the role of translator. Light waves impinge on the eyes and we experience colour. Sound waves impinge on the ear and we experience pitch. Without the sensory apparatus we would have no contact with the external world. There remains, however, the question of the nature of this translation.

A number of philosophers and psychologists have conceived of the translation process as an essentially passive one, completely determined by the physical properties of the stimulus and by the structure of the receptors and sensory nervous system. They conceive of our sensory apparatus as working somewhat like a high-speed translation device. Physical impingements are looked up and the proper experiential attribute is read out. This conception has led to arguments as to how much of this dictionary is present at birth and how much is the product of our learning history. One reason for the popularity of the passive recording view of perception is the immediacy and 'givenness' of our experience. Our experiences are immediate and they feel direct. These feelings led to the belief that the translation process must be automatic and built in.

The primary argument against that position stems from the fact that our experience of the world is highly selective. If we passively translated and recorded stimuli, our world would be a jumble of experiences; while you were reading a book, you would also be aware of the pressure of your clothes on your body and of all the sounds around you. Actually, from a myriad of impinging stimuli, we are aware of only certain objects and certain attributes of the objects. Anyone who has asked two different persons to describe the same scene has been struck by the fact that they often describe it very differently; each selects different events and different attributes of the events. Given this phenomenon, we must be more than passive translators. In fact, we must be active processors of information. The world is not merely revealed to us; rather, we play an active role in the creation of our experiences.

Let us take an example from the research of Robert W. Leeper to illustrate our point (Leeper, 1935). The stimulus he used was an ambiguous picture which can be seen as either an old hag or an attractive young woman (Figure 13.1a). Continued inspection of the picture usually permits an observer to see first one and then the other. Leeper had the original picture redrawn so that one version emphasized the young woman (b) and another emphasized the old hag (c). Subjects who had been exposed to one or the other of these redrawings found themselves 'locked in' on that view when the original ambiguous picture was presented. One hundred per cent of the subjects who had had prior experience with the version emphasizing the young woman saw only the young woman when first looking at the same ambiguous picture. The subjects had been given a set to process the input stimuli in a certain way, and they created a structure consistent with that set. Although our experiences are both immediate and structured, extremely complex participation by the organism, including the active selection and processing of stimulus impingements, is involved in their creation.

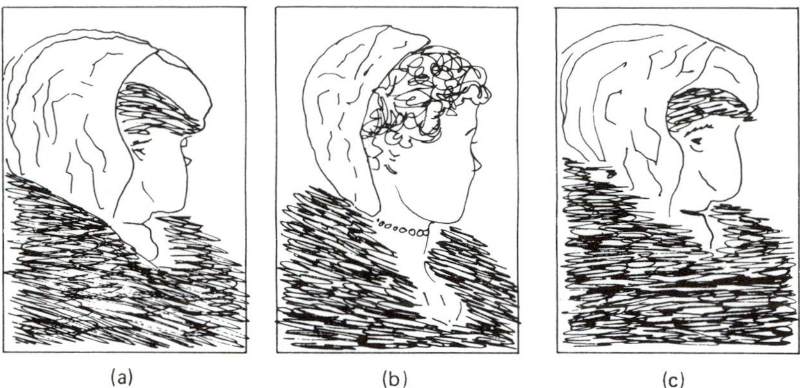

Figure 13.1

One of the most salient features of the person's participation in structuring his experiential world can be described as a categorizing process. He extracts stimuli from the world and forces them into a set of categories. We have here a powerful example of the effects of linguistic coding on the structuring of experience. The subjects in Leeper's experiment did not see a complex pattern of light and dark nor even 'a person' (a possible category); they saw an old hag or a young woman. The categories we use are derived from our past history and are dependent on our language and our cultural background. Some of these categories are markedly ubiquitous and well agreed on by perceivers. Classification of objects according to the attributes of size and shape seems obvious, but some persons may employ different sets of categories. For example, they may perceive in terms of colour and softness. Moreover, there are occasions when all of us change categories in perceiving objects. Instead of size and colour, we may see things in terms of function: the large blue pen and the small green pencil are suddenly similar when we want only to jot down a telephone number. Whatever the nature of the categories we use, they play an important role in the processing of information.

We have begun with the experiential fact that our perceptions are both structured and organized. This structure is immediate and appears to be given by the world of objects. We have argued that a causal analysis of the situation clearly indicates that structured perceptions are the outcome of the organism's engaging in active processing of information, which includes the translation of physical impingements to nerve impulses and the active selection and categorizing of the inputs.

Our world of experience has stability

When we open our eyes and look at a scene, we are not overwhelmed with constant shifts in the picture as our eyes and our attention wander. There is a certain enduring aspect to our experience. We select certain facets of the situation and stick with them. Check this statement against your own experience with the ambiguous picture in Figure 13.1. If it was like the experience of most people, the first organization of the picture, whether it was the old hag or the young woman continued to demand your attention. It was hard to 'find' the other one. You made various attempts to shift the focus of attention by blinking your eyes or by concentrating on a certain part of the picture, but those stratagems did not always work. Although stability in a case of this kind may frustrate us to such an extent that it deserves to be given a different and more pejorative label – rigidity – the example demonstrates that we do *not* experience a world of chaotic instability.

The most obvious example of the maintenance of stability in our experience has been termed *the constancies* in perception. Constancy phenomena have been most carefully described in regard to the perception of size, colour, shape, and brightness. Let us consider an example. You are sitting in a chair in your living room. Another person walks into the room, moves over to a table by the window, picks up a magazine, and then goes across the room to sit down and read it. What are the successive visual-stimulus events impinging on your retina and your successive experiences? Every time the person moves closer to you, the impingement, or *proximal stimulus*, gets larger; in fact, if he moves from 20 feet away to 10 feet away, the height of the image on your eyes doubles in size. The opposite occurs as he moves away from you because the size of the retinal image is inversely proportional to a distance of the object from you. Furthermore, when the person moves near the window, more light is available and more light is reflected to the retina. Yet your perception does not fit this description of the stimulus events. While the person is moving about the room, you experience him as remaining relatively constant in size and brightness. In spite of dramatic alterations in the proximal stimulus, you experience a stable world. Given this discrepancy between proximal-stimulus events and experience, the organism must actively process information to produce the stability in his world of experience.

Psychologists are not in total agreement as to how this information-processing takes place, but certain general characteristics of the organism's contribution are apparent. The organism seems to seek *invariance*; that is, he perceives as constant those aspects of the physical world which are most enduring, e.g. size and shape, even though the information he has about them may change radically. The perceived invariance seems to depend on the ability of the organism to combine information from different sources, and to result from the application of equations which define proximal stimulation as a joint function of the distal stimulus (the object) and environmental mediating factors, such as distance and incident illumination. For example, our person moving about the room is always the same height, say six feet. The height of the retinal image, on the other hand, varies, but it is always a constant direct function of his height and an inverse function of his distance from the observer. An invariant function exists:

$$\text{Proximal size} = K \times \frac{\text{Distal size}}{\text{Distance}}$$

Figure 13.2 illustrates the relationships. Note that K is the distance from the lens to the retina, which is assumed to be constant. The invariant relationship allows the formula to be 'solved' by the perceiver; e.g., knowing retinal size and estimating distance, one can arrive at an estimate of the size of the object. By applying this invariant relationship to a particular case, the perceiver can account for variation in proximal size and perceive the object as of a constant size, as he knows from other experiences it must be. Finding invariance by applying

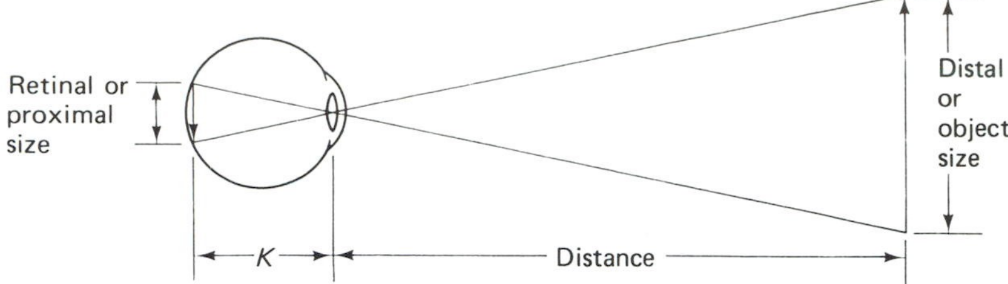

Figure 13.2

relationships such as the above requires the processing of considerable amounts of information, including the physical size of the object, distance, and illumination; and the information-processing involved in this kind of perceptual act must be quite complicated.

Let us think of the perceptual act as a complex form of problem-solving, the goal of which is to create a stability in which our perceptions bear some relationship to external events. We can then draw an analogy between perceptual problem-solving and scientific problem-solving. Just as the scientist attempts to reduce a complex jumble of events to a small set of variables which serve as a set of determining conditions for a particular event, so we search out the invariant aspects of a situation to produce stable perceptions. The scientist searches for invariance in order to understand and to predict the world; we as perceivers also seek to understand and to predict the world in order that we may behave in it to our advantage. In other words, the perceptual act can be said to generate a prediction that we can use as a basis for action. The goal in both cases is predictability of the environment, and the means to the goal is the specification of causal relationships.

Our world of experience is meaningful

The connotation of 'meaningful' here is that structured and stable events are not isolated from one another but appear to be related in some orderly fashion over time. Both structure and stability are probably necessary for meaning to exist. It is so common for the world of experience to make sense to us that the most powerful way to point out the importance of the phenomenon is to suggest trying to conceive of a world that does not make sense. Events would follow each other with no apparent causal relationships. Almost every event would entail surprise. Nothing would seem familiar. The general experience would be one of chaos. Such a state of affairs is so alien to our everyday-life experience that it is extremely difficult to imagine. Our experiences usually *are* meaningful in that they are structured and they are stable; they are related in the sense that they seem familiar, but particularly in the sense that events have implications for one another.

We must look at the organism as an active processor of input stimuli who categorizes stimulus events and relates them to both past and present events. One property of the organism as an information processor is that he possesses a linguistic coding system which possesses a set of implicative relationships. The impinging stimuli provide the raw material; the organism, with the aid of language, produces the meaning. The organism exists in time: he has a past and he anticipates the future. Past experience, language, and present motivational state or goals for the future influence our perceptions of the present. Our past learning has a significant influence on perception, but it operates within a framework of purposive activity. The experience-derived rules we apply are selected by the purposes we are trying to accomplish. The perceptual process is an achievement by the organism, and perception would not exist without active problem-solving on the part of the perceiver. Our perceptions do have meaning, they do make sense; and meaning and sense derive from both our own past experiences and our present purposes. Without the presence of meaning and sense as active, organizing agents, perception, as we know it, would not exist.

14

Social communication

Stuart Sigman

From Sigman, Stuart 1987: *A perspective on social communication.* New York: Lexington Books, 4–13.

Golding and Murdock (1978) suggest that no theory of communication can be forthcoming without a commensurate theory of society and of the place of communication processes in society. Birdwhistell similarly writes, 'Research on human communication as a systematic and structured organization could not be initiated until we had some idea about the organization of society itself' (1970, 72). The organic approach to communication takes as its starting point society as an 'objective facticity' (Berger, 1963), as an analytic abstraction as real as any other utilized by science (Lundberg, 1939; Mandelbaum, 1973). In Durkheim's (1938) words, society is an entity *sui generis*, and the behavioural performances of society members are conditioned by regularities that exist at the social level (O'Neill [1973] provides essays on the converse position, methodological individualism). This social level is supraindividual and multi-generational, comprising a history and pattern of behaviour transcending any individual's biography, yet organizing and integrating the behaviour of society members (cf. Kroeber, 1963).

Social communication theory, which is the particular variant of the organic model to be developed in this book (for an alternative, see Watzlawick, Beavin, and Jackson, 1967), is based in part on considerations of society as both a structural entity and a supraindividual behavioural process. Communication in this view is seen not as an individual-level phenomenon, but as a societal-level one. More specifically, communication is conceived of as a process that functions to integrate and support the diverse components of society (or of selected subgroups), of which separate actors are but one of many. Birdwhistell (1970) defines communication as the dynamic or processual aspect of social structure, and as that behavioural organization which facilitates orderly multisensorial interaction. Scheflen similarly writes, 'Communication can be defined . . . as the integrated system of behaviour that mediates, regulates, sustains and makes possible human relationships. We can, therefore, think of communication as the mechanism of social organization.' (1965, 26). Communication is considered to be a means by which the biosocial interdependencies of species members are maintained (cf. Lundberg, 1939; LaBarre, 1954), and a means by which their behaviour is predictably ordered.

Social communication theory suggests that communication not be reductively defined as a process through which individual cognitions are exchanged, or as a process of information transmission between isolates (for example, 'senders' and 'receivers'). While it is certainly the case that individuals transmit messages to others about internal states and often do so intentionally, it is suggested that interpersonal (or interindividual) messages do not typify or exhaust the human capacity and the group requirement for communication. Pittenger, Hockett, and Danehy write in this regard:

> It is not really useful to think of individuals as the units out of which groups and societies are constructed; it is more fruitful to think of an individual [or the dyad] as the limiting case of a group when, for the moment, there is no one else around. It is treacherously misleading to think of language and other communication systems as cloaks donned by the ego when it ventures into the interpersonal world: rather, we think of ego (or 'mind') as arising from the internalization of interpersonal communicative processes. (1960, 223)

Just as individual persons are moments in society, so interpersonal behaviour is a moment in social communication.

Communication thus appears as a concomitant of all social life and as an exigency of social survival in general, not primarily as a consequence of either individual motivation or initiative. Indeed, from a social communication perspective, motivation and initiative are socially programmed features of mind, that is, delimited options for the appearance of consciousness, emotional display, and verbal reference to consciousness and conduct (cf. Blum and McHugh, 1971; Hochschild, 1983; Denzin, 1984). For example, Heritage and Watson's (1979) analysis of one structural element of conversation, namely, 'formulations', demonstrates that personality attributions or assessments require an understanding not of the presumed 'internal' workings of people but of *standard and non-standard courses of action* available to society members. 'Humourless', 'unsympathetic', 'hostile', and so forth are evidenced in interactants' adherence and non-adherence to specific sequential relationships between and among behaviour units. Personalities do not 'cause' persons to act in certain ways; certain ways of acting provide materials for the construction of social identities and are subject to social regulation and definition. Goffman's focus on the behaviour of interaction is influential here: 'I assume that the proper study of interaction is not the individual and his psychology, but rather the syntactical relations among the acts of different persons mutually present to one another' (1967, 2). In other words, emphasis is placed on the organizing patterns (rules) of behaviour and not on traits of persons.

It follows from this discussion that communication can be defined as a process of information handling – including activities of production, dissemination, reception, and storage – within a social system (cf. Wilden, 1979). This process provides for members' behavioural predictability and ensures societal (group) continuity. Communication is the means by which social reality is created, lived through, sustained, and/or altered. Rather than a process whereby information about an external, 'real' world is shared *by* individuals, it is the mechanism whereby information is used to construct reality *for* individuals (cf. Berger and Luckmann, 1967; Carey, 1975; Pearce and Cronen, 1980). As implied above, it also serves to constitute group membership, to create the social boundaries of persons. Communication establishes meaningful distinctions between and among persons, objects, and behaviours, and defines the structure and goals of interactional events. In this sense, communication is the active or dynamic aspect of social reality, although this does not mean that all persons having membership in a group share identical vantage points on social reality. Rather, as described in greater detail below, there is a supraindividual patterning to the related yet separate social realities, for example, knowledge states and communication rules, which group members are permitted to access. Thus, the suggestion that communication and social reality are related is not intended to mean that individuals, at particular moments of interaction, *construct* or *create* social reality. Instead, social communication proposes that interactants' behaviour serves to *recreate* and *invoke* the historically prior and continuing social reality (cf. McCall and Simmons, 1978).

Communication involves a dynamic structure that allows or prohibits various orders of information or message flow.[1] Research emphasis within social communication is directed toward the continuities and regularities of this information flow over time, rather than toward discrete transmission moments. Focus is placed on the socially constructed limitations or rules governing message flow, and on the functioning of these limitations for the social group under consideration.[2] In this manner, the traditional disciplinary concern for speaker/hearers' message production and reception abilities – that is, 'output' and 'input' trait variables – gives way to analyses of the semiotic codes and the social contexts that sustain *continuous* information flow within society. Communication defines and organizes the spatiotemporal features of interactional events; the sensorial contact among persons 'forcing' either focused or unfocused behaviour (Goffman, 1963); the membership and behavioural requirements sustaining each interactional event; and so on.

It is the sense of communication as a continuous social phenomenon that the two axioms 'one cannot not communicate' (Watzlawick, Beavin and Jackson, 1967) and 'nothing never happens' (Smith and Trager, cited by Pittenger, Hockett, and Danehy, 1960, 234) take on meaning. The social environment is in a constant state of messagefulness, although selected channels and participants may be momentarily 'silent'. Communication as a totality takes place and endures even though certain channels are not actively employed at particular moments. Moreover, one cannot not communicate because one indeed does not communicate; rather, *one partially contributes behaviour to the supraindividual process of communication.*

It should be stressed that by postulating communication as a social process, organic communication theorists do not intend for the word *social* to imply simply individuals in association, in collectivities, in the physical and interactional co-presence of others. Rather, communication processes are considered *societal* phenomena, prerequisites for the continuity, integration and adaptation of the social system. This social system comprises interdependent members and patterns of behaviour that transcend the individual (Durkheim, 1933, 1938; Linton, 1936, 1940; Kroeber, 1963; Radcliffe-Brown, 1965; Sorokin, 1947).

For example, communication serves to articulate and sustain the processual aspects of the social division of labour. This division of labour functions to allocate responsibility for the group's process and to integrate the diverse activities constitutive of the social group. Individual contributions to the division of labour are not necessarily identical or equivalent. Aberle and his colleagues write of the significance of communication for the division of social labour thus: 'Communication is indispensable if socialization and role differentiation are to function effectively' (Aberle *et al.*, 1950, 106). Kemper similarly notes, 'If actors are engaged in coaction to complete a task, there is inevitably communication between them. . . . Thus, the arrangement of actors in the division of labour coincides in an important sense with the pattern of communication between them' (1972, 743). The behaviour displayed in particular interactional episodes draws upon codes, or integrated sets of rules, which enable the participants to signal (1) their places in the social hierarchy; (2) the amount and type of work responsibility they may be expected to contribute to various activities and contexts; and (3) the integration of their behaviour with that found in other episodes and on the part of other group members. Joos describes one set of such signals for recruiting persons on behalf of the division of labour as follows:

> The community's survival depends on cooperation; and adequate cooperation depends on recognizing the more and less responsible types of persons around us. We need to identify the natural burden-bearers of the community so that we can give them the responsibility which is heaviest of all: we make them responsible for cooperation itself. (1967, 14–15)

Individuals contribute unequally and incompletely to the behaviour that is constitutive of particular interactional scenes and events. Reciprocally, there is a differential distribution of rules and resources for conduct across societal members (see Poole and McPhee, 1983).

Communication is not a single, temporally linear process; a number of subsidiary processes, temporal laminates, and behavioural consequences can be discerned. For example, Erickson and Shultz (1982) recognize two aspects of the social organization of time: *kairos*, the appropriate or right time for a particular action; and *chronos*, the duration of an activity that is mechanically measurable via clock time. Both orders of time contribute to the regulation and structuring of communication.

The multilevel organization of communication can also be seen in terms of multiple message functions. Birdwhistell, for example, distinguishes two dynamic features of messages, which he labels 'integrational' and 'new informational' communication.

> 'Integrational' communication involves such interaction as invokes common past experiences and is related to the initiation, maintenance or severance of interaction. 'New informational', while symbolically consistent with and made up of past experience, involves the inclusion of information not held in common by the communicants. (1952, 3–4)

In a more recent formulation, Birdwhistell (1970) extends the notion of integrational communication processes to comprise regulation of interaction, maintenance of systemic operations, and cross-referencing of particular message units to those contexts providing for their comprehensibility (see also Scheflen, 1968). Lasswell suggests the recognition of three functional processes (see also Jakobson 1960 and Halliday 1978):

> Our analysis of communication will deal with the specializations that carry on certain functions, of which the following may be clearly distinguished: (1) the surveillance of the environment; (2) the correlation of the parts of society in responding to the environment; (3) the transmission of the social heritage from one generation to the next. (1971, 85)

Of special note here is the idea that communication serves multiple social and interactional ends and that a given unit of behaviour can be seen to fulfil several such functions. In addition, there is an awareness of, and an emphasis on, messages of system stability, continuity, monitoring and/or adaptation, rather than on those associated solely with person change.

The social communication position calling for studies of messages of continuity and predictability is designed to contrast with the mechanical model's apparent emphasis (or overemphasis) on novel information transmissions (Thomas, 1980). I have elsewhere written that within contemporary discourse and interaction analysis, the mechanical approach fails to account for the many messages (or message units) which seemingly function to signal the development, stability, and continuity of the participants' social relationships and group affiliations (Sigman, 1983a). Moreover, I suggest that the examination of the verbalizations exchanged within a single interactional episode for some 'internal' organizing principle essentially precludes consideration of the *continuities of information across episodes* that form the full pattern of the social information system.

It should be noted that from a social system perspective, a behavioural function is not always, nor necessarily, seen as isomorphic with an actor's intentions for producing a particular unit of behaviour (see Radcliffe-Brown, 1965; Merton, 1968; Scheflen, 1973; Sigman, 1980; Giddens, 1984). The usefulness of a distinction between group-level and individual-level (psychological) functions when studying face-to-face interaction is summarized in the following:

> We need not explain the regularities observed in social interaction by reverting to the interests or motives of the individual. Instead, the system of interaction can be treated as having a structure of its own, *sui generis*. . . . Moreover, the structure of the interaction that results may bear little relationship to the motives with which persons entered the situation. (Aldrich, 1972, 171)

In contrast, many definitions of communication explicitly concern themselves with intentional message transmission (cf. Fisher, 1978; Scott, 1977) or delimit that part of social behaviour which is communicative on the basis of presumed intentions (cf. Cushman, 1980). Nevertheless, from the social communication perspective outlined above, an actor's ostensible motive for performing a particular behaviour does not exhaust the regularity of that behaviour or the richness of its contribution to an interactional event and to the social system at large. Although intentionality as a component of some communicative activity cannot be denied, communication analysts should not be limited to this aspect of behaviour. Kockelmans writes that there is a legitimate 'distinction between the meaning a phenomenon has for a society and the meaning the same phenomenon may have for a particular individual who finds himself in this or that particular situation' (1975, 76).

That particular groups define communication in terms of actor intentionality, or differentially judge behaviour as either intentionally or unintentionally performed, is not the same issue as suggesting that the communication theorist examine all examples of socially patterned behaviour. A distinction can be drawn in this manner between a general and culture-specific definition of communication, between a broad social-scientific approach to the study of communication and a narrower native-delimited conception of what does and does not count as communication. Behaviour for which individuals have intentions, that is, for which socially regulated and defined intentions are available and accountable, is not to be accorded different status for research purposes from that behaviour for which no ostensible (verbalizable) intentions are forthcoming. The ranges of acceptable and unacceptable intentions for behaviour are worthy of systematic study (cf. Gergen, 1982; Harré, Clarke and DeCarlo, 1985).

Thus, this position does not deny or overlook social actors' possession and use of knowledge, regarding their conduct and its spatiotemporal location (see Giddens, 1984), but it does situate such knowledge as part of the group's larger system of communication. The very notions of motive, intentionality, individual responsibility, and so on must be seen as semiotic tokens in a more encompassing network of actions and meanings. In this regard, non-Western cultural data remind us that not all groups account for persons' acts in terms of internal psychological states. Rosen writes, for example:

> A person's inner state [in Morocco] is largely irrelevant to an account of events in the world: since motive and intent are discernible in words and deeds, there is no felt need to discuss a person's interior state directly. (1984, 169)
>
> Narrative accounts – whether as conversational exposition, oral history or popular storytelling – work on the assumption that no man acts without contexts; therefore to reveal a person in a variety of circumstances is to reveal him as a social person. (1984, 171)

The more encompassing network of meanings, for example, the vocabularies of motives and the conditions under which persons are socially expected to be responsible for their behaviour (see Gerth and Mills, 1953), is the proper domain of social communication investigations.

The above discussion should not be taken to mean that social communication theorists assume a unilateral correlation between a unit of behaviour and its function. Rather, each functional aspect of the communication process is said to represent only a heuristic abstraction from the total stream of behaviour; communication events and the constituent behaviour 'partials' (see below) may serve numerous functions. Social communication analysts observe such multifunctionality of behaviour in at least three interrelated ways. First, communication behaviour appears to be a hierarchically structured process (cf. Pike, 1967).[3] As such, the contribution or function of communication units on one level of structure may not be the same as on other levels. A unit's functioning may be apparent for some (or all) of the levels considered. As Pittenger, Hockett, and Danehy write, 'In theory, the relative importance of a single small event can be assayed by observing how far its effects 'reverberate' up through more and more inclusive larger events of which it is a constituent' (1960, 250). Moreover, social behaviour is so structured as to involve multiples of hierarchies, and hierarchically arranged hierarchies; the same unit of behaviour may thus be part of, and function within, one or more hierarchies.

A second aspect of multifunctionality concerns the idea that, as noted above, the function and meaning of behaviour for individual participants may not be isomorphic with the function of behaviour for the social group (or for particular subcomponents within a society or group). This again is a consideration of level, although it attends to the relationship among social actors, social systems, and their constitutive behaviour, not to the hierarchical patterning of the behaviour itself. A further consideration here is that the same behaviour may have different consequences (and different message values) for individuals, depending upon where in the social group they occupy statuses. For example, I have previously discussed the different group membership requirements that nursing-home staff members and residents may have for an entering patient; the same behaviour exhibited by a newcomer may be interpreted by one group as a sign of 'health' and 'alertness' while seen by others as a sign of 'mental unfitness' and 'disease', depending on their institutional location and perspective (Sigman, 1982; cf. Annandale, 1985).

Finally, multiple codes constrain the form and contextual placement of participants' behaviour; each such code provides for a meaning or function component of the behaviour (cf. Frentz and Farrell, 1976; Halliday, 1978), and possibly more than one. A simple metaphor may be useful in clarifying this point. A behavioural form in its processing may be thought of as passing through a series of filters or generators before it is finally produced (performed) by a social actor; each filter, representing a particular social-behavioural code, shapes the behaviour according to its own specifications and in interaction with other codes. Interactional behaviour may be assumed to be conditioned by *inter alia* phonological, morphological, discursive and interpersonal regulative principles; the behaviour so produced provides information derived from each one of these semiotic constraints (cf. Halliday, 1978). For example, an utterance such as a greeting – more specifically, the particular form a greeting takes at any one time – may be conditioned by rules for sound production (partially embedding a sociolinguistic message regarding the speaker's socioeconomic background, among other things); social politeness (potentially transmitting information about the speaker's attitude toward the recipient); lexical choice (potentially including information on the speaker's level of education); and so on. In this manner, the multiple codes enable a greeting to serve numerous functions in interaction.

The notion of multifunctionality can be related to the notion of behaviour 'partials' (cf. Birdwhistell, 1970; Wilden, 1979). Individuals only partially possess or evidence all the rules constraining the totality of socially patterned behaviour that constitutes the communication system, and not all individuals share the indentical or fully overlapping codes for conduct (see Hymes, 1974; Bernstein, 1975).[4] Interactional events are composed of members' rule-governed behaviour, yet the behaviour of any individual interactant is an incomplete contribution to these events (cf. Mead, 1934; Birdwhistell, 1970). Interactional events require multiparticipant coordination in much the same way that, for example, a formal dinner party is comprised of, and emerges from, the non-overlapping rules and behaviour units of the domestic, who knows to 'serve from the left and take from the right', and the guest, who knows only to signal completion of one course and readiness for the next. In accordance with the division of labour theory noted above, group members must have available to them (1) behaviour for signalling their knowledge states and places within the overall social system, and (2) routines for coordinating their behaviour partials with those of their fellows.

Poole, Seibold and McPhee write, 'Structures cannot be reduced to "cognitive maps" in individual actors' heads' (1985, 77). Nwoye's (1985) discussion of courtship among the Igbo of Nigeria presents a good illustration of the relevance of the concept of structural partials to the ethnographic study of continuous social communication. In the African group studied by Nwoye, courtship is a family affair proceeding from the man's having asked for the girl's hand in marriage: 'The young man is joined by his parents, his relatives, and friends who start treating and behaving toward the girl as his "wife" ' (1985, 188–9). As a communication event, courtship in this group is spread across numerous interactions and is constituted by the partial contributions of a number of individuals (see also Rosen, 1984). The construction of an individual's social identity is likewise illuminated by the concept of interaction partials: 'Evidence of this possession [of a self] is thoroughly a product of joint ceremonial labour, the part expressed through the individual's demeanour being no more significant than the part conveyed by others through their deferential treatment toward him' (Goffman, 1967, 85).

The notion of partials leads to the methodological proposition that the analytic boundaries of a unit of behaviour be seen beyond any individual group member's body; a unit of communication behaviour may be constituted by the separate yet related contributions of several members. As Scheflen remarks, 'A unit is not necessarily performed by one individual. A given unit, usually performed by one person, may be performed by several interactants on some occasions: e.g., one speaker may start a statement, another finish it' (1965, 20). (See Heritage and Watson's [1979] discussion of the collaborations involved in formulations of conversational topics.) Similarly, behaviour performed by an individual may serve as a partial contribution to a group-level function; forms by several individuals, which complement each other or are integrated with each other, may in combination serve this particular function more completely. Discourse units separated by space and time, that is, occurring in what the social actors consider to be discrete events, can still be judged to constitute a single, albeit complex, message unit (see Sigman, 1983). Birdwhistell (1970) suggests that a communication event may last as briefly as a phoneme or as long as a generation or two. Pearce and Cronen similarly write, 'The patterns of human action that exist in a marriage, a formal organization, or a nation may be extremely complex and take years or generations to emerge' (1980, 161). Leach (1976) notes the message relatedness of symbols separated by many years, for example, as in the case of veils of marriage and of widowhood.

In addition, certain of the partials contributed to the constitution of particular interactional events are only indirectly offered by persons then present in the scene. Rapoport's (1982) discussion of the semiotics of the built environment indicates the importance of various architectural and physical elements for the meaningful construction of interactional events. In such cases, it is reductionistic and simplistic to suggest that the individual architect is communicating *to* the actors, or that the environment is merely a surround ('context') for behaviour; rather, the architect and his/her design can be better seen as meaningful, albeit partial, units of the total scene's structure and composition.

Finally, any one interactional moment is a partial of the larger continuous system of information flow; communication systems, as noted above, are continuous, even though the 'active' and partial message displays of particular individuals may not be.

These observations have implications for the appropriate size of analytic units and the 'ethnographic present' (time framework) selected by researchers for their analyses (see Birdwhistell, 1977). Rather than look at the separate behaviours of various interactants in terms of their apparent role as either 'stimulus' or 'response,' social communication theory highlights the larger scene or event that is accomplished and contributed to by the behaviours. The behaviours of individual interactants can be seen as the partial constituents of communication episodes requiring varying degrees of integration and co-ordination in adherence to the event 'programs' (Scheflen, 1968). When studying interactional events as units of communication constituted by the partial contributions of multiple participants – as in the Igbo case just described – there are several patterns to be explored. The social communication investigator considers (1) the complete array of behaviours constitutive of the event; (2) the assignment of persons, for example, through recruitment and/or self-selection, to the different phases and subsidiary behaviour units of the event; (3) signals available to the participants for co-ordinating their contributions and indicating the progression of the event; and (4) differential outcomes and evaluations of repeated communication events, based on the varying assignments of persons to behavioural contributions.

15

Stereotypes

Walter Lippmann

From Lippmann, Walter 1922: *Public opinion*. London: George Allen and Unwin, 79–94.

I

Each of us lives and works on a small part of the earth's surface, moves in a small circle, and of these acquaintances knows only a few intimately. Of any public event that has wide effects we see at best only a phase and an aspect. This is as true of the eminent insiders who draft treaties, make laws, and issue orders, as it is of those who have treaties framed for them, laws promulgated to them, orders given at them. Inevitably our opinions cover a bigger space, a longer reach of time, a greater number of things, than we can directly observe. They have, therefore, to be pieced together out of what others have reported and what we can imagine.

Yet even the eyewitness does not bring back a naïve picture of the scene.[1] For experience seems to show that he himself brings something to the scene which later he takes away from it, that oftener than not what he imagines to be the account of an event is really a transfiguration of it. Few facts in consciousness seem to be merely given. Most facts in consciousness seem to be partly made. A report is the joint product of the knower and known, in which the rôle of the observer is always selective and usually creative. The facts we see depend on where we are placed, and the habits of our eyes.

An unfamiliar scene is like the baby's world,'one great, blooming, buzzing confusion.'[2] This is the way, says Mr. John Dewey,[3] that any new thing strikes an adult, so far as the thing is really new and strange. 'Foreign languages that we do not understand always seem jibberings, babblings, in which it is impossible to fix a definite, clear-cut, individualized group of sounds. The countryman in the crowded street, the landlubber at sea, the ignoramus in sport at a contest between experts in a complicated game, are further instances. Put an inexperienced man in a factory, and at first the work seems to him a meaningless medley. All strangers of another race proverbially look alike to the visiting stranger. Only gross differences of size or colour are perceived by an outsider in a flock of sheep, each of which is perfectly individualized to the shepherd. A diffusive blur and an indiscriminately shifting suction characterize what we do not understand. The problem of the acquisition of meaning by things, or (stated in another way) of forming habits of simple apprehension, is thus the problem of introducing (1) *definiteness* and *distinction* and (2) *consistency* or *stability* of meaning into what is otherwise vague and wavering.'

But the kind of definiteness and consistency introduced depends upon who introduces them. In a later passage[4] Dewey gives an example of how differently an experienced layman and a chemist might define the word metal. 'Smoothness, hardness, glossiness, and brilliancy, heavy weight for its size . . . the serviceable properties of capacity for being hammered and pulled without breaking, of being softened by heat and hardened by cold, of retaining the shape and form given, of resistance to pressure and decay, would probably be included' in the layman's definition. But the chemist would likely as not ignore these aesthetic and utilitarian qualities, and define a metal as 'any chemical element that enters into combination with oxygen so as to form a base.'

For the most part we do not first see, and then define, we define first and then see. In the great blooming, buzzing confusion of the outer world we pick out what our culture has already defined for us, and we tend to perceive that which we have picked out in the form stereotyped for us by our culture. Of the great men who assembled at Paris to settle the affairs of mankind, how many were there who were able to see much of the Europe about them, rather than their commitments about Europe? Could anyone have penetrated the mind of M. Clemenceau, would he have found these images of the Europe of 1919, or a great sediment of stereotyped ideas accumulated and hardened in a long and pugnacious existence? Did he see the Germans of 1919, or the German type as he had learned to see it since 1871? He saw the type, and among the reports that came to him from Germany, he took to heart those reports, and, it seems, those only, which fitted the type that was in his mind. If a junker blustered, that was an authentic German; if a labour leader confessed the guilt of the empire, he was not an authentic German.

At a Congress of Psychology in Göttingen an interesting experiment was made with a crowd of presumably trained observers.[5]

> 'Not far from the hall in which the Congress was sitting there was a public fête with a masked ball. Suddenly the door of the hall was thrown open and a clown rushed in madly pursued by a negro, revolver in hand. They stopped in the middle of the room fighting; the clown fell, the negro leapt upon him, fired, and then both rushed out of the hall. The whole incident hardly lasted twenty seconds.
>
> 'The President asked those present to write immediately a report since there was sure to be a judicial inquiry. Forty reports were sent in. Only one had less than 20 per cent of mistakes in regard to the principal facts; fourteen had 20 per cent to 40 per cent of mistakes; twelve from 40 per cent to 50 per cent; thirteen more than 50 per cent. Moreover in twenty-four accounts 10 per cent of the details were pure inventions and this proportion was exceeded in ten accounts and diminished in six. Briefly a quarter of the accounts were false.
>
> 'It goes without saying that the whole scene had been arranged and even photographed in advance. The ten false reports may then be relegated to the category of tales and legends; twenty-four accounts are half legendary, and six have a value approximating to exact evidence.'

Thus out of forty trained observers writing a responsible account of a scene that had just happened before their eyes, more than a majority saw a scene that had not taken place. What then did they see? One would suppose it was easier to tell what had occurred, than to invent something which had not occurred. They saw their stereotype of such a brawl. All of them had in the course of their lives acquired a series of images of brawls, and these images flickered before their eyes. In one man these images displaced less than 20 per cent of the actual scene, in thirteen men more than half. In thirty-four out of the forty observers the stereotypes pre-empted at least one-tenth of the scene.

A distinguished art critic has said[6] that 'what with the almost numberless shapes assumed by an object. . . . What with our insensitiveness and inattention, things scarcely would have for us features and outlines so determined and clear that we could recall them at will, but for the stereotyped shapes art has lent them.' The truth is even broader than that, for the stereotyped shapes lent to the world come not merely from art, in the sense of painting and sculpture and literature, but from our moral codes and our social philosophies and our political agitations as well. Substitute in the following passage of Mr. Berenson's the words 'politics,' 'business,' and 'society,' for the word 'art' and the sentences will be no less true: '. . . unless years devoted to the study of all schools of art have taught us also to see with our own eyes, we soon fall into the habit of moulding whatever we look at into the forms borrowed from the one art with which we are acquainted. There is our standard of artistic reality. Let anyone give us shapes and colours which we cannot instantly match in our paltry stock of hackneyed forms and tints, and we shake our heads at his failure to reproduce things as we know they certainly are, or we accuse him of insincerity.'

Mr. Berenson speaks of our displeasure when a painter 'does not visualize objects exactly as we do,' and of the difficulty of appreciating the art of the Middle Ages because since then 'our manner of visualizing forms has changed in a thousand ways.'[7] He goes on to show how in regard to the human figure we have been taught to see what we do see. 'Created by Donatello and Masaccio, and sanctioned by the Humanists, the new canon of the human figure, the new cast of features . . . presented to the ruling classes of that time the type of human being most likely to win the day in the combat of human forces. . . Who had the power to break through this new standard of vision and, out of the chaos of things, to select shapes more definitely expressive of reality than those fixed by men of genius? No one had such power. People had perforce to see things in that way and in no other, and to see only the shapes depicted, to love only the ideals presented. . . .'[8]

II

If we cannot fully understand the acts of other people, until we know what they think they know, then in order to do justice we have to appraise not only the information which has been at their disposal, but the minds through which they have filtered it. For the accepted types, the current patterns, the standard versions, intercept information on its way to consciousness. Americanization, for example, is superficially at least the substitution of American for European stereotypes. Thus the peasant who might see his landlord as if he were the lord of the manor, his employer as he saw the local magnate, is taught by Americanization to see the landlord and employer according to American standards. This constitutes a change of mind, which is, in effect, when the innoculation succeeds, a change of vision. His eye sees differently. One kindly gentlewoman has confessed that the stereotypes are of such overweening importance, that when hers are not indulged, she at least is unable to accept the brotherhood of man and the fatherhood of God: 'we are strangely affected by the clothes we wear. Garments create a mental and social atmosphere. What can be hoped for the Americanism of a man who insists on employing a London tailor? One's very food affects his Americanism. What kind of American consciousness can grow in the atmosphere of sauerkraut and Limburger cheese? Or what can you expect of the Americanism of the man whose breath always reeks of garlic?'[9]

This lady might well have been the patron of a pageant which a friend of mine once attended. It was called the Melting Pot, and it was given on the Fourth of July in an automobile town where many foreign-born workers are employed. In the centre of the baseball park at second base stood a huge wooden and canvas pot. There were flights of steps up to the rim on two sides. After the audience had settled itself, and the band had played, a procession came through an opening at one side of the field. It was made up of men of all the foreign nationalities employed in the factories. They wore their native costumes, they were singing their national songs; they danced their folk dances, and carried the banners of all Europe. The master of ceremonies was the principal of the grade school dressed as Uncle Sam. He led them to the pot. He directed them up the steps to the rim, and inside. He called them out again on the other side. They came, dressed in derby hats, coats, pants, vest, stiff collar and polka-dot tie, undoubtedly, said my friend, each with an Eversharp pencil in his pocket, and all singing the Star-Spangled Banner.

To the promoters of this pageant, and probably to most of the actors, it seemed as if they had managed to express the most intimate difficulty to friendly association between the older peoples of America and the newer. The contradiction of their stereotypes interfered with the full recognition of their common humanity. The people who change their names know this. They mean to change themselves, and the attitude of strangers toward them.

There is, of course, some connection between the scene outside and the mind through which we watch it, just as there are some long-haired men and short-haired women in radical gatherings. But to the hurried observer a slight connection is enough. If there are two bobbed heads and four beards in the audience, it will be a bobbed and bearded audience to the reporter who knows beforehand that such gatherings are composed of people with these tastes in the management of their hair. There is a connection between our vision and the facts, but it is often a strange connection. A man has rarely looked at a landscape, let us say, except to examine its possibilities for division into building lots, but he has seen a number of landscapes hanging in the parlour. And from them he has learned to think of a landscape as a rosy sunset, or as a country road with a church steeple and a silver moon. One day he goes to the country, and for hours he does not see a single landscape. Then the sun goes down looking rosy. At once he recognizes a landscape and exclaims that it is beautiful. But two days later, when he tries to recall what he saw, the odds are that he will remember chiefly some landscape in a parlour.

Unless he has been drunk or dreaming or insane he did see a sunset, but he saw in it, and above all remembers from it, more of what the oil painting taught him to observe, than what an impressionist painter, for example, or a cultivated Japanese would have seen and taken away with him. And the Japanese and the painter in turn will have seen and remembered more of the form they had learned, unless they happen to be the very rare people who find fresh sight for mankind. In untrained observation we pick recognizable signs out of the environment. The signs stand for ideas, and these ideas we fill out with our stock of images. We do not so much see this man and that sunset; rather we notice that the thing is man or sunset, and then see chiefly what our mind is already full of on those subjects.

III

There is economy in this. For the attempt to see all things freshly and in detail, rather than as types and generalities, is exhausting, and among busy affairs practically out of the question. In a circle of friends, and in relation to close associates or competitors, there is no shortcut through, and no substitute for, an individualized understanding. Those who we love and admire most are the men and women whose consciousness is peopled thickly with persons rather than with types, who know us rather than the classification into which we might fit. For even without phrasing it to ourselves, we feel intuitively that all classification is in relation to some purpose not necessarily our own; that between two human beings no association has final dignity in which each does not take the other as an end in himself. There is a taint on any contact between two people which does not affirm as an axiom the personal inviolability of both.

But modern life is hurried and multifarious, above all physical distance separates men who are often in vital contact with each other, such as employer and employee, official and voter. There is neither time nor opportunity for intimate acquaintance. Instead we notice a trait which marks a well known type, and fill in the rest of the picture by means of the stereotypes we carry about in our heads. He is an agitator. That much we notice, or are told. Well, an agitator is this sort of person, and so *he* is this sort of person. He is an intellectual. He is a plutocrat. He is a foreigner. He is a 'South European.' He is from Back Bay. He is a Harvard Man. How different from the statement: he is a Yale Man. He is a regular fellow. He is a West Pointer. He is an old army sergeant. He is a Greenwich Villager: what don't we know about him then, and about her? He is an international banker. He is from Main Street.

The subtlest and most pervasive of all influences are those which create and maintain the repertory of stereotypes. We are told about the world before we see it. We imagine most things before we experience them. And those preconceptions, unless education has made us acutely aware, govern deeply the whole process of perception. They mark out certain objects as familiar or strange, emphasizing the difference, so that the slightly familiar is seen as very familiar, and the somewhat strange as sharply alien. They are aroused by small signs, which may vary from a true index to a vague analogy. Aroused, they flood fresh vision with older images, and project into the world what has been resurrected in memory. Were there no practical uniformities in the environment, there would be no economy and only error in the human habit of accepting foresight for sight. But there are uniformities sufficiently accurate, and the need of economizing attention is so inevitable, that the abandonment of all stereotypes for a wholly innocent approach to experience would impoverish human life.

What matters is the character of the stereotypes, and the gullibility with which we employ them. And these in the end depend upon those inclusive patterns which constitute our philosophy of life. If in that philosophy we assume that the world is codified according to a code which we possess, we are likely to make our reports of what is going on describe a world run by our code. But if our philosophy tells us that each man is only a small part of the world, that his intelligence catches at best only phases and aspects in a coarse net of ideas, then, when we use our stereotypes, we tend to know that they are only stereotypes, to hold them lightly, to modify them gladly. We tend, also, to realize more and more clearly when our ideas started, where they started, how they came to us, why we accepted them. All useful history is antiseptic in this fashion. It enables us to know what fairytale, what school book, what tradition, what novel, play, picture, phrase, planted one preconception in this mind, another in that mind.

IV

Those who wish to censor art do not at least underestimate this influence. They generally misunderstand it, and almost always they are absurdly bent on preventing other people from discovering anything not sanctioned by them. But at any rate, like Plato in his argument about the poets, they feel vaguely that the types acquired through fiction tend to be imposed on reality. Thus there can be little doubt that the moving picture is steadily building up imagery which is then evoked by the words people read in their newspapers. In the whole experience of the race there has been no aid to visualization comparable to the cinema. If a Florentine wished to visualize the saints, he could go to the frescoes in his church, where he might see a vision of saints standardized for his time by Giotto. If an Athenian wished to visualize the gods he went to the temples. But the number of objects which were pictured was not great. And in the East, where the spirit of the second commandment was widely accepted, the portraiture of concrete things was even more meagre, and for that reason perhaps the faculty of practical decision was by so much reduced. In the western world, however, during the last few centuries there has been an enormous increase in the volume and scope of secular description, the word picture, the narrative, the illustrated narrative, and finally the moving picture and, perhaps, the talking picture.

Photographs have the kind of authority over imagination to-day, which the printed word had yesterday, and the spoken word before that. They seem utterly real. They come, we imagine, directly to us without human meddling, and they are the most effortless food for the mind conceivable. Any description in words, or even any inert picture, requires an effort of memory before a picture exists in the mind. But on the screen the whole process of observing, describing, reporting, and then imagining, has been accomplished for you. Without more trouble than is needed to stay awake the result which your imagination is always aiming at is reeled off on the screen. The shadowy idea becomes vivid; your hazy notion, let us say, of the Klu Klux Klan, thanks to Mr. Griffiths, takes vivid shape when you see the Birth of a Nation. Historically it may be the wrong shape, morally it may be a pernicious shape, but it is a shape, and I doubt whether anyone who has seen the film and does not know more about the Klu Klux Klan than Mr. Griffiths, will ever hear the name again without seeing those white horsemen.

V

And so when we speak of the mind of a group of people, of the French mind, the militarist mind, the bolshevik mind, we are liable to serious confusion unless we agree to separate the instinctive equipment from the stereotypes, the patterns, and the formulæ which play so decisive a part in building up the mental world to which the native character is adapted and responds. Failure to make this distinction accounts for oceans of loose talk about collective minds, national souls, and race psychology. To be sure a stereotype may be so consistently and authoritatively transmitted in each generation from parent to child that it seems almost like a biological fact. In some respects, we may indeed have become, as Mr. Wallas says,[10] biologically parasitic upon our social heritage. But certainly there is not the least scientific evidence which would enable anyone to argue that men are born with the political habits of the country in which they are born. In so far as political habits are alike in a nation, the first places to look for an explanation are the nursery, the school, the church, not in that limbo inhabited by Group Minds and National Souls. Until you have thoroughly failed to see tradition being handed on from parents, teachers, priests, and uncles, it is a solecism of the worst order to ascribe political differences to the germ plasm.

It is possible to generalize tentatively and with a decent humility about comparative differences within the same category of education and experience. Yet even this is a tricky enterprise. For almost no two experiences are exactly alike, not even of two children in the same household. The older son never does have the experience of being the younger. And therefore, until we are able to discount the differences in nurture, we must withhold judgment about differences of nature. As well judge the productivity of two soils by comparing their yield before you know which is in Labrador and which in Iowa, whether they have been cultivated and enriched, exhausted, or allowed to run wild.

16

They saw a game: a case study

Albert H. Hastorf and Hadley Cantril

From Hastorf, Albert H. and Cantril, Hadley 1954: They saw a game: a case study. *Journal of Abnormal and Social Psychology* 49, 129–34.

On a brisk Saturday afternoon, November 23, 1951, the Dartmouth football team played Princeton in Princeton's Palmer Stadium. It was the last game of the season for both teams and of rather special significance because the Princeton team had won all its games so far and one of its players, Kazmaier, was receiving All-American mention and had just appeared as the cover man on *Time* magazine, and was playing his last game.

A few minutes after the opening kick-off, it became apparent that the game was going to be a rough one. The referees were kept busy blowing their whistles and penalizing both sides. In the second quarter, Princeton's star left the game with a broken nose. In the third quarter, a Dartmouth player was taken off the field with a broken leg. Tempers flared both during and after the game. The official statistics of the game, which Princeton won, showed that Dartmouth was penalized 70 yards, Princeton 25, not counting more than a few plays in which both sides were penalized.

Needless to say, accusations soon began to fly. The game immediately became a matter of concern to players, students, coaches, and the administrative officials of the two institutions, as well as to alumni and the general public who had not seen the game but had become sensitive to the problem of big-time football through the recent exposures of subsidized players, commercialism, etc. Discussion of the game continued for several weeks.

One of the contributing factors to the extended discussion of the game was the extensive space given to it by both campus and metropolitan newspapers. An indication of the fervour with which the discussions were carried on is shown by a few excerpts from the campus dailies.

For example, on November 27 (four days after the game), the *Daily Princetonian* (Princeton's student newspaper) said:

> This observer has never seen quite such a disgusting exhibition of so-called 'sport.' Both teams were guilty but the blame must be laid primarily on Dartmouth's doorstep. Princeton, obviously the better team, had no reason to rough up Dartmouth. Looking at the situation rationally, we don't see why the Indians should make a deliberate attempt to cripple Dick Kazmaier or any other Princeton player. The Dartmouth psychology, however, is not rational itself.

The November 30th edition of the *Princeton Alumni Weekly* said:

> But certain memories of what occurred will not be easily erased. Into the record books will go in indelible fashion the fact that the last game of Dick Kazmaier's career was cut short by more than half when he was forced out with a broken nose and a mild concussion, sustained from a tackle that came well after he had thrown a pass.
>
> This second-period development was followed by a third quarter outbreak of roughness that was climaxed when a Dartmouth player deliberately kicked Brad Glass in the ribs while the latter was on his back. Throughout the often unpleasant afternoon, there was undeniable evidence that the losers' tactics were the result of an actual style of play, and reports on other games they have played this season substantiate this.

Dartmouth students were 'seeing' an entirely different version of the game through the editorial eyes of the *Dartmouth* (Dartmouth's undergraduate newspaper). For example, on November 27 the *Dartmouth* said:

> However, the Dartmouth-Princeton game set the stage for the other type of dirty football. A type which may be termed as an unjustifiable accusation.
>
> Dick Kazmaier was injured early in the game. Kazmaier was the star, an All-American. Other stars have been injured before, but Kazmaier had been built to represent a Princeton idol. When an idol is hurt there is only one recourse – the tag of dirty football. So what did the Tiger Coach Charley Caldwell do? He announced to the world that the Big Green had been out to extinguish the Princeton star. His purpose was achieved.
>
> After this incident, Caldwell instilled the old see-what-they-did-go-get-them attitude into his players. His talk got results. Gene Howard and Jim Miller were both injured. Both had dropped back to pass, had passed, and were standing unprotected in the backfield. Result: one bad leg and one leg broken.
>
> The game was rough and did get a bit out of hand in the third quarter. Yet most of the roughing penalties were called against Princeton while Dartmouth received more of the illegal-use-of-the-hands variety.

On November 28 the *Dartmouth* said:

> Dick Kazmaier of Princeton admittedly is an unusually able football player. Many Dartmouth men traveled to Princeton, not expecting to win – only hoping to see an All-American in action. Dick Kazmaier was hurt in the second period, and played only a token part in the remainder of the game. For this, spectators were sorry.
>
> But there were no such feelings for Dick Kazmaier's health. Medical authorities have confirmed that as a relatively unprotected passing and running star in a contact sport, he is quite liable to injury. Also, his particular injuries – a broken nose and slight concussion – were no more serious than is experienced almost any day in any football practice, where there is no more serious stake than playing the following Saturday. Up to the Princeton game, Dartmouth players suffered about 10 known nose fractures and face injuries, not to mention several slight concussions.
>
> Did Princeton players feel so badly about losing their star? They shouldn't have. During the past undefeated campaign they stopped several individual stars by a concentrated effort, including such mainstays as Frank Hauff of Navy, Glenn Adams of Pennsylvania and Rocco Calvo of Cornell.
>
> In other words, the same brand of football condemned by the *Prince* – that of stopping the big man – is practiced quite successfully by the Tigers.

Basically, then, there was disagreement as to what had happened during the 'game.' Hence we took the opportunity presented by the occasion to make a 'real life' study of a perceptual problem.[1]

Procedure

Two steps were involved in gathering data. The first consisted of answers to a questionnaire designed to get reactions to the game and to learn something of the climate of opinion in each institution. This questionnaire was administered a week after the game to both Dartmouth and Princeton undergraduates who were taking introductory and intermediate psychology courses.

The second step consisted of showing the same motion picture of the game to a sample of undergraduates in each school and having them check on another questionnaire, as they watched the film, any infraction of the rules they saw and whether these infractions were 'mild' or 'flagrant.'[2] At Dartmouth, members of two fraternities were asked to view the film on December 7; at Princeton, members of two undergraduate clubs saw the film early in January.

The answers to both questionnaires were carefully coded and transferred to punch cards.[3]

Results

Table 16.1 shows the questions which received different replies from the two student populations on the first questionnaire.

Questions asking if the students had friends on the team, if they had ever played football themselves, if they felt they knew the rules of the game well, etc. showed no differences in either school and no relation to answers given to other questions. This is not surprising since the students in both schools come from essentially the same type of educational, economic, and ethnic background.

Summarizing the data of Tables 16.1 and 16.2, we find a marked contrast between the two student groups.

Nearly all *Princeton* students judged the game as 'rough and dirty' – not one of them thought it 'clean and fair.' And almost nine-tenths of them thought the other side started the rough play. By and large they felt that the charges they understood were being made were true; most of them felt the charges were made in order to avoid similar situations in the future.

When Princeton students looked at the movie of the game, they saw the Dartmouth team make over twice as many infractions as their own team made. And they saw the Dartmouth team make over twice as many infractions as were seen by Dartmouth students. When Princeton students judged these infractions as 'flagrant' or 'mild,' the ratio was about two 'flagrant' to one 'mild' on the Dartmouth team, and about one 'flagrant' to three 'mild' on the Princeton team.

As for the *Dartmouth* students, while the plurality of answers fell in the 'rough and dirty' category, over one-tenth thought the game was 'clean and fair' and over a third introduced their own category of 'rough and fair' to describe the action. Although a third of the Dartmouth students felt that Dartmouth was to blame for starting the rough play, the majority of Dartmouth students thought both sides were to blame. By and large, Dartmouth men felt that the charges they understood were being made were not true, and most of them thought the reason for the charges was Princeton's concern for its football star.

When Dartmouth students looked at the movie of the game they saw both teams make about the same number of infractions. And they saw their own team make only half the number of infractions the Princeton students saw them make.

The ratio of 'flagrant' to 'mild' infractions was about one to one when Dartmouth students judged the Dartmouth team, and about one 'flagrant' to two 'mild' when Dartmouth students judged infractions made by the Princeton team.

It should be noted that Dartmouth and Princeton students were thinking of different charges in judging their validity and in assigning reasons as to why the charges were made. It should also be noted that whether or not students were spectators of the game in the stadium made little difference in their responses.

Interpretation: the nature of a social event[4]

It seems clear that the 'game' actually was many different games and that each version of the events that transpired was just as 'real' to a particular person as other versions were to other people. A consideration of the experiential phenomena that constitute a 'football game' for the spectator may help us both to account for the results obtained and illustrate something of the nature of any social event.

Like any other complex social occurrence, a 'football game' consists of a whole host of happenings. Many different events are occurring simultaneously. Furthermore, each happening is a link in a chain of happenings, so that one follows another in sequence. The 'football game,' as well as other complex social situations, consists of a whole matrix of events. In the game situation, this matrix of events consists of the actions of all the players, together with the behaviour of the referees and linesmen, the action on the sidelines, in the grandstands, over the loud-speaker, etc.

Of crucial importance is the fact that an 'occurrence' on the football field or in any other social situation does not become an experiential 'event' unless and until some significance is given to it: an 'occurrence' becomes an *'event'* only when the happening has significance. And a happening generally has significance only if it reactivates learned significances already registered in what we have called a person's assumptive form-world (Table 16.1).

Hence the particular occurrences that different people experienced in the football game were a limited series of events from the total matrix of events *potentially* available to them. People experienced those occurrences that reactivated significances they brought to the occasion; they failed to experience those occurrences which did not reactivate past significances. We do not need to introduce 'attention' as an 'intervening third' (to paraphrase James on memory) to account for the selectivity of the experiential process.

In this particular study, one of the most interesting examples of this phenomenon was a telegram sent to an officer of Dartmouth College by a member of a Dartmouth alumni group in the Midwest. He had viewed the film which had been shipped to his alumni group from Princeton after its use with Princeton students, who saw, as we noted, an average of over nine infractions by Dartmouth players during the game. The alumnus, who couldn't see the infractions he had heard publicized, wired:

> Preview of Princeton movies indicates considerable cutting of important part please wire explanation and possibly air mail missing part before showing scheduled for January 25 we have splicing equipment.

The 'same' sensory impingements emanating from the football field, transmitted through the visual mechanism to the brain, also obviously gave rise to different experiences in different people. The significances assumed by different happenings for different people depend in large part on the purposes people bring to the occasion and the assumptions they have of other people involved. This was amusingly pointed out by the New York *Herald Tribune's* sports columnist, Red Smith, in describing a prize fight between Chico Vejar and Carmine Fiore in his column of December 21, 1951. Among other things, he wrote:

> You see, Steve Ellis is the proprietor of Chico Vejar, who is a highly desirable tract of Stamford, Conn., welterweight. Steve is also a radio announcer. Ordinarily there is no conflict between Ellis the Brain and Ellis the Voice because Steve is an uncommonly substantial lump of meat who can support both halves of a split personality and give away weight on each end without missing it.
>
> This time, though, the two Ellises met head-on, with a sickening, rending crash. Steve the Manager sat at ringside in the guise of Steve the Announcer broadcasting a dispassionate, unbiased, objective report of Chico's adventures in the ring . . . .
>
> Clear as mountain water, his words came through, winning big for Chico. Winning? Hell, Steve was slaughtering poor Fiore.
>
> Watching and listening, you could see what a valiant effort the reporter was making to remain cool and detached. At the same time you had an illustration of the old, established truth that when anybody with a preference watches a fight, he sees only what he prefers to see.
>
> That is always so. That is why, after any fight that doesn't end in a clean knockout, there always are at least a few hoots when the decision is announced. A guy from, say, Billy Graham's neighborhood goes to see Billy fight and he watches Graham all the time. He sees all the punches Billy throws, and hardly any of the punches Billy catches. So it was with Steve.
>
> 'Fiore feints with a left,' he would say, honestly believing that Fiore hadn't caught Chico full on the chops. 'Fiore's knees buckle,' he said, 'and Chico backs away.' Steve didn't see the hook that had driven Chico back. . . .

In brief, the data here indicate that there is no such 'thing' as a 'game' existing 'out there' in its own right which people merely 'observe.' The 'game' 'exists' for a person and is experienced by him only in so far as certain happenings have significances in terms of his purpose. Out of all the occurrences going on in the environment, a person selects those that have some significance for him from his own egocentric position in the total matrix.

Obviously in the case of a football game, the value of the experience of watching the game is enhanced if the purpose of 'your' team is accomplished, that is, if the happening of the desired consequence is experienced – i.e., if your team wins. But the value attribute of the experience can, of course, be spoiled if the desire to win crowds out behaviour we value and have come to call sportsmanlike.

The sharing of significances provides the links except for which a 'social' event would not be experienced and would not exist for anyone.

A 'football game' would be impossible except for the rules of the game which we bring to the situation and which enable us to share with others the significances of various happenings. These rules make possible a certain repeatability of events such as first downs, touchdowns, etc. If a person is unfamiliar with the rules of the game, the behaviour he sees lacks repeatability and consistent significance and hence 'doesn't make sense.'

And only because there is the possibility of repetition is there the possibility that a happening has a significance. For example, the balls used in games are designed to give a high degree of repeatability. While a football is about the only ball used in games which is not a sphere, the shape of the modern football has apparently evolved in order to achieve a higher degree of accuracy and speed in forward passing than would be obtained with a spherical ball, thus increasing the repeatability of an important phase of the game.

The rules of a football game, like laws, rituals, customs, and mores, are registered and preserved forms of sequential significances enabling people to share the significances of occurrences. The sharing of sequential significances which have value for us provides the links that operationally make social events possible. They are analogous to the forces of attraction that hold parts of an atom together, keeping each part from following its individual, independent course.

From this point of view it is inaccurate and misleading to say that different people have different 'attitudes' concerning the same 'thing.' For the 'thing' simply is *not* the same for different people whether the 'thing' is a football game, a presidential candidate, Communism, or spinach. We do not simply 'react to' a happening or to some impingement from the environment in a determined way (except in behaviour that has become reflexive or habitual). We behave according to what we bring to the occasion, and what each of us brings to the occasion is more or less unique. And except for these significances which we bring to the occasion, the happenings around us would be meaningless occurrences, would be 'inconsequential.'

From the transactional view, an attitude is not a predisposition to react in a certain way to an occurrence or stimulus 'out there' that exists in its own right with certain fixed characteristics which we 'colour' according to our predisposition (Table 16.2). That is, a subject does not simply 'react to' an 'object'. An attitude would rather seem to be a complex of registered significances reactivated by some stimulus which assumes its own particular significance for us in terms of our purposes. That is, the object as experienced would not exist for us except for the reactivated aspects of the form-world which provide particular significance to the hieroglyphics of sensory impingements.

TABLE 16.1 Data from First Questionnaire

Question	Dartmouth Students (N = 163) %	Princeton Students (N = 161) %
1 Did you happen to see the actual game between Dartmouth and Princeton in Palmer Stadium this year?		
Yes	33	71
No	67	29
2 Have you seen a movie of the game or seen it on television?		
Yes, movie	33	2
Yes, television	0	1
No, neither	67	97

3 (Asked of those who answered 'yes' to either or both of above questions.) From your observations of what went on at the game, do you believe the game was clean and fairly played, or that it was unnecessarily rough and dirty?		
Clean and fair	6	0
Rough and dirty	24	69
Rough and fair*	25	2
No answer	45	29
4 (Asked of those who answered 'no' on both of the first questions.) From what you have heard and read about the game, do you feel it was clean and fairly played, or that it was unnecessarily rough and dirty?		
Clean and fair	7	0
Rough and dirty	18	24
Rough and fair*	14	1
Don't know	6	4
No answer	55	71
(Combined answers to questions 3 and 4 above)		
Clean and fair	13	0
Rough and dirty	42	93
Rough and fair*	39	3
Don't know	6	4
5 From what you saw in the game or the movies, or from what you have read, which team do you feel started the rough play?		
Dartmouth started it	36	86
Princeton started it	2	0
Both started it	53	11
Neither	6	1
No answer	3	2
6 What is your understanding of the charges being made?**		
Dartmouth tried to get Kazmaier	71	47
Dartmouth intentionally dirty	52	44
Dartmouth unnecessarily rough	8	35
7 Do you feel there is any truth to these charges?		
Yes	10	55
No	57	4

Partly	29	35
Don't know	4	6
8 Why do you think the charges were made?		
Injury to Princeton star	70	23
To prevent repetition	2	46
No answer	28	31

*This answer was not included on the checklist but was written in by the percentage of students indicated.

**Replies do not add to 100% since more than one charge could be given.

TABLE 16.2 Data from Second Questionnaire Checked While Seeing Film

Group	Number	Total Number of Infractions Checked Against			
		Dartmouth Team		Princeton Team	
		Mean	*SD**	Mean	*SD**
Dartmouth students	48	4.3**	2.7	4.4	2.8
Princeton students	49	9.8**	5.7	4.2	3.5

*Standard Deviation

**Significant at the .1 level

17

Introduction to the presentation of self in everyday life

Erving Goffman

From Goffman, Erving 1959: *The presentation of self in everyday life.* New York: Doubleday Anchor. Extract taken from Penguin books edition: Harmondsworth 1969, 14–27.

When an individual enters the presence of others, they commonly seek to acquire information about him or to bring into play information about him already possessed. They will be interested in his general socio-economic status, his conception of self, his attitude towards them, his competence, his trustworthiness, etc. Although some of this information seems to be sought almost as an end in itself, there are usually quite practical reasons for acquiring it. Information about the individual helps to define the situation, enabling others to know in advance what he will expect of them and what they may expect of him. Informed in these ways, the others will know how best to act in order to call forth a desired response from him.

For those present, many sources of information become accessible and many carriers (or 'sign-vehicles') become available for conveying this information. If unacquainted with the individual, observers can glean clues from his conduct and appearance which allow them to apply their previous experience with individuals roughly similar to the one before them or, more important, to apply untested stereotypes to him. They can also assume from past experience that only individuals of a particular kind are likely to be found in a given social setting. They can rely on what the individual says about himself or on documentary evidence he provides as to who and what he is. If they know, or know of, the individual by virtue of experience prior to the interaction, they can rely on assumptions as to the persistence and generality of psychological traits as a means of predicting his present and future behaviour.

However, during the period in which the individual is in the immediate presence of the others, few events may occur which directly provide the others with the conclusive information they will need if they are to direct wisely their own activity. Many crucial facts lie beyond the time and place of interaction or lie concealed within it. For example, the 'true' or 'real' attitudes, beliefs, and emotions of the individual can be ascertained only indirectly, through his avowals or through what appears to be involuntary expressive behaviour. Similarly, if the individual offers the others a product or service, they will often find that during the interaction there will be no time and place immediately available for eating the pudding that the proof can be found in. They will be forced to accept some events as conventional or natural signs of something not directly available to the senses. In Ichheiser's terms,[1] the individual will have to act so that he intentionally or unintentionally *expresses* himself, and the others will in turn have to be *impressed* in some way by him.

The expressiveness of the individual (and therefore his capacity to give impressions) appears to involve two radically different kinds of sign activity: the expression that he *gives*, and the expression that he *gives off*. The first involves verbal symbols or their substitutes which he uses admittedly and solely to convey the information that he and the others are known to attach to these symbols. This is communication in the traditional and narrow sense. The second involves a wide range of action that others can treat as symptomatic of the actor, the expectation being that the action was performed for reasons other than the information conveyed in this way. As we shall have to see, this distinction has an only initial validity. The individual does of course intentionally convey misinformation by means of both of these types of communication, the first involving deceit, the second feigning.

Taking communication in both its narrow and broad sense, one finds that when the individual is in the immediate presence of others, his activity will have a promissory character. The others are likely to find that they must accept the individual on faith, offering him a just return while he is present before them in exchange for something whose true value will not be established until after he has left their presence. (Of course, the others also live by inference in their dealings with the physical world, but it is only in the world of social interaction that the objects about which they make inferences will purposely facilitate and hinder this inferential process.) The security that they justifiably feel in making inferences about the individual will vary, of course, depending on such factors as the amount of information they already possess about him, but no amount of such past evidence can entirely obviate the necessity of acting on the basis of inferences. As William I. Thomas suggested:

> It is also highly important for us to realize that we do not as a matter of fact lead our lives, make our decisions, and reach our goals in everyday life either statistically or scientifically. We live by inference. I am, let us say, your guest. You do not know, you cannot determine scientifically, that I will not steal your money or your spoons. But inferentially I will not, and inferentially you have me as a guest.[2]

Let us now turn from the others to the point of view of the individual who presents himself before them. He may wish them to think highly of him, or to think that he thinks highly of them, or to perceive how in fact he feels towards them, or to obtain no clear-cut impression; he may wish to ensure sufficient harmony so that the interaction can be sustained, or to defraud, get rid of, confuse, mislead, antagonize, or insult them. Regardless of the particular objective which the individual has in mind and of his motive for having his objective, it will be in his interests to control the conduct of the others, especially their responsive treatment of him.[3] This control is achieved largely by influencing the definition of the situation which the others come to formulate, and he can influence this definition by expressing himself in such a way as to give them the kind of impression that will lead them to act voluntarily in accordance with his own plan. Thus, when an individual appears in the presence of others, there will usually be some reason for him to mobilize his activity so that it will convey an impression to others which it is in his interests to convey.

Since a girl's dormitory mates will glean evidence of her popularity from the calls she receives on the phone, we can suspect that some girls will arrange for calls to be made, and Willard Waller's finding can be anticipated:

> It has been reported by many observers that a girl who is called to the telephone in the dormitories will often allow herself to be called several times, in order to give all the other girls ample opportunity to hear her paged.[4]

Of the two kinds of communication – expressions given and expressions given off – this report will be primarily concerned with the latter, with the more theatrical and contextual kind, the non-verbal, presumably unintentional kind, whether this communication be purposely engineered or not. As an example of what we must try to examine, I would like to cite at length a novelistic incident in which Preedy, a vacationing Englishman, makes his first appearance on the beach of his summer hotel in Spain:

> But in any case he took care to avoid catching anyone's eye. First of all, he had to make it clear to these potential companions of his holiday that they were of no concern to him whatsoever. He stared through them, round them, over them – eyes lost in space. The beach might have been empty. If by chance a ball was thrown his way, he looked surprised; then let a smile of amusement lighten his face (Kindly Preedy), looked round dazed to see that there *were* people on the beach, tossed it back with a smile to himself and not a smile *at* the people, and then resumed carelessly his nonchalant survey of space.
>
> But it was time to institute a little parade, the parade of the Ideal Preedy. By devious handlings he gave any who wanted to look a chance to see the title of his book – a Spanish translation of Homer, classic thus, but daring, cosmopolitan too – and then gathered together his beach-wrap and bag into a neat sand-resistant pile (Methodical and Sensible Preedy), rose slowly to stretch at ease his huge frame (Big-Cat Preedy), and tossed aside his sandals (Carefree Preedy, after all).
>
> The marriage of Preedy and the sea! There were alternative rituals. The first involved the stroll that turns into a run and a dive straight into the water, thereafter smoothing into a strong splashless crawl toward the horizon. But of course not really to the horizon. Quite suddenly he would turn on to his back and thrash great white splashes with his legs, somehow thus showing that he could have swum further had he wanted to, and then would stand up a quarter out of the water for all to see who it was.
>
> The alternative course was simpler, it avoided the cold-water shock and it avoided the risk of appearing too high-spirited. The point was to appear to be so used to the sea, the Mediterranean, and this particular beach, that one might as well be in the sea as out of it. It involved a slow stroll down and into the edge of the water – not even noticing his toes were wet, land and water all the same to *him!* – with his eyes up at the sky gravely surveying portents, invisible to others, of the weather (Local Fisherman Preedy).[5]

The novelist means us to see that Preedy is improperly concerned with the extensive impressions he feels his sheer bodily action is giving off to those around him. We can malign Preedy further by assuming that he has acted merely in order to give a particular impression, that this is a false impression, and that the others present receive either no impression at all, or, worse still, the impression that Preedy is affectedly trying to cause them to receive this particular impression. But the important point for us here is that the kind of impression Preedy thinks he is making is in fact the kind of impression that the others correctly and incorrectly glean from someone in their midst.

I have said that when an individual appears before others his actions will influence the definition of the situation which they come to have. Sometimes the individual will act in a thoroughly calculating manner, expressing himself in a given way solely in order to give the kind of impression to others that is likely to evoke from them a specific response he is concerned to obtain. Sometimes the individual will be calculating in his activity but be relatively unaware that this is the case. Sometimes he will intentionally and consciously express himself in a particular way, but chiefly because the tradition of his group or social status require this kind of expression and not because of any particular response (other than vague acceptance or approval) that is likely to be evoked from those impressed by the expression. Sometimes the traditions of an individual's role will lead him to give a well-designed impression of a particular kind and yet he may be neither consciously nor unconsciously disposed to create such an impression. The others, in their turn, may be suitably impressed by the individual's efforts to convey something, or may misunderstand the situation and come to conclusions that are warranted neither by the individual's intent nor by the facts. In any case, in so far as the others act as if the individual had conveyed a particular impression, we may take a functional or pragmatic view and say that the individual has 'effectively' projected a given definition of the situation and 'effectively' fostered the understanding that a given state of affairs obtains.

There is one aspect of the others' response that bears special comment here. Knowing that the individual is likely to present himself in a light that is favourable to him, the others may divide what they witness into two parts: a part that is relatively easy for the individual to manipulate at will, being chiefly his verbal assertions, and a part in regard to which he seems to have little concern or control, being chiefly derived from the expressions he gives off. The others may then use what are considered to be the ungovernable aspects of his expressive behaviour as a check upon the validity of what is conveyed by the governable aspects. In this a fundamental asymmetry is demonstrated in the communication process, the individual presumably being aware of only one stream of his communication, the witnesses of this stream and one other. For example, in Shetland Isle one crofter's wife, in serving native dishes to a visitor from the mainland of Britain, would listen with a polite smile to his polite claims of liking what he was eating; at the same time she would take note of the rapidity with which the visitor lifted his fork or spoon to his mouth, the eagerness with which he passed food into his mouth, and the gusto expressed in chewing the food, using these signs as a check on the stated feelings of the eater. The same woman, in order to discover what one acquaintance (A) 'actually' thought of another acquaintance (B), would wait until B was in the presence of A but engaged in conversation with still another person (C). She would then covertly examine the facial expressions of A as he regarded B in conversation with C. Not being in conversation with B, and not being directly observed by him, A would sometimes relax usual constraints and tactful deceptions, and freely express what he was 'actually' feeling about B. This Shetlander, in short, would observe the unobserved observer.

Now given the fact that others are likely to check up on the more controllable aspects of behaviour by means of the less controllable, one can expect that sometimes the individual will try to exploit this very possibility, guiding the impression he makes through behaviour felt to be reliably informing.[6] For example, in gaining admission to a tight social circle, the participant observer may not only wear an accepting look while listening to an informant, but may also be careful to wear the same look when observing the informant talking to others; observers of the observer will then not as easily discover where he actually stands. A specific illustration may be cited from Shetland Isle. When a neighbour dropped in to have a cup of tea, he would ordinarily wear at least a hint of an expectant warm smile as he passed through the door into the cottage. Since lack of physical obstructions outside the cottage and lack of light within it usually made it possible to observe the visitor unobserved as he approached the house, islanders sometimes took pleasure in watching the visitor drop whatever expression he was manifesting and replace it with a sociable one just before reaching the door. However, some visitors, in appreciating that this examination was occurring, would blindly adopt a social face a long distance from the house, thus ensuring the projection of a constant image.

This kind of control upon the part of the individual reinstates the symmetry of the communication process, and sets the stage for a kind of information game – a potentially infinite cycle of concealment, discovery, false revelation, and rediscovery. It should be added that since the others are likely to be relatively unsuspicious of the presumably unguided aspect of the individual's conduct, he can gain much by controlling it. The others of course may sense that the individual is manipulating the presumably spontaneous aspects of his behaviour, and seek in this very act of manipulation some shading of conduct that the individual has not managed to control. This again provides a check upon the individual's behaviour, this time his presumably uncalculated behaviour, thus re-establishing the asymmetry of the communication process. Here I would like only to add the suggestion that the arts of piercing an individual's effort at calculated unintentionality seem better developed than our capacity to manipulate our own behaviour, so that regardless of how many steps have occurred in the information game, the witness is likely to have the advantage over the actor, and the initial asymmetry of the communication process is likely to be retained.

When we allow that the individual projects a definition of the situation when he appears before others, we must also see that the others, however passive their role may seem to be, will themselves effectively project a definition of the situation by virtue of their response to the individual and by virtue of any lines of action they initiate to him. Ordinarily the definitions of the situation projected by the several different participants are sufficiently attuned to one another, so that open contradiction will not occur. I do not mean that there will be the kind of consensus that arises when each individual present candidly expresses what he really feels and honestly agrees with the expressed feelings of the others present. This kind of harmony is an optimistic ideal and in any case not necessary for the smooth working of society. Rather, each participant is expected to suppress his immediate heartfelt feelings, conveying a view of the situation which he feels the others will be able to find at least temporarily acceptable. The maintenance of this surface of agreement, this veneer of consensus, is facilitated by each participant concealing his own wants behind statements which assert values to which everyone present feels obliged to give lip service. Further, there is usually a kind of division of definitional labour. Each participant is allowed to establish the tentative official rule regarding matters which are vital to him but not immediately important to others, e.g. the rationalizations and justifications by which he accounts for his past activity. In exchange for this courtesy he remains silent or noncommittal on matters important to others but not immediately important to him. We have then a kind of interactional *modus vivendi*. Together the participants contribute to a single overall definition of the situation which involves not so much a real agreement as to what exists but rather a real agreement as to whose claims concerning what issues will be temporarily honoured. Real agreement will also exist concerning the desirability of avoiding an open conflict of definitions of the situation.[7] I will refer to this level of agreement as a 'working consensus'. It is to be understood that the working consensus established in one interaction setting will be quite different in content from the working consensus established in a different type of setting. Thus, between two friends at lunch, a reciprocal show of affection, respect, and concern for the other is maintained. In service occupations, on the other hand, the specialist often maintains an image of disinterested involvement in the problem of the client, while the client responds with a show of respect for the competence and integrity of the specialist. Regardless of such differences in content, however, the general form of these working arrangements is the same.

In noting the tendency for a participant to accept the definitional claims made by the others present, we can appreciate the crucial importance of the information that the individual *initially* possesses or acquires concerning his fellow participants, for it is on the basis of this initial information that the individual starts to define the situation and starts to build up lines of responsive action. The individual's initial projection commits him to what he is proposing to be and requires him to drop all pretences of being other things. As the interaction among the participants progresses, additions and modifications in this initial informational state will of course occur, but it is essential that these later developments be related without contradiction to, and even built up from, the initial positions taken by the several participants. It would seem that an individual can more easily make a choice as to what line of treatment to demand from and extend to the others present at the beginning of an encounter than he can alter the line of treatment that is being pursued once the interaction is under way.

In everyday life, of course, there is a clear understanding that first impressions are important. Thus, the work adjustment of those in service occupations will often hinge upon a capacity to seize and hold the initiative in the service relation, a capacity that will require subtle aggressiveness on the part of the server when he is of lower socio-economic status than his client. W.F. Whyte suggests the waitress as an example.

> The first point that stands out is that the waitress who bears up under pressure does not simply respond to her customers. She acts with some skill to control their behaviour. The first question to ask when we look at the customer relationship is, 'Does the waitress get the jump on the customer, or does the customer get the jump on the waitress?' The skilled waitress realizes the crucial nature of this question. . . .
>
> The skilled waitress tackles the customer with confidence and without hesitation. For example, she may find that a new customer has seated himself before she could clear off the dirty dishes and change the cloth. He is now leaning on the table studying the menu. She greets him, says, 'May I change the cover, please?' and, without waiting for an answer, takes his menu away from him so that he moves back from the table, and she goes about her work. The relationship is handled politely but firmly, and there is never any question as to who is in charge.[8]

When the interaction that is initiated by 'first impressions' is itself merely the initial interaction is an extended series of interactions involving the same participants, we speak of 'getting off on the right foot' and feel that it is crucial that we do so. Thus, one learns that some teachers take the following view:

> You can't ever let them get the upper hand on you or you're through. So I start out tough. The first day I get a new class in, I let them know who's boss. . . . You've got to start off tough, then you can ease up as you go along. If you start out easy-going, when you try to get tough, they'll just look at you and laugh.[9]

Similarly, attendants in mental institutions may feel that if the new patient is sharply put in his place the first day on the ward and made to see who is boss, much future difficulty will be prevented.[10]

Given the fact that the individual effectively projects a definition of the situation when he enters the presence of others, we can assume that events may occur within the interaction which contradict, discredit, or otherwise throw doubt upon this projection. When these disruptive events occur, the interaction itself may come to a confused and embarrassed halt. Some of the assumptions upon which the responses of the participants had been predicted become untenable, and the participants find themselves lodged in an interaction for which the situation has been wrongly defined and is now no longer defined. At such moments the individual whose presentation has been discredited may feel ashamed while the others present may feel hostile, and all the participants may come to feel ill at ease, nonplussed, out of countenance, embarrassed, experiencing the kind of anomaly that is generated when the minute social system of face-to-face interaction breaks down.

In stressing the fact that the initial definition of the situation projected by an individual tends to provide a plan for the cooperative activity that follows – in stressing this action point of view – we must not overlook the crucial fact that any projected definition of the situation also has a distinctive moral character. It is this moral character of projections that will chiefly concern us in this report. Society is organized on the principle that any individual who possesses certain social characteristics has a moral right to expect that others will value and treat him in an appropriate way. Connected with this principle is a second, namely that an individual who implicitly or explicitly signifies that he has certain social characteristics ought in fact to be what he claims he is. In consequence, when an individual projects a definition of the situation and thereby makes an implicit or explicit claim to be a person of a particular kind, he automatically exerts a moral demand upon the others, obliging them to value and treat him in the manner that persons of this kind have a right to expect. He also implicitly forgoes all claims to be things he does not appear to be[11] and hence forgoes the treatment that would be appropriate for such individuals. The others find, then, that the individual has informed them as to what is and as to what they *ought* to see as the 'is'.

One cannot judge the importance of definitional disruptions by the frequency with which they occur, for apparently they would occur more frequently were not constant precautions taken. We find that preventive practices are constantly employed to avoid these embarrassments and that corrective practices are constantly employed to compensate for discrediting occurrences that have not been successfully avoided. When the individual employs these strategies and tactics to protect his own projections, we may refer to them as 'defensive practices'; when a participant employs them to save the definition of the situation projected by another, we speak of 'protective practices' or 'tact'. Together, defensive and protective practices comprise the techniques employed to safeguard the impression fostered by an individual during his presence before others. It should be added that while we may be ready to see that no fostered impression would survive if defensive practices were not employed, we are less ready perhaps to see that few impressions could survive if those who received the impression did not exert tact in their reception of it.

In addition to the fact that precautions are taken to prevent disruption of projected definitions, we may also note that an intense interest in these disruptions comes to play a significant role in the social life of the group. Practical jokes and social games are played in which embarrassments which are to be taken unseriously are purposely engineered.[12] Fantasies are created in which devastating exposures occur. Anecdotes from the past – real, embroidered, or fictitious – are told and retold, detailed disruptions which occurred, almost occurred, or occurred and were admirably resolved. There seems to be no grouping which does not have a ready supply of these games, reveries, and cautionary tales, to be used as a source of humour, a catharsis for anxieties, and a sanction for inducing individuals to be modest in their claims and reasonable in their projected expectations. The individual may tell himself through dreams of getting into impossible positions. Families tell of a time a guest got his dates mixed and arrived when neither the house nor anyone in it was ready for him. Journalists tell of times when an all-too-meaningful misprint occurred, and the paper's assumption of objectivity or decorum was humorously discredited. Public servants tell of times a client ridiculously misunderstood form instructions, giving answers which implied an unanticipated and bizarre definition of the situation.[13] Seamen, whose home away from home is rigorously he-man, tell stories of coming back home and inadvertently asking mother to 'pass the fucking butter'.[14] Diplomats tell of the time a near-sighted queen asked a republican ambassador about the health of his king.[15]

To summarize, then, I assume that when an individual appears before others he will have many motives for trying to control the impression they receive of the situation. This report is concerned with some of the common techniques that persons employ to sustain such impressions and with some of the common contingencies associated with the employment of these techniques. The specific content of any activity presented by the individual participant, or the role it plays in the interdependent activities of an on-going social system, will not be at issue; I shall be concerned only with the participants' dramaturgical problems of presenting the activity before others. The issues dealt with by stage-craft and stage management are sometimes trivial but they are quite general; they seem to occur everywhere in social life, providing a clear-cut dimension for formal sociological analysis.

18

Mass communication and para-social interaction

Donald Horton and R. Richard Wohl

From Horton, Donald and Wohl, R. Richard 1956: Mass communication and para-social interaction: observations on intimacy at a distance. *Psychiatry* 19, 215–23. Two final sections ('Extreme para-sociability' and 'The image as artifact') are omitted.

One of the striking characteristics of the new mass media – radio, television, and the movies – is that they give the illusion of face-to-face relationship with the performer. The conditions of response to the performer are analogous to those in a primary group. The most remote and illustrious men are met *as if* they were in the circle of one's peers; the same is true of a character in a story who comes to life in these media in an especially vivid and arresting way. We propose to call this seeming face-to-face relationship between spectator and performer a *para-social relationship*.

In television, especially, the image which is presented makes available nuances of appearance and gesture to which ordinary social perception is attentive and to which interaction is cued. Sometimes the 'actor' – whether he is playing himself or performing in a fictional role – is seen engaged with others; but often he faces the spectator, uses the mode of direct address, talks as if he were conversing personally and privately. The audience, for its part, responds with something more than mere running observation; it is, as it were, subtly insinuated into the programme's action and internal social relationships and, by dint of this kind of staging, is ambiguously transformed into a group which observes and participates in the show by turns. The more the performer seems to adjust his performance to the supposed response of the audience, the more the audience tends to make the response anticipated. This simulacrum of conversational give and take may be called *para-social interaction*.

Para-social relations may be governed by little or no sense of obligation, effort, or responsibility on the part of the spectator. He is free to withdraw at any moment. If he remains involved, these para-social relations provide a framework within which much may be added by fantasy. But these are differences of degree, not of kind, from what may be termed the ortho-social. The crucial difference in experience obviously lies in the lack of effective reciprocity, and this the audience cannot normally conceal from itself. To be sure, the audience is free to choose among the relationships offered, but it cannot create new ones. The interaction, characteristically, is one-sided, non-dialectical, controlled by the performer, and not susceptible of mutual development. There are, of course, ways in which the spectators can make their feelings known to the performers and the technicians who design the programmes, but these lie outside the para-social interaction itself. Whoever finds the experience unsatisfying has only the option to withdraw.

What we have said so far forcibly recalls the theatre as an ambiguous meeting ground on which real people play out the roles of fictional characters. For a brief interval, the fictional takes precedence over the actual, as the actor becomes identified with the fictional role in the magic of the theatre. This glamorous confusion of identities is temporary: the worlds of fact and fiction meet only for the moment. And the actor, when he takes his bows at the end of the performance, crosses back over the threshold into the matter-of-fact world.

Radio and television, however – and in what follows we shall speak primarily of television – are hospitable to both these worlds in continuous interplay. They are alternately public platforms and theatres, extending the para-social relationship now to leading people of the world of affairs, now to fictional characters, sometimes even to puppets anthropomorphically transformed into 'personalities', and, finally, to theatrical stars who appear in their capacities as real celebrities. But of particular interest is the creation by these media of a new type of performer: quizmasters, announcers, 'interviewers' in a new 'show-business' world – in brief, a special category of 'personalities' whose existence is a function of the media themselves. These 'personalities,' usually, are not prominent in any of the social spheres beyond the media.[1] They exist for their audience only in the para-social relation. Lacking an appropriate name for these performers, we shall call them *personae.*

The role of the persona

The persona is the typical and indigenous figure of the social scene presented by radio and television. To say that he is familiar and intimate is to use pale and feeble language for the pervasiveness and closeness with which multitudes feel his presence. The spectacular fact about such personae is that they can claim and achieve an intimacy with what are literally crowds of strangers, and this intimacy, even if it is an imitation and a shadow of what is ordinarily meant by that word, is extremely influential with, and satisfying for, the great numbers who willingly receive it and share in it. They 'know' such a persona in somewhat the same way they know their chosen friends: through direct observation and interpretation of his appearance, his gestures and voice, his conversation and conduct in a variety of situations. Indeed, those who make up his audience are invited, by designed informality, to make precisely these evaluations – to consider that they are involved in a face-to-face exchange rather than in passive observation. When the television camera pans down on a performer, the illusion is strong that he is enhancing the presumed intimacy by literally coming closer. But the persona's image, while partial, contrived, and penetrated by illusion, is no fantasy or dream; his performance is an objectively perceptible action in which the viewer is implicated imaginatively, but which he does not imagine.

The persona offers, above all, a continuing relationship. His appearance is a regular and dependable event, to be counted on, planned for, and integrated into the routines of daily life. His devotees 'live with him' and share the small episodes of his public life – and to some extent even of his private life away from the show. Indeed, their continued association with him acquires a history, and the accumulation of shared past experiences gives additional meaning to the present performance. This bond is symbolized by allusions that lack meaning for the casual observer and appear occult to the outsider. In time, the devotee – the 'fan' – comes to believe that he 'knows' the persona more intimately and profoundly than others do; that he 'understands' his character and appreciates his values and motives.[2] Such an accumulation of knowledge and intensification of loyalty, however, appears to be a kind of growth without development, for the one-sided nature of the connection precludes a progressive and mutual reformulation of its values and aims.[3]

The persona may be considered by his audience as a friend, counsellor, comforter, and model; but, unlike real associates, he has the peculiar virtue of being standardized according to the 'formula' for his character and performance which he and his managers have worked out and embodied in an appropriate 'production format'. Thus his character and pattern of action remain basically unchanged in a world of otherwise disturbing change. The persona is ordinarily predictable, and gives his adherents no unpleasant surprises. In their association with him there are no problems of understanding or empathy too great to be solved. Typically, there are no challenges to a spectator's self – to his ability to take the reciprocal part in the performance that is assigned to him – that cannot be met comfortably. This reliable sameness is only approximated, and then only in the short run, by the figures of fiction. On television, Groucho is always sharp; Godfrey is always warm-hearted.

The bond of intimacy

It is an unvarying characteristic of these 'personality' programmes that the greatest pains are taken by the persona to create an illusion of intimacy. We call it an illusion because the relationship between the persona and any member of his audience is inevitably one-sided, and reciprocity between the two can only be suggested. There are several principal strategies for achieving this illusion of intimacy.

Most characteristic is the attempt of the persona to duplicate the gestures, conversational style, and milieu of an informal face-to-face gathering. This accounts, in great measure, for the casualness with which even the formalities of programme scheduling are treated. The spectator is encouraged to gain the impression that what is taking place on the programme gains a momentum of its own in the very process of being enacted. Thus, Steve Allen is always pointing out to his audience that 'we never know what is going to happen on this show.' In addition, the persona tries to maintain a flow of small talk which gives the impression that he is responding to and sustaining the contributions of an invisible interlocutor. Dave Garroway, who has mastered his style to perfection, has described how he stumbled on the device in his early days in radio.

> Most talk on the radio in those days was formal and usually a little stiff. But I just rambled along, saying whatever came into my mind. I was introspective. I tried to pretend that I was chatting with a friend over a highball late in the evening. . . . Then – and later – I consciously tried to talk to the listener as an individual, to make each listener feel that he knew me and I knew him. It seemed to work pretty well then and later. I know that strangers often stop me on the street today, call me Dave and seem to feel that we are old friends who know all about each other.[4]

In addition to creating an appropriate tone and patter, the persona tries as far as possible to eradicate, or at least to blur, the line which divides him and his show, as a formal performance, from the audience both in the studio and at home. The most usual way of achieving this ambiguity is for the persona to treat his supporting cast as a group of close intimates. Thus all the members of the cast will be addressed by their first names, or by special nicknames, to emphasize intimacy. They very quickly develop, or have imputed to them, stylized character traits which, as members of the supporting cast, they will indulge in and exploit regularly in programme after programme. The member of the audience, therefore, not only accumulates an historical picture of 'the kinds of people they really are,' but tends to believe that this fellowship includes him by extension. As a matter of fact, all members of the programme who are visible to the audience will be drawn into this by-play to suggest this ramification of intimacy.

Furthermore, the persona may try to step out of the particular format of his show and literally blend with the audience. Most usually, the persona leaves the stage and mingles with the studio audience in a question-and-answer exchange. In some few cases, and particularly on the Steve Allen show, this device has been carried a step further. Thus Allen has managed to blend even with the home audience by the manoeuver of training a television camera on the street outside the studio and, in effect, suspending his own show and converting all the world outside into a stage. Allen, his supporting cast, and the audience, both at home and in the studio, watch together what transpires on the street – the persona and his spectators symbolically united as one big audience. In this way, Allen erases for the moment the line which separates persona and spectator.

In addition to the management of relationships between the persona and performers, and between him and his audience, the technical devices of the media themselves are exploited to create illusions of intimacy.

> For example [Dave Garroway explains in this connection], we developed the 'subjective-camera' idea, which was simply making the camera be the eyes of the audience. In one scene the camera – that's you, the viewer – approached the door of a dentist's office, saw a sign that the dentist was out to lunch, sat down nervously in the waiting room. The dentist returned and beckoned to the camera, which went in and sat in the big chair. 'Open wide,' the dentist said, poking a huge, wicked-looking drill at the camera. There was a roar as the drill was turned on, sparks flew and the camera vibrated and the viewers got a magnified version of sitting in the dentist's chair – except that it didn't hurt.[5]

All these devices are indulged in not only to lure the attention of the audience, and to create the easy impression that there is a kind of participation open to them in the programme itself, but also to highlight the chief values stressed in such 'personality' shows. These are sociability, easy affability, friendship, and close contact – briefly, all the values associated with free access to and easy participation in pleasant social interaction in primary groups. Because the relationship between persona and audience is one-sided and cannot be developed mutually, very nearly the whole burden of creating a plausible imitation of intimacy is thrown on the persona and on the show of which he is the pivot. If he is successful in initiating an intimacy which his audience can believe in, then the audience may help him maintain it by fan mail and by the various other kinds of support which can be provided indirectly to buttress his actions.

The role of the audience

At one extreme, the 'personality' programme is like a drama in having a cast of characters, which includes the persona, his professional supporting cast, non-professional contestants and interviewees, and the studio audience. At the other extreme, the persona addresses his entire performance to the home audience with undisturbed intimacy. In the dramatic type of programme, the participation of the spectator involves, we presume, the same taking of successive roles and deeper empathic involvements in the leading roles which occurs in any observed social interaction.[6] It is possible that the spectator's 'collaborative expectancy'[7] may assume the more profound form of identification with one or more of the performers. But such identification can hardly be more than intermittent. The 'personality' programme, unlike the theatrical drama, does not demand or even permit the aesthetic illusion – that loss of situational reference and self-consciousness in which the audience not only accepts the symbol as reality, but fully assimilates the symbolic role. The persona and his staff maintain the para-social relationship, continually referring to and addressing the home audience as a third party to the programme; and such references remind the spectator of his own independent identity. The only illusion maintained is that of directness and immediacy of participation.

When the persona appears alone, in apparent face-to-face interaction with the home viewer, the latter is still more likely to maintain his own identity without interruption, for he is called upon to make appropriate responses which are complementary to those of the persona. This 'answering' role is, to a degree, voluntary and independent. In it, the spectator retains control over the content of his participation rather than surrendering control through identification with others, as he does when absorbed in watching a drama or movie.

This independence is relative, however, in a twofold sense: First, it is relative in the profound sense that the very act of entering into any interaction with another involves *some* adaptation to the other's perspectives, if communication is to be achieved at all. And, second, in the present case it is relative because the role of the persona is enacted in such a way, or is of such a character, that an *appropriate* answering role is specified by implication and suggestion. The persona's performance, therefore, is open-ended, calling for a rather specific answering role to give it closure.[8]

The general outlines of the appropriate audience role are perceived intuitively from familiarity with the common cultural patterns on which the role of the persona is constructed. These roles are chiefly derived from the primary relations of friendship and the family, characterized by intimacy, sympathy, and sociability. The audience is expected to accept the situation defined by the programme format as credible, and to concede as 'natural' the rules and conventions governing the actions performed and the values realized. It should play the role of the loved one to the persona's lover; the admiring dependent to his father-surrogate; the earnest citizen to his fearless opponent of political evils. It is expected to benefit by his wisdom, reflect on his advice, sympathize with him in his difficulties, forgive his mistakes, buy the products that he recommends, and keep his sponsor informed of the esteem in which he is held.

Other attitudes than compliance in the assigned role are, of course, possible. One may reject, take an analytical stance, perhaps even find a cynical amusement in refusing the offered gambit and playing some other role not implied in the script, or view the proceedings with detached curiosity or hostility. But such attitudes as these are, usually, for the one-time viewer. The faithful audience is one that can accept the gambit offered; and the functions of the programme for this audience are served not by the mere perception of it, but by the role-enactment that completes it.

The coaching of audience attitudes

Just how the situation should be defined by the audience, what to expect of the persona, what attitudes to take toward him, what to 'do' as a participant in the programme is not left entirely to the common experience and intuitions of the audience. Numerous devices are used in a deliberate 'coaching of attitudes,' to use Kenneth Burke's phrase.[9] The typical programme format calls for a studio audience to provide a situation of face-to-face interaction for the persona, and exemplifies to the home audience an enthusiastic and 'correct' response. The more interaction occurs, the more clearly is demonstrated the kind of man the persona is, the values to be shared in association with him, and the kind of support to give him. A similar model of appropriate response may be supplied by the professional assistants who, though technically performers, act in a subordinate and deferential reciprocal relation toward the persona. The audience is schooled in correct responses to the persona by a variety of other means as well. Other personae may be invited as guests, for example, who play up to the host in exemplary fashion; or persons drawn from the audience may be manoeuvred into fulfilling this function. And, in a more direct and literal fashion, reading excerpts from fan-mail may serve the purpose.

Beyond the coaching of specific attitudes towards personae, a general propaganda on their behalf flows from the performers themselves, their press agents, and the mass communication industry. Its major theme is that the performer should be loved and admired. Every attempt possible is made to strengthen the illusion of reciprocity and rapport in order to offset the inherent impersonality of the media themselves. The jargon of show business teems with special terms for the mysterious ingredients of such rapport: ideally, a performer should have 'heart,' should be 'sincere';[10] his performance should be 'real' and 'warm.'[11] The publicity campaigns built around successful performers continually emphasize the sympathetic image which, it is hoped, the audience is perceiving and developing.[12]

The audience, in its turn, is expected to contribute to the illusion by believing in it, and by rewarding the persona's 'sincerity' with 'loyalty.' The audience is entreated to assume a sense of personal obligation to the performer, to help him in his struggle for 'success' if he is 'on the way up,' or to maintain his success if he has already won it. 'Success' in show business is itself a theme which is prominently exploited in this kind of propaganda. It forms the basis of many movies; it appears often in the patter of the leading comedians and in the exhortations of MC's; it dominates the so-called amateur hours and talent shows; and it is subject to frequent comment in interviews with 'show people.'[13]

Conditions of acceptance of the para-social role by the audience

The acceptance by the audience of the role offered by the programme involves acceptance of the explicit and implicit terms which define the situation and the action to be carried out in the programme. Unless the spectator understands these terms, the role performances of the participants are meaningless to him; and unless he accepts them, he cannot 'enter into' the performance himself. But beyond this, the spectator must be able to play the part demanded of him; and this raises the question of the compatibility between his normal self – as a system of role-patterns and self-conceptions with their implicated norms and values – and the kind of self postulated by the programme scheme and the actions of the persona. In short, one may conjecture that the probability of rejection of the proffered role will be greater the less closely the spectator 'fits' the role prescription.

To accept the gambit without the necessary personality 'qualifications' is to invite increasing dissatisfaction and alienation – which the student of the media can overcome only by a deliberate, imaginative effort to take the postulated role. The persona himself takes the role of his projected audience in the interpretation of his own actions, often with the aid of cues provided by a studio audience. He builds his performance on a cumulative structure of assumptions about their response, and so postulates – more or less consciously – the complex of attitudes to which his own actions are adapted. A spectator who fails to make the anticipated responses will find himself further and further removed from the base-line of common understanding.[14] One would expect the 'error' to be cumulative, and eventually to be carried, perhaps, to the point at which the spectator is forced to resign in confusion, disgust, anger, or boredom. If a significant portion of the audience fails in this way, the persona's 'error in role-taking'[15] has to be corrected with the aid of audience research, 'programme doctors,' and other aids. But, obviously, the intended adjustment is to some average or typical spectator, and cannot take too much account of deviants.

The simplest example of such a failure to fulfill the role prescription would be the case of an intellectual discussion in which the audience is presumed to have certain basic knowledge and the ability to follow the development of the argument. Those who cannot meet these requirements find the discussion progressively less comprehensible. A similar progressive alienation probably occurs when children attempt to follow an adult programme or movie. One observes them absorbed in the opening scenes, but gradually losing interest as the developing action leaves them behind. Another such situation might be found in the growing confusion and restiveness of some audiences watching foreign movies or 'high-brow' drama. Such resistance is also manifested when some members of an audience are asked to take the opposite-sex role – the woman's perspective is rejected more commonly by men than vice versa – or when audiences refuse to accept empathically the roles of outcasts or those of racial or cultural minorities whom they consider inferior.[16]

It should be observed that merely witnessing a programme is not evidence that a spectator has played the required part. Having made the initial commitment, he may 'string along' with it at a low level of empathy but reject it retrospectively. The experience does not end with the programme itself. On the contrary, it may be only after it has ended that it is submitted to intellectual analysis and integrated into, or rejected by, the self; this occurs especially in those discussions which the spectator may undertake with other people in which favourable or unfavourable consensual interpretations and judgments are arrived at. It is important to enter a qualification at this point. The suspension of immediate judgment is probably more complete in the viewing of the dramatic programme where there is an aesthetic illusion to be accepted, than in the more self-conscious viewing of 'personality' programmes.

Values of the para-social role for the audience

What para-social roles are acceptable to the spectator and what benefits their enactment has for him would seem to be related to the systems of patterned roles and social situations in which he is involved in his everyday life. The values of a para-social role may be related, for example, to the demands being made upon the spectator for achievement in certain statuses. Such demands, to pursue this instance further, may be manifested in the expectations of others, or they may be self-demands, with the concomitant emergence of more or less satisfactory self-conceptions. The enactment of a para-social role may therefore constitute an exploration and development of new role possibilities, as in the experimental phases of actual, or aspired to, social mobility.[17] It may offer a recapitulation of roles no longer played – roles which, perhaps, are no longer possible. The audience is diversified in terms of life-stages, as well as by other social and cultural characteristics; thus, what for youth may be the anticipatory enactment of roles to be assumed in the future may be, for older persons, a reliving and re-evaluation of the actual or imagined past.

The enacted role may be an idealized version of an everyday performance – a 'successful' para-social approximation of an ideal pattern, not often, perhaps never, achieved in real life. Here the contribution of the persona may be to hold up a magic mirror to his followers, playing his reciprocal part more skilfully and ideally than do the partners of the real world. So Liberace, for example, outdoes the ordinary husband in gentle understanding, or Nancy Berg outdoes the ordinary wife in amorous complaisance. Thus, the spectator may be enabled to play his part suavely and completely in imagination as he is unable to do in actuality.

If we have emphasized the opportunities offered for playing a vicarious or actual role, it is because we regard this as the key operation in the spectator's activity, and the chief avenue of the programme's meaning for him. This is not to overlook the fact that every social role is reciprocal to the social roles of others, and that it is as important to learn to understand, to decipher, and to anticipate their conduct as it is to manage one's own. The function of the mass media, and of the programmes we have been discussing, is also the exemplification of the patterns of conduct one needs to understand and cope with in others as well as of those patterns which one must apply to one's self. Thus the spectator is instructed variously in the behaviours of the opposite sex, of people of higher and lower status, of people in particular occupations and professions. In a quantitative sense, by reason of the sheer volume of such instruction, this may be the most important aspect of the para-social experience, if only because each person's roles are relatively few, while those of the others in his social worlds are very numerous. In this culture, it is evident that to be prepared to meet all the exigencies of a changing social situation, no matter how limited it may be, could

– and often does – require a great stream of plays and stories, advice columns and social how-to-do-it books. What, after all, is soap opera but an interminable exploration of the contingencies to be met with in 'home life?'[18]

In addition to the possibilities we have already mentioned, the media present opportunities for the playing of roles to which the spectator has – or feels he has – a legitimate claim, but for which he finds no opportunity in his social environment. This function of the para-social then can properly be called compensatory, inasmuch as it provides the socially and psychologically isolated with a chance to enjoy the elixir of sociability. The 'personality' programme – in contrast to the drama – is especially designed to provide occasion for good-natured joking and teasing, praising and admiring, gossiping and telling anecdotes, in which the values of friendship and intimacy are stressed.

It is typical of the 'personality' programmes that ordinary people are shown being treated, for the moment, as persons of consequence. In the interviews of non-professional contestants, the subject may be praised for having children – whether few or many does not matter; he may be flattered on his youthful appearance; and he is likely to be honoured the more – with applause from the studio audience – the longer he has been 'successfully' married. There is even applause, and a consequent heightening of ceremony and importance for the person being interviewed, at mention of the town he lives in. In all of this, the values realized for the subject are those of a harmonious, successful participation in one's appointed place in the social order. The subject is represented as someone secure in the affections and respect of others, and he probably senses the experience as a gratifying reassurance of social solidarity and self-confidence. For the audience, in the studio and at home, it is a model of appropriate role performance – as husband, wife, mother, as 'attractive' middle age, 'remarkably youthful' old age, and the like. It is, furthermore, a demonstration of the fundamental generosity and good will of all concerned, including, of course, the commercial sponsor.[19] But unlike a similar exemplification of happy sociability in a play or a novel, the television or radio programme is real; that is to say, it is enveloped in the continuing reassurances and gratifications of objective responses. For instance there may be telephone calls to 'outside' contestants, the receipt and acknowledgement of requests from the home audience, and so on. Almost every member of the home audience is left with the comfortable feeling that he too, if he wished, could appropriately take part in this healing ceremony.

19

Holbein's *The Ambassadors*

John Berger

From Berger, John 1972: *Ways of seeing.* Harmondsworth: Penguin/London: British Broadcasting Corporation, 89–91, 94–7.

Note: In order to focus on the author's discussion of one painting, two sections of comment and one illustration are omitted.

Holbein's painting of *The Ambassadors* (1533) stands at the beginning of the tradition and, as often happens with a work at the opening of a new period, its character is undisguised. The way it is painted shows what it is about. How is it painted?

It is painted with great skill to create the illusion in the spectator that he is looking at real objects and materials. We pointed out in the first essay that

the sense of touch was like a restricted, static sense of sight. Every square inch of the surface of this painting, whilst remaining purely visual, appeals to, importunes, the sense of touch. The eye moves from fur to silk to metal to wood to velvet to marble to paper to felt, and each time what the eye perceives is already translated, within the painting itself, into the language of tactile sensation. The two men have a certain presence and there are many objects which symbolize ideas, but it is the materials, the stuff, by which the men are surrounded and clothed which dominate the painting.

Except for the faces and hands, there is not a surface in this picture which does not make one aware of how it has been elaborately worked over – by weavers, embroiderers, carpet-makers, goldsmiths, leather workers, mosaic-makers, furriers, tailors, jewellers – and of how this working-over and the resulting richness of each surface has been finally worked-over and reproduced by Holbein the painter.

This emphasis and the skill that lay behind it was to remain a constant of the tradition of oil painting.

Works of art in earlier traditions celebrated wealth. But wealth was then a symbol of a fixed social or divine order. Oil painting celebrated a new kind of wealth – which was dynamic and which found its only sanction in the supreme buying power of money. Thus painting itself had to be able to demonstrate the desirability of what money could buy. And the visual desirability of what can be bought lies in its tangibility, in how it will reward the touch, the hand, of the owner.

In the foreground of Holbein's *Ambassadors* there is a mysterious, slanting, oval form. This represents a highly distorted skull: a skull as it might be seen in a distorting mirror. There are several theories about how it was painted and why the ambassadors wanted it put there. But all agree that it was a kind of memento mori: a play on the medieval idea of using a skull as a continual reminder of the presence of death. What is significant for our argument is that the skull is painted in a (literally) quite different optic from everything else in the picture. If the skull had been painted like the rest, its metaphysical implication would have disappeared; it would have become an object like everything else, a mere part of a mere skeleton of a man who happened to be dead.

This was a problem which persisted throughout the tradition. When metaphysical symbols are introduced (and later there were painters who, for instance, introduced realistic skulls as symbols of death), their symbolism is usually made unconvincing or unnatural by the unequivocal, static materialism of the painting-method.

[. . .]

Let us now return to the two ambassadors, to their presence as men. This will mean reading the painting differently: not at the level of what it shows within its frame, but at the level of what it refers to outside it.

The two men are confident and formal; as between each other they are relaxed. But how do they look at the painter – or at us? There is in their gaze and their stance a curious lack of expectation of any recognition. It is as though in principle their worth cannot be recognized by others. They look as though they are looking at something of which they are not part. At something which surrounds them but from which they wish to exclude themselves. At the best it may be a crowd honouring them; at the worst, intruders.

What were the relations of such men with the rest of the world?

The painted objects on the shelves between them were intended to supply – to the few who could read the allusions – a certain amount of information about their position in the world. Four centuries later we can interpret this information according to our own perspective.

The scientific instruments on the top shelf were for navigation. This was the time when the ocean trade routes were being opened up for the slave trade and for the traffic which was to siphon the riches from other continents into Europe, and later supply the capital for the take-off of the Industrial Revolution.

In 1519 Magellan had set out, with the backing of Charles V, to sail round the world. He and an astronomer friend, with whom he had planned the voyage, arranged with the Spanish court that they personally were to keep twenty per cent of the profits made, and the right to run the government of any land they conquered.

The globe on the bottom shelf is a new one which charts this recent voyage of Magellan's. Holbein has added to the globe the name of the estate in France which belonged to the ambassador on the left. Beside the globe are a book of arithmetic, a hymn book and a lute. To colonize a land it was necessary to convert its people to Christianity and accounting, and thus to prove to them that European civilization was the most advanced in the world. Its art included.

[. . .]

How directly or not the two ambassadors were involved in the first colonizing ventures is not particularly important, for what we are concerned with here is a stance towards the world; and this was general to a whole class. The two ambassadors belonged to a class who were convinced that the world was there to furnish their residence in it. In its extreme form this conviction was confirmed by the relations being set up between colonial conqueror and the colonized.

These relations between conqueror and colonized tended to be self-perpetuating. The sight of the other confirmed each in his inhuman estimate of himself. The circularity of the relationship can be seen in the following diagram – as also the

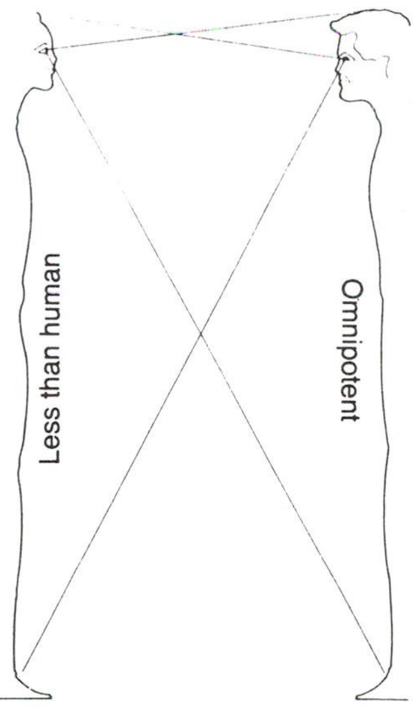

mutual solitude. The way in which each sees the other confirms his own view of himself.

The gaze of the ambassadors is both aloof and wary. They expect no reciprocity. They wish the image of their presence to impress others with their vigilance and their distance. The presence of kings and emperors had once impressed in a similar way, but their images had been comparatively impersonal. What is new and disconcerting here is the *individualized presence* which needs to suggest *distance*. Individualism finally posits equality. Yet equality must be made inconceivable.

20

The flower dream

Sigmund Freud

From Freud, Sigmund 1976: *The interpretation of dreams.* Translated by James Strachey, edited by James Strachey and Alan Tyson, revised by Angela Richards. Harmondsworth: Penguin Books, 493–8.

The naïve dreams of healthy people actually often contain a much simpler, more perspicuous and more characteristic symbolism than those of neurotics; for in the latter, as a result of the more powerful working of the censorship and of the consequently more far-reaching dream-distortion, the symbolism may be obscure and hard to interpret. The dream recorded below will serve to illustrate this fact. It was dreamt by a girl who is not neurotic but is of a somewhat prudish and reserved character. In the course of conversation with her I learnt that she was engaged, but that there were some difficulties in the way of her marriage which were likely to lead to its postponement. Of her own accord she told me the following dream.

' *"I arrange the centre of a table with flowers for a birthday."** In reply to a question she told me that in the dream she seemed to be in her own home (where she was not at present living) and had "a feeling of happiness".

' "Popular" symbolism made it possible for me to translate the dream unaided. It was an expression of her bridal wishes: the table with its floral centre-piece symbolized herself and her genitals; she represented her wishes for the future as fulfilled, for her thoughts were already occupied with the birth of a baby; so her marriage lay a long way behind her.

'I pointed out to her that *"the 'centre' of a table"* was an unusual expression (which she admitted), but I could not of course question her further directly on that point. I carefully avoided suggesting the meaning of the symbols to her, and merely asked what came into her head in connection with the separate parts of the dream. In the course of the analysis her reserve gave place to an evident interest in the interpretation and to an openness made possible by the seriousness of the conversation.

'When I asked what flowers they had been, her first reply was *"expensive flowers; one has to pay for them"*, and then that they had been *"lilies of the valley, violets and pinks or carnations"*. I assumed that the word "lily" appeared in the dream in its popular sense as a symbol of chastity; she confirmed this assumption, for her association to "lily" was *"purity"*. *"Valley"* is a frequent female symbol in

*In the present analysis all the material printed in italics occurs in English in the original, exactly as here reproduced.

dreams; so that the chance combination of the two symbols in the English name of the flower was used in the dream-symbolism to stress the preciousness of her virginity – *"expensive flowers, one has to pay for them"* – and to express her expectation that her husband would know how to appreciate its value. The phrase *"expensive flowers*, etc.", as will be seen, had a different meaning in the case of each of the three flower-symbols.

"Violets" was ostensibly quite asexual; but, very boldly, as it seemed to me, I thought I could trace a secret meaning for the word in an unconscious link with the French word *"viol"* ["rape"]. To my surprise the dreamer gave as an association the English word *"violate"*. The dream had made use of the great chance similarity between the words *"violet"* and *"violate"* – the difference in their pronunciation lies merely in the different stress upon their final syllables – in order to express "in the language of flowers" the dreamer's thoughts on the violence of defloration (another term that employs flower symbolism) and possibly also a masochistic trait in her character. A pretty instance of the 'verbal bridges'* crossed by the paths leading to the unconscious. The words *"one has to pay for them"* signified having to pay with her life for being a wife and a mother.

'In connection with *"pinks"*, which she went on to call *"carnations"*, I thought of the connection between that word and "carnal". But the dreamer's association to it was *"colour"*. She added that *"carnations"* were the flowers which her *fiancé* gave her frequently and in great numbers. At the end of her remarks she suddenly confessed of her own accord that she had not told the truth: what had occurred to her had not been *"colour"* but *"incarnation"* – the word I had expected. Incidentally *"colour"* itself was not a very remote association, but was determined by the meaning of *"carnation"* (flesh-colour) – was determined, that is, by the same complex. This lack of straightforwardness showed that it was at this point that resistance was greatest, and corresponded to the fact that this was where the symbolism was most clear and that the struggle between libido and its repression was at its most intense in relation to this phallic theme. The dreamer's comment to the effect that her *fiancé* frequently gave her flowers of that kind was an indication not only of the double sense of the word *"carnations"* but also of their phallic meaning in the dream. The gift of flowers, an exciting factor of the dream derived from her current life, was used to express an exchange of sexual gifts: she was making a gift of her virginity and expected a full emotional and sexual life in return for it. At this point, too, the words *"expensive flowers, one has to pay for them"* must have had what was no doubt literally a financial meaning. – Thus the flower symbolism in this dream included virginal femininity, masculinity and an allusion to defloration by violence. It is worth pointing out in this connection that sexual flower symbolism, which, indeed, occurs very commonly in other connections, symbolizes the human organs of sex by blossoms, which are the sexual organs of plants. It may perhaps be true in general that gifts of flowers between lovers have this unconscious meaning.

'The birthday for which she was preparing in the dream meant, no doubt, the birth of a baby. She was identifying herself with her *fiancé*, and was representing

*Freud here refers the reader to the following footnote from earlier in the volume: *Footnote added* 1909: See my volume on jokes (1905a) [especially the later part of Chapter VI] and the use of 'verbal bridges' in the solution of neurotic symptoms. [See, e.g., the synthesis of Dora's first dream at the end of Section II of Freud, 1905b (where the term 'switch-words' is also used), and the solution of the 'Rat Man's' rat-obsession in Section I (G) of Freud, 1909.] (Comment in square brackets added by Angela Richards.)

him as "arranging" her for a birth – that is, as copulating with her. The latent thought may have run: "If I were he, I wouldn't wait – I would deflower my *fiancée* without asking her leave – I would use violence." This was indicated by the word *"violate"*, and in this way the sadistic component of the libido found expression.

'In a deeper layer of the dream, the phrase *"I arrange . . ."* must no doubt have an auto-erotic, this is to say, an infantile, significance.

'The dreamer also revealed an awareness, which was only possible to her in a dream, of her physical deficiency: she saw herself like a table, without projections, and on that account laid all the more emphasis on the preciousness of the *"centre"* – on another occasion she used the words, *"a centre-piece of flowers"* – that is to say, on her virginity. The horizontal attribute of a table must also have contributed something to the symbol.

'The concentration of the dream should be observed: there was nothing superfluous in it, every word was a symbol.

'Later on the dreamer produced an addendum to the dream: *"I decorate the flowers with green crinkled paper."* She added that it was *"fancy paper"* of the sort used for covering common flower-pots. She went on: *"to hide untidy things, whatever was to be seen, which was not pretty to the eye; there is a gap, a little space in the flowers. The paper looks like velvet or moss."* – To *"decorate"* she gave the association *"decorum"*, as I had expected. She said the green colour predominated, and her association to it was *"hope"* – another link with pregnancy. – In this part of the dream the chief factor was not identification with a man; ideas of shame and self-relevation came to the fore. She was making herself beautiful for him and was admitting physical defects which she felt ashamed of and was trying to correct. Her associations *"velvet"* and *"moss"* were a clear indication of a reference to pubic hair.

'This dream, then, gave expression to thoughts of which the girl was scarcely aware in her waking life – thoughts concerned with sensual love and its organs. She was being "arranged for a birthday" – that is, she was being copulated with. The fear of being deflowered was finding expression, and perhaps, too, ideas of pleasurable suffering. She admitted her physical deficiencies to herself and over-compensated for them by an over-valuation of her virginity. Her shame put forward as an excuse for the signs of sensuality the fact that its purpose was the production of a baby. Material considerations, too, alien to a lover's mind, found their way to expression. The affect attaching to this simple dream – a feeling of happiness – indicated that powerful emotional complexes had found satisfaction in it.'

Ferenczi (1917)* had justly pointed out that the meaning of symbols and the significance of dreams can be arrived at with particular ease from the dreams of precisely those people who are uninitiated into psycho-analysis.

*This paragraph was added in 1919.

Section IV

Media Form and Cultural Process

Most Communication Studies programmes include a substantial amount of work on 'mass communication' or 'the media'. Everyone easily understands these terms to include television, radio and the press, but the inclusion of cinema, popular music or publishing is a matter of variation according to institutional usage. The term 'mass communication' has its origins in a tradition of sociological inquiry whereas the idea of 'media' tends to suggest an increased awareness of communicative form itself and of arts-related issues. Both terms have potential drawbacks. 'Mass communication' can easily slip into a suggestion that, rather than mass production and distribution, it is communication to 'a mass' which is involved, with a consequent distortion and denigration of viewers and readers. 'The media' can be used to suggest that what is involved is essentially or exclusively 'mediation', radically understating the role of press and broadcasting in originating items of public communication.

However, both terms classify as separate and worthy of special study those kinds of communication which are, to a greater or lesser extent, *industrialized* and thus able to operate on a larger scale than the primary communicative activities of, for instance, conversation, letter-writing and group talk. In 'mass communication' very few people are senders, whereas quite often millions of them are receivers. So the communication is less directly interactive than in much primary activity and this is likely to remain the case, despite the advent of new video and information technologies which increase the scope for interactivity. Nevertheless, we should not assume that the receivers are essentially passive simply because they can't send anything straight back to the television studio or newspaper offices. This view (often resulting from seeing the audience rather than the system of communication as 'mass') begs a lot of questions about the knowledge and critical intelligence of viewers and readers.

There are good reasons for academic interest in the media. For a start, the media are widely regarded as absolutely central to the politics and organization of modern societies; among other things they are a primary means of actually creating a 'public' of citizens, informed about what is going on nationally and internationally. There is continuing debate about the degree of *power* which the media exercise in the performance of this role and the debate quickly becomes one with several aspects – linguistic, psychological, cultural, organizational and economic.

In many societies, advertising provides the revenue for much media activity, thus linking the media system into the market economy in a direct and often controversial way. Then there is the question of Government regulation – how far does the State use the media in an attempt to organise consent for its policies? Or how far do the media perform a public, civic duty by maintaining a 'space' for the flow of information and for discussion, including criticism of government?

To believe that the media are in many respects 'powerful' (the assumption of most people in modern societies) does not require us to think either that media agencies themselves are routinely bent on persuasion and manipulation or that viewers, listeners and readers are susceptible to some kind of mind-control. All we have to do is reflect on the way in which political and social information now circulates in society and the extent to which we draw on the images, accounts and entertainments of the media in making sense of our own daily lives. Without disparaging the intelligence of the average citizen, it can be said that the media now contribute extensively to the make-up of modern consciousness – what we think and feel about ourselves and about our relations with others across the differences of nationality, race, gender, social class and age. In doing this, it is the full range of media output which is involved – the popular songs, the adverts, the situation comedies and the fiction alongside the journalism and the documentaries.

We can sum this up by using three influential terms in media research. In addition to performing a *gate-keeping* function (allowing certain things through while ignoring or filtering out others) and an *agenda-setting* function (establishing *what* is to be talked about locally or nationally), the media also have a *reality-defining* effect. That is, they provide us with many of the bits and pieces out of which we fashion our sense of the 'ways things are'.

Another, if related, reason for studying the media is that their communicative range and richness attracts attention as part of the analysis of contemporary culture. Media products are made in most cases by small groups of production staff who hope to use the full, available range of technologies and skills to appeal to large, perhaps vast, groupings who are 'addressed' individually (as viewer, listener or reader). The result is culturally both dynamic and complex.

But how to study the media? Given the variety of academic specialisms which have developed an interest, it's not surprising that there has been a lack of orthodoxy here. The broadest division has been between those approaches drawn from social science and those drawn from the arts subjects. The former have emphasised the economic and social character of the media whilst the latter have paid particular attention to media content and style. The idea of the media being studied as a component of 'culture' tries to bring both these approaches together but it would be wrong to suggest that this has been so successful that no tensions now exist. However, even since the last edition of this book, a further convergence and integration of ideas and methods has occurred, often via intensive debate. Work on policy, on production, on output and on audiences has increasingly shown itself to be aware of other aspects of the media process and, overall, a far stronger sense of *meaning-making* as the cultural hub of media activity has emerged.

At the time of writing, still the most rapidly developing work in this area is audience research, particularly of the kind that has come to be called 'reception study'. This work proceeds from the observation that meanings do not somehow exist *in* texts (as inherent properties) but are the product of interpretative conventions being applied to the images, sounds and words out of which texts are made. When these interpretative conventions differ, the text will mean different things to different viewers or readers – the degree and kind of difference being dependent on just what has varied in the interpretative system (e.g. it could be the connotations 'received' from a particular visual symbol in the news or a whole way of responding to, say, the fictional depiction of violence). Audience research of this kind, carried out alongside the close study of media output, opens up interesting questions about the range of interpretative resources upon which people draw in 'making sense' of the media and highlights the different levels of activity involved in sense-making. When added to the 'reality-defining' idea, it also gives a new dimension to inquiry into the processes of media 'influence' as an interconnection of signification (how the form of the communication is organized) with subjectivity (how the reader/viewer's consciousness is engaged and made generative of meaning).

So the argument about media power is one which is conducted both at the level of international media economics (who owns what?) and at the level of social psychology (how do people produce knowledge from viewing and reading?) as well as at several levels in between.

In the extracts which follow we have, as in previous editions, concentrated on television. This is primarily because television, with its dominance of national media systems, has attracted a far greater amount of research attention than either radio or the press in recent years and has (consequently) been the area in which a number of important theoretical and analytic advances have been made.

The cohesion of the items in this section is provided by a shared concern with television as involving the cultural construction of meanings. By what practices, according to what professional conventions and audience expectations do TV meanings get made? And what are the consequences of having a society in which one medium and its characteristics have become such a focus for political and cultural expression? This way of engaging with preliminary questions – placing 'form' within 'process' and the media within the more general culture – allows us to do two things. First of all, to develop our understanding of the distinctive features of the different media as modes of communicative action. Secondly, to connect our studies to broader questions of social organisation, including those other kinds of human communication which are discussed earlier in this volume.

Our selection opens with a piece by the distinguished social theorist **Anthony Giddens**, in which he offers an analysis of some of the defining conditions of 'modernity' – that very broad set of economic, political and social circumstances which, during the middle and late twentieth century, has become established across many parts of the world. Giddens is particularly interested in the cultural characteristics of 'modern' life and he calls attention to the new relations of space and of time which have emerged. He moves on to consider how new systems for 'the mediation of experience' are at the centre of modernity and its distinctive problems of cultural and social identity. His account provides a provocative context for considering just how decisive the role of media is in the transition from pre-modern to modern culture and how closely their activities have become woven into the fabric of contemporary experience.

Following this general account of the social conditions within which contemporary media operate, our second item is drawn from the classic media sociology of the 1950s. By taking one extended, 'live' television broadcast as a case-study, **Lang** and **Lang** raise questions about the nature of the 'unique perspective' on reality which television offers to its viewers. Their analysis concerns the methods of visual and verbal depiction which television uses and the consequence which this usage may have for the nature of public knowledge and political judgement. Despite being more than forty years old, the study is intriguing not just as a document about television in its first phase of development but as a clear and provocative inquiry into issues which are still on the agenda of international media research. Students should find it useful to compare the situation which the Langs describe to that of today, both in terms of TV's development and the political assumptions which surround it.

In our third extract, **John Ellis** also concerns himself with how television form relates to its social character. He compares television with cinema in a way which brings out some of its special qualities as a medium. In particular, Ellis wants to explore the different ways in which television is 'domestic'. He has something to say, too, about how the facility of direct address (by which people appear to 'look out from the screen' and speak directly to the viewer) regulates the relationship between programme material and audience. It's worth testing out his propositions against your own viewing experience and viewing habits before considering further their pertinence for our understanding of television's political and social function. The argument starts with questions of visual and aural form and then moves on through presumed modes of reception to consider some of the implications of a 'television culture'.

Ericson, Baranek and **Chan** consider television as a medium for journalism. They take production conventions as a primary topic, following these through into an account of textual form, and particularly, of the relation of visuals to speech. Connections with work contained in the earlier part of this volume are clear as Ericson *et al* discuss the kinds of viewing relationships which news teams try to establish and the various ways in which teams seeks to 'naturalize' their selective and constructive activity through the artful linking of discrete sections of report by speech 'bridges', by visual continuity and by narrative pace. The authors usefully cite the seminal studies of the Glasgow University Media Group in considering how news-making is open to being shaped by undeclared assumptions and values (by ideological factors).

Corner and **Richardson** also engage with the processes of television journalism. Their study takes as its focus the account of a documentary programme (about the consequences of unemployment for the way of life of a community) and asks questions about the way in which viewers made sense of particular sequences. This involves close attention to the visual and verbal structure of the programme in a way which is similar to the analytic approach of Ericson and his colleagues. But it also involves listening carefully to the accounts of respondents who, having viewed and listened, are asked to recall what they saw and heard and to evaluate this. Such an investigation is able to plot the movements and variations of interpretation in some detail, as our extract indicates.

The same broad objective – further understanding of viewer interpretation – is to be found in **Ien Ang**'s study of responses to the popular drama series *Dallas*. Ang asks questions about the enjoyment which viewers derive from the programme and about the assessments they make of its 'realism'. Her sample is drawn from letters which she invited by advertising in the (Dutch) press. She attempts to reconstruct the 'Dallas experience' through a close and comparative study of what the letters said. This allows her not only to develop some ideas about the different kinds of 'realism' which TV can be seen to provide, but about both the pleasures and the ideological power of popular drama.

'Television' usually means the broadcast output of professional production groups, but increasingly the sale of camcorders is allowing people to become TV producers not just TV consumers. To what use do people put the newly domesticized video technology? What conventions do they follow in recording aspects of their lives, including recording images of themselves? In a fascinating article, **Lawrence Vale** looks at how the camcorder has entered family life and personal space. The cultural precedent of the camera (and the less widely popular home movie camera) acts as a marker here, but the flexibility and instantaneity of home video and its interconnections with professional TV raise new questions about, as Vale puts it, the 'personalization' of TV and the 'capturing' of the real. It is worth comparing this suggestive study with Lang and Lang's comments on the 'unique perspective' of the broadcast medium.

Our final piece brings together the communicative conventions of a distinctive and important kind of television – 'commercials' – and the conventions of music. **Hroar Klempe** looks at the way in which the principle of repetition works in the use of music for creating mood, regulating narrative continuity and reinforcing a positive 'message'. Ranging across a number of examples, he focuses upon *Coca-Cola* advertising and its deployment of myth around its product. Analysis of music was neglected in much early Communication Studies, apart from occasional attention to the lyrics of popular song. However, popular music study is now a developing field of cultural inquiry and Klempe's piece shows just how illuminating it can be when, as here, it can locate itself at the intersection point of economic, visual and linguistic issues as well as pursuing its own specialist concerns.

Further reading

Over the last few years, the number of books offering introductions to aspects of the mass media has massively increased. Below, we indicate a brief selection of some of the most useful texts for students pursuing independent reading.

Corner, John (ed.) 1991: *Popular television in Britain: studies in cultural history*. London. British Film Institute.
A collection of articles on the development of British television which pays particular attention to the emergence of generic styles, the relationship between technology and aesthetics and the social context of programming. It includes work on situation comedy, quiz shows, music programmes, documentaries, drama, films on television and the popular television audience.

Curran, James and Gurevitch, Michael (eds.) 1991: *Mass media and society*. London: Edward Arnold.
Articles on the current state of research into the media's political and social character. Some pieces focus on theoretical and methodological issues, others concentrate on a particular area of media activity and offer an 'overview' of it. Generally a high standard of contribution, organised throughout by the editors' awareness of radical changes both in the nature of international media and in approaches to study.

Curran, James and Seaton, Jean 1991: *Power without responsibility: the press and broadcasting in Britain*. London: Routledge.
An excellent, if necessarily rather 'potted', account of the major historical

developments in British press and broadcasting together with chapters on media research, on changing structures of funding and regulation and on recent thinking about the media and democracy. Good on detail, it is a clear primer which also has uses as a basic reference work.

Garnham, Nicholas 1990: *Capitalism and communication*. London: Sage.

A collection of articles putting forward a view of international media from a 'political economy' perspective. Garnham presents an account of the media as primarily a sphere of commerce, linked sensitively to state policies. He is particularly alert to the impact of new telecommunication technologies and combines theory with detailed, substantive study in order to offer what is finally a global analysis.

Scannell, Paddy (ed.) 1991: *Broadcast talk*. London: Sage.

This is the first volume of studies to focus attention on the language used by broadcasters. In doing so, its contributors bring together sociological and linguistic methods of analysis to explore the different kinds of 'professional speech' heard on television and radio. The formal features of microphone talk, its development within the genres of broadcasting and the social relationships it projects to its audience are all given attention.

Schlesinger, Philip 1991: *Media, state and nation*. London: Sage.

'National identity' is one of the central issues in the media analysis of the 1990s. This book contains a number of studies in political sociology which set national questions within the contexts both of official state activities and of broader cultural shifts. Schlesinger is particularly concerned with the political use of violence and the effects which mediations of this have upon perceptions of nationality and security.

Seiter, E. *et al* (eds.) 1989: *Remote control: television, audiences and cultural power*. London: Routledge.

Audience research continues to be a major area of development. The articles in this volume offer an authoritative and stimulating account of the major findings to date and also point to new directions. The volume includes a number of chapters which report in detail on particular research projects, while other authors reflect more generally on principal themes and problems.

Tulloch, John 1990: *Television drama*. London: Routledge.

A critical survey of the whole area of television drama, looking at its distinctive character, its varieties of sub-genres and at both the productional and textual questions which it raises. Drawing on both literary and sociological ideas, Tulloch succeeds in covering a lot of ground whilst providing selective detail and a good continuity of argument.

21

The contours of high modernity

Anthony Giddens

From Giddens, A. 1991: *Modernity and self-identity*. Oxford: Polity, 14–27.

Modernity: some general considerations

I use the term 'modernity' in a very general sense, to refer to the institutions and modes of behaviour established first of all in post-feudal Europe, but which in the twentieth century increasingly have become world-historical in their impact. 'Modernity' can be understood as roughly equivalent to 'the industrialised world', so long as it be recognised that industrialism is not its only institutional dimension.[1] I take industrialism to refer to the social relations implied in the widespread use of material power and machinery in production processes. As such, it is one institutional axis of modernity. A second dimension is capitalism, where this term means a system of commodity production involving both competitive product markets and the commodification of labour power. Each of these can be distinguished analytically from the institutions of surveillance, the basis of the massive increase in organisational power associated with the emergence of modern social life. Surveillance refers to the supervisory control of subject populations, whether this control takes the form of 'visible' supervision in Foucault's sense, or the use of information to coordinate social activities. This dimension can in turn be separated from control of the means of violence in the context of the 'industrialisation of war'. Modernity ushers in an era of 'total war', in which the potential destructive power of weaponry, signalled above all by the existence of nuclear armaments, becomes immense.

Modernity produces certain distinct social forms, of which the most prominent is the nation-state. A banal observation, of course, until one remembers the established tendency of sociology to concentrate on 'society' as its designated subject-matter. The sociologist's 'society', applied to the period of modernity at any rate, is a nation-state, but this is usually a covert equation rather than an explicitly theorised one. As a sociopolitical entity the nation-state contrasts in a fundamental way with most types of traditional order. It develops only as part of a wider nation-state system (which today has become global in character), has very specific forms of territoriality and surveillance capabilities, and monopolises effective control over the means of violence.[2] In the literature of international relations, nation-states are often treated as 'actors' – as 'agents' rather than 'structures' – and there is a definite justification for this. For modern states are reflexively monitored systems which, even if they do not 'act' in the strict sense of the term, follow coordinated policies and plans on a geopolitical scale. As such, they are a prime example of a more general feature of modernity, the rise of the *organisation*. What distinguishes modern organisations is not so much their size, or their bureaucratic character, as the concentrated reflexive monitoring they both permit and entail. Who says modernity says not just organisations, but organisation – the regularised control of social relations across indefinite time-space distances.

Modern institutions are in various key respects *discontinuous* with the gamut of pre-modern cultures and ways of life. One of the most obvious characteristics separating the modern era from any other period preceding it is modernity's extreme dynamism. The modern world is a 'runaway world': not only is the *pace* of social change much faster than in any prior system, so also is its *scope*, and the *profoundness* with which it affects pre-existing social practices and modes of behaviour.[3]

What explains the peculiarly dynamic character of modern social life? Three main elements, or sets of elements, are involved. The first is what I call the *separation of time and space*. All cultures, of course, have possessed modes of time-reckoning of one form or another, as well as ways of situating themselves spatially. There is no society in which individuals do not have a sense of future, present and past. Every culture has some form of standardised spatial markers which designate a special awareness of place. In pre-modern settings, however, time and space were connected *through* the situatedness of place.

Larger pre-modern cultures developed more formal methods for the calculation of time and the ordering of space – such as calendars and (by modern standards) crude maps. Indeed, these were the prerequisites for the 'distancing' across time and space which the emergence of more extensive forms of social system presupposed. But in pre-modern eras, for the bulk of the population, and for most of the ordinary activities of day-to-day life, time and space remained essentially linked through place. 'When' markers were connected not just to the 'where' of social conduct, but to the substance of that conduct itself.

The separation of time from space involved above all the development of an 'empty' dimension of time, the main lever which also pulled space away from place. The invention and diffusion of the mechanical clock is usually seen – rightly – as the prime expression of this process, but it is important not to interpret this phenomenon in too superficial a way. The widespread use of mechanical timing devices facilitated, but also presumed, deeply structured changes in the tissue of everyday life – changes which could not only be local, but were inevitably universalising. A world that has a universal dating system, and globally standardised time zones, as ours does today, is socially and experientially different from all pre-modern eras. The global map, in which there is no privileging of place (a universal projection), is the correlate symbol to the clock in the 'emptying' of space. It is not just a mode of portraying 'what has always been there' – the geography of the earth – but is constitutive of quite basic transformations in social relations.

The emptying out of time and space is in no sense a unilinear development, but proceeds dialectically. Many forms of 'lived time' are possible in social settings structured through the separation of time and space. Moreover, the severance of time from space does not mean that these henceforth become mutually alien aspects of human social organisation. On the contrary: it provides the very basis for their recombination in ways that coordinate social activities without necessary reference to the particularities of place. The organisations, and organisation, so characteristic of modernity are inconceivable without the reintegration of separated time and space. Modern social organisation presumes the precise coordination of the actions of many human beings physically absent from one another; the 'when' of these actions is directly connected to the 'where', but not, as in pre-modern epochs, via the mediation of place.

We can all sense how fundamental the separation of time from space is for the massive dynamism that modernity introduces into human social affairs. The phenomenon universalises that 'use of history to make history' so intrinsic to the processes which drive modern social life away from the hold of tradition. Such historicity becomes global in form with the creation of a standardised 'past' and a universally applicable 'future': a date such as the 'year 2000' becomes a recognisable marker for the whole of humanity.

The process of the emptying of time and space is crucial for the second major influence on modernity's dynamism, the *disembedding* of social institutions. I choose the metaphor of disembedding in deliberate opposition to the concept of 'differentiation' sometimes adopted by sociologists as a means of contrasting pre-modern with modern social systems. Differentiation carries the imagery of the progressive separation of functions, such that modes of activity organised in a diffuse fashion in pre-modern societies become more specialised and precise with the advent of modernity. No doubt this idea has some validity, but it fails to capture an essential element of the nature and impact of modern institutions – the 'lifting out' of social relations from local contexts and their rearticulation across indefinite tracts of time-space. This 'lifting out' is exactly what I mean by disembedding, which is the key to the tremendous acceleration in time-space distanciation which modernity introduces.

Disembedding mechanisms are of two types, which I term 'symbolic tokens' and 'expert systems'. Taken together, I refer to these as *abstract systems*. Symbolic tokens are media of exchange which have standard value, and thus are interchangable across a plurality of contexts. The prime example, and the most pervasively important, is money. Although the larger forms of pre-modern social system have all developed monetary exchange of one form or another, money economy becomes vastly more sophisticated and abstract with the emergence and maturation of modernity. Money brackets time (because it is a means of credit) and space (since standardised value allows transactions between a multiplicity of individuals who never physically meet one another). Expert systems bracket time and space through deploying modes of technical knowledge which have validity independent of the practitioners and clients who make use of them. Such systems penetrate virtually all aspects of social life in conditions of modernity – in respect of the food we eat, the medicines we take, the buildings we inhabit, the forms of transport we use and a multiplicity of other phenomena. Expert systems are not confined to areas of technological expertise. They extend to social relations themselves and to the intimacies of the self. The doctor, counsellor and therapist are as central to the expert systems of modernity as the scientist, technician or engineer.

Both types of expert system depend in an essential way on *trust*. Trust is different from the form of confidence which Georg Simmel called the 'weak inductive knowledge' involved in formal transactions.[4] Some decisions in life are based on inductive inferences from past trends, or from past experience believed in some way to be dependable for the present. This kind of confidence may be an element in trust, but it is not sufficient in itself to define a trust relation. Trust presumes a leap to commitment, a quality of 'faith' which is irreducible. It is specifically related to absence in time and space, as well as to ignorance. We have no need to trust someone who is constantly in view and whose activities can be directly monitored. Thus, for example, jobs which are monotonous or unpleasant, and poorly paid, in which the motivation to perform the work conscientiously is weak, are usually 'low-trust' positions. 'High-trust' posts are those carried out largely outside the presence of management or supervisory staff.[5] Similarly, there is no requirement of trust when a technical system is more or less completely known to a particular individual. In respect of expert systems, trust brackets the limited technical knowledge which most people possess about coded information which routinely affects their lives.

Trust, of varying sorts and levels, underlies a host of day-to-day decisions that all of us take in the course of orienting our activities. But trusting is not by any means always the result of consciously taken decisions: more often, it is a generalised attitude of mind that underlies those decisions, something which has its roots in the connection between trust and personality development. We *can* make the decision to trust, a phenomenon which is common because of the third underlying element of modernity (already mentioned, but also further discussed below): its intrinsic reflexivity. But the faith which trust implies also tends to resist such calculative decision-making.

Attitudes of trust, in relation to specific situations, persons or systems, and on a more generalised level, are directly connected to the psychological *security* of individuals and groups. Trust and security, risk and danger: these exist in various historically unique conjunctions in conditions of modernity. The disembedding mechanisms, for example, purchase wide arenas of relative security in daily social activity. People living in the industrialised countries, and to some extent elsewhere today, are generally protected from some of the hazards routinely faced in pre-modern times – such as those emanating from inclement nature. On the other hand, new risks and dangers are created through the disembedding mechanisms themselves, and these may be local or global. Foodstuffs purchased with artificial ingredients may have toxic characteristics absent from more traditional foods; environmental hazards might threaten the ecosystems of the earth as a whole.

Modernity is essentially a post-traditional order. The transformation of time and space, coupled with the disembedding mechanisms, propel social life away from the hold of pre-established precepts or practices. This is the context of the thoroughgoing *reflexivity* which is the third major influence on the dynamism of modern institutions. The reflexivity of modernity has to be distinguished from the reflexive monitoring of action intrinsic to all human activity. Modernity's reflexivity refers to the susceptibility of most aspects of social activity, and material relations with nature, to chronic revision in the light of new information or knowledge. Such information or knowledge is not incidental to modern institutions, but constitutive of them – a complicated phenomenon, because many possibilities of reflection about reflexivity exist in modern social conditions. The social sciences play a basic role in the reflexivity of modernity: they do not simply 'accumulate knowledge' in the way in which the natural sciences may do.

Separation of time and space: the condition for the articulation of social relations across wide spans of time-space, up to and including global systems.

Disembedding mechanisms: consist of symbolic tokens and expert systems (these together = abstract systems). Disembedding mechanisms separate interaction from the particularities of locales.

Institutional reflexivity: the regularised use of knowledge about circumstances of social life as a constitutive element in its organisation and transformation.

Figure 21·1 The dynamism of modernity

In respect both of social and natural scientific knowledge, the reflexivity of modernity turns out to confound the expectations of Enlightenment thought – although it is the very product of that thought. The original progenitors of modern science and philosophy believed themselves to be preparing the way for securely founded knowledge of the social and natural worlds: the claims of reason were due to overcome the dogmas of tradition, offering a sense of certitude in place of the arbitrary character of habit and custom. But the reflexivity of modernity actually undermines the certainty of knowledge, even in the core domains of natural science. Science depends, not on the inductive accumulation of proofs, but on the methodological principle of doubt. No matter how cherished, and apparently well established, a given scientific tenet might be, it is open to revision – or might have to be discarded altogether – in the light of new ideas or findings. The integral relation between modernity and radical doubt is an issue which, once exposed to view, is not only disturbing to philosophers but is *existentially troubling* for ordinary individuals.

The local, the global and the transformation of day-to-day life

The globalising tendencies of modernity are inherent in the dynamic influences just sketched out. The reorganising of time and space, disembedding mechanisms and the reflexivity of modernity all presume universalising properties that explain the expansionist, coruscating nature of modern social life in its encounters with traditionally established practices. The globalisation of social activity which modernity has served to bring about is in some ways a process of the development of genuinely worldwide ties – such as those involved in the global nation-state system or the international division of labour. However, in a general way, the concept of globalisation is best understood as expressing fundamental aspect of time-space distanciation. Globalisation concerns the intersection of presence and absence, the interlacing of social events and social relations 'at distance' with local contextualities. We should grasp the global spread of modernity in terms of an ongoing relation between distanciation and the chronic mutability of local circumstances and local engagements. Like each of the other processes mentioned above, globalisation has to be understood as a dialectical phenomenon, in which events at one pole of a distanciated relation often produce divergent or even contrary occurrences at another. The *dialectic of the local and global* is a basic emphasis of the arguments employed [here].

Globalisation means that, in respect of the consequences of at least some disembedding mechanisms, no one can 'opt out' of the transformations brought about by modernity: this is so, for example, in respect of the global risks of nuclear war or of ecological catastrophe. Many other aspects of modern institutions, including those operating on the small scale, affect people living in more traditional settings, outside the most strongly 'developed' portions of the world. In those developed sectors, however, the connecting of the local and global has been tied to a profound set of transmutations in the nature of day-to-day life.

We can understand these transmutations directly in terms of the impact of disembedding mechanisms, which act to deskill many aspects of daily activities. Such deskilling is *not* simply a process where everyday knowledge is appropriated by experts or technical specialists (since very often there are imponderable or hotly disputed features of their fields of expertise); and it is not only a one-way process, because specialist information, as part of the reflexivity of modernity, is in one form or another constantly reappropriated by lay actors. These observations apply to the writings of sociologists as much as to any other specialists: it has been seen that the findings of books such as *Second Chances* [an investigation of divorce and remarriage] are likely to filter back into the milieux in which people take decisions about relationships, marriage and divorce. Trust in disembedding mechanisms is not confined to laypeople, because no one can be an expert about more than a tiny part of the diverse aspects of modern social life conditioned by abstract systems. Everyone living in conditions of modernity is affected by a multitude of abstract systems, and can at best process only superficial knowledge of their technicalities.

Awareness of the frailties and limits of abstract systems is not confined to technical specialists. Few individuals sustain an unswerving trust in the systems of technical knowledge that impinge on them, and everyone, whether consciously or not, selects among the competing possibilities of action that such systems (or disengagement from them) provide. Trust often merges with pragmatic acceptance: it is a sort of 'effort-bargain' that the individual makes with the institutions of modernity. Various attitudes of scepticism or antagonism towards abstract systems may coexist with a taken-for-granted confidence in others. For example, a person may go to great lengths to avoid eating foods that contain additives, but if that individual does not grow everything he or she eats, trust must necessarily be invested in the purveyors of 'natural foods' to provide superior products. Someone might turn towards holistic medicine after becoming disenchanted with the orthodox medical profession, but of course this is a transfer of faith. A sufferer from an illness might be so sceptical of the claims of all forms of expertise in healing that she avoids contact with medical practitioners altogether, no matter how the illness progresses. But even a person who effected a radical disengagement of this type would find it virtually impossible to escape altogether from the impact of systems of medicine and medical research, since these influence many aspects of the 'knowledge environment' as well as concrete elements of day-to-day life. For instance, they affect the regulations governing the production of foodstuffs – whether these be 'artificial' or 'natural' in character.

The mediation of experience

Virtually all human experience is mediated – through socialisation and in particular the acquisition of language. Language and memory are intrinsically connected, both on the level of individual recall and that of the institutionalisation of collective experience.[6] For human life, language is the prime and original means of time-space distanciation, elevating human activity beyond the immediacy of the experience of animals.[7] Language, as Lévi-Strauss says, is a time machine, which permits the re-enactment of social practices across the generations, while also making possible the differentiation of past, present and future.[8] The spoken word is a medium, a trace, whose evanescence in time and space is compatible with the preservation of meaning across time-space distances because of human mastery of language's structural characteristics. Orality and tradition are inevitably related closely to one another. As Walter Ong puts it in his study of speaking and writing, oral cultures 'have a heavy investment in the past, which registers in their highly conservative institutions and in their verbal performances and poetic processes, which are formulaic, relatively invariable, calculated to preserve the hard-won knowledge garnered out of past experience which, since there is no writing to record it, would otherwise slip away'.[9]

Although Lévi-Strauss and others have skilfully explored the relation between writing and the emergence of 'hot', dynamic social systems, only Innis and, following him, McLuhan, have theorised the impact of media on social development in a sophisticated fashion, especially in relation to the emergence of modernity.[10] Both authors emphasise the connections between dominant kinds of media and time-space transformations. The degree to which a medium serves to alter time-space relations does not depend primarily on the content or the 'messages' it carries, but on its form and reproducibility. Innis points out, for example, that the introduction of papyrus as a medium for the inscribing of writing greatly extended the scope of administrative systems because it was so much easier to transport, store and reproduce than previously used materials.

Modernity is inseparable from its 'own' media: the printed text and, subsequently, the electronic signal. The development and expansion of modern institutions were directly bound up with the tremendous increase in the mediation of experience which these communication forms brought in their train. When books were produced by hand, readership was sequential: the book had to pass from one person to another. The books and texts of pre-modern civilisations remained substantially geared to the transmission of tradition, and were almost always essentially 'classical' in character. Printed materials cross space as easily as time because they can be distributed to many readers more or less simultaneously.[11] Only half a century after the appearance of Gutenberg's bible, hundreds of printing shops had sprung up in cities all over Europe. Today the printed word remains at the core of modernity and its global networks. Practically every known language of humankind has been set down in print, and even in those societies where levels of literacy are low, printed materials and the ability to produce and interpret them are indispensable means of administrative and social coordination. It has been calculated that, on a global level, the amount of printed materials produced has doubled every fifteen years since the days of Gutenberg.[12]

Printing was one of the main influences upon the rise of the early modern state, and other antecedent institutions of modernity, but when we look to the origins of high modernity it is the increasingly intertwined development of mass printed media and electronic communication that is important. The emergence of mass circulation printed materials is customarily thought of as belonging to an era prior to that of electronic messages – particularly by McLuhan, who radically set off one against the other. In terms of sheer temporal succession, it is true that the prime example of mass printed material – the newspaper – came into being about a century before the advent of television. Yet it is quite misleading to see one merely as a phase prior to the emergence of the other; from early on electronic communication has been vital to the development of mass printed media. Although the invention of the telegraph came some while after the first flourishing of dailies and periodicals, it was fundamental to what we now know as the newspaper and indeed to the very concept of 'news' itself. Telephone and radio communication further expanded this connection.

The early newspapers (and a whole diversity of other magazines and periodicals) played a major role in completing the separation of space from place, but this process only became a global phenomenon because of the integration of printed and electronic media. This is easily demonstrated by reference to the development of the modern newspaper. Thus Susan Brooker-Gross has examined changes in the time-space 'reach' of newspapers. She found that typical news items in an American paper from the mid-nineteenth century, before the diffusion of the telegraph, differed both from those of the early 1800s, and from those produced subsequently. The news items reported stories from cities some way distant in the US, but these did not have the immediacy the reader is used to with the newspapers of today.[13]

Prior to the coming of the telegraph, Brooker-Gross showed, news stories described events that were close at hand and recent; the further away a particular happening, the more it would appear at a very late date. News from far came in the form of what she calls 'geographic bundling'. Materials from Europe, for example, literally came in packages from the ship, and would be presented in the form in which they were found: 'a ship arrived from London, and here is the news it brought.' In other words, channels of communication, and the pressures of time-space differences, directly shaped the presentation of the printed news-pages. Following the introduction of the telegraph, and then the telephone and other electronic media, it was the event that increasingly became the determining factor governing inclusion – rather than the place in which it occurred. Most news media preserve a sense of 'privileged place' in respect of their own position – giving a bias towards local news – but only against the backcloth of the pre-eminence of the event.[14]

The visual images which television, films and videos present no doubt create textures of mediated experience which are unavailable through the printed word. Yet, like newspapers, magazines, periodicals and printed materials of other sorts, these media are as much an expression of the disembedding, globalising tendencies of modernity as they are the instruments of such tendencies. As modalities of reorganising time and space, the similarities between printed and electronic media are more important than their differences in the constituting of modern institutions. This is so in respect of two basic features of mediated experience in conditions of modernity. One is the *collage effect.* Once the event has become more or less completely dominant over location, media presentation takes the form of the juxtaposition of stories and items which share nothing in common other than that they are 'timely' and consequential. The newspaper page and the television programme guide are equally significant examples of the collage effect. Does this effect mark the disappearance of narratives and even perhaps the severence of signs from their referents, as some have suggested?[15] Surely not. A collage is by definition not a narrative; but the coexistence of different items in mass media does not represent a chaotic jumble of signs. Rather, the separate 'stories' which are displayed alongside one another express orderings of consequentiality typical of a transformed time-space environment from which the hold of place has largely evaporated. They do not, of course, add up to a single narrative, but they depend on, and also in some ways express, unities of thought and consciousness.

Characteristic of mediated experience in modern times is a second major feature: the *intrusion of distant events into everyday consciousness*, which is in some substantial part organised in terms of awareness of them. Many of the events reported on the news, for instance, might be experienced by the individual as external and remote; but many equally enter routinely into everyday activity. Familiarity generated by mediated experience might perhaps quite often produce feelings of 'reality inversion': the real object and event, when encountered, seem to have a less concrete existence than their media representation. Moreover many experiences that might be rare in day-to-day life (such as direct contact with death and the dying) are encountered routinely in media representations; confrontation with the real phenomena themselves is psychologically problematic. In conditions of modernity, in sum, the media do not mirror realities but in some part form them; but this does not mean that we should draw the conclusion that the media have created an autonomous realm of 'hyperreality' where the sign or image is everything.

It has become commonplace to claim that modernity fragments, dissociates. Some have even presumed that such fragmentation marks the emergence of a novel phase of social development beyond modernity – a postmodern era. Yet the unifying features of modern institutions are just as central to modernity – especially in the phase of high modernity – as the disaggregating ones. The 'emptying' of time and space set in motion processes that established a single 'world' where none existed previously. In the majority of pre-modern cultures, including those of medieval Europe, time and space merged with domains of the gods and spirits as well as with the 'privileging of place'.[16] Taken overall, the many diverse modes of culture and consciousness characteristic of pre-modern 'world systems' formed a genuinely fragmented array of human social communities. By contrast, late modernity produces a situation in which humankind in some respects becomes a 'we', facing problems and opportunities where there are no 'others'.

22

The unique perspective of television and its effect: a pilot study

Kurt Lang and Gladys Engel Lang

From the *American Sociological Review* Vol 18(1), 1953, 3–12.

Note: In 1951, during the Korean War, President Truman dismissed General Douglas MacArthur of his duties as Commander-in-Chief of US forces, following disagreements over policy and White House disquiet about the extent to which the General was threatening the prerogative of the president in policy-making. MacArthur returned to the United States amidst great publicity and popular demonstrations of support. A number of major cities welcomed him on civic visits, including Chicago. The study reprinted below concerns itself with the TV coverage of this visit and its political implications.

This paper aims to investigate a public event as viewed over television or, to put it differently, to study in the context of public life, an event transmitted over video. The concern is not with the effects of television on individual persons, irrespective of the spread of this effect. Our assumption is, on the contrary, that the effect of exposure to TV broadcasting of public events cannot be measured most successfully in isolation. For the influence on one person is communicated to others, until the significance attached to the video event overshadows the 'true' picture of the event, namely the impression obtained by someone physically present at the scene of the event. The experience of spectators may not be disseminated at all or may be discounted as the biased version of a specially interested participant. Or, again, the spectator's interpretation of his own experience may be reinterpreted when he finds the event in which he participated discussed by friends, newspapermen, and radio commentators. If the significance of the event is magnified, even casual spectatorship assumes importance. The fact of having 'been there' is to be remembered – not so much because the event, in itself, has left an impression, but because the event has been recorded by others. At the opposite extreme, privately significant experiences, unless revived in subsequent interpersonal relations, soon recede into the deeper layers of memory.

By taking MacArthur Day in Chicago,[1] as it was experienced by millions of spectators and video viewers, we have attempted to study an event transmitted over video. The basis of this report is the contrast between the actually recorded experience of participant observers on the scene, on the one hand, and the picture which a video viewer received by way of the television screen, and the way in which the event was interpreted, magnified, and took on added significance on the other. The contrast between these two perspectives from which the larger social environment can be viewed and 'known,' forms the starting point for the assessment of a particular effect of television in structuring public events.

The research design

The present research was undertaken as an exploration in collective behaviour.[2] the design of the communications analysis differs significantly from most studies of content analysis. The usual process of inferring effect from content and validating the effect by means of interviews with an audience and control group is reversed. A generally apparent effect, i.e. the 'landslide effect' of national indignation at MacArthur's abrupt dismissal and the impression of enthusiastic support, bordering on 'mass hysteria,' given to him, was used to make inferences on given aspects of the television content. The concern was with the picture disseminated, especially as it bore on the political atmosphere. To explain how people could have a false imagery (the implication of participant observational data), it was necessary to show how their perspective of the larger political environment was limited and how the occasion of Chicago's welcome to MacArthur, an event mediately known already, was given a particular structure. The concern is how the picture of the events was shaped by selection, emphasis, and suggested inferences which fitted into the already existing pattern of expectations.

The content analysis was therefore focused on two aspects – the selections made by the camera and their structuring of the event in terms of foreground and background, and the explanation and interpretations of televised events given by commentators and persons interviewed by them. Moreover, each monitor was instructed to give his impression of what was happening, on the basis of the picture and information received by way of television. The monitors' interpretations and subjective impressions were separately recorded. They served as a check that the structure inferred from the two operations of 'objective' analysis of content were, in fact, legitimate inferences.[3] At the same time, utilizing the categories of the objective analysis, the devices by which the event was structured could be isolated, and the specific ways in which television reportage differed from the combined observations could be determined.

Thirty-one participant observers took part in the study. They were spatially distributed to allow for the maximum coverage of all the important phases of the day's activities, i.e., no important vantage point of spectatorship was neglected. Since the events were temporarily distributed, many observers took more than one station, so that coverage was actually based on more than 31 perspectives. Thus the sampling error inherent in individual participant observation or unplanned mass-observation was greatly reduced. Observers could witness the arrival at Midway Airport and still arrive in the Loop area long before the scheduled time for the parade. Reports were received from 43 points of observation.

Volunteers received instruction sheets which drew their attention to principles of observation[4] and details to be carefully recorded. Among these was the directive to take careful note of any activity indicating possible influences of the televising of the event upon the behaviour of spectators, e.g., actions specifically addressed to the cameras, indications that events were staged with an eye towards transmission over television, and the like.

Summary of findings

The pattern of expectations.

The mass-observation concentrated on discerning the psychological structure of the unfolding event in terms of present and subsequent anticipations. Certainly the crowd which turned out for the MacArthur Day celebration was far from a casual collection of individuals: the members *intended* to be witnesses to this 'unusual event.' One may call these intentions specific attitudes, emergent acts, expectations, or predispositions. Whatever the label, materials on these patterns of expectations were taken from two sources: (1) all statements of spectators recorded in the observer reports which could be interpreted as indicative of such expectations (coded in terms of the inferences therein); (2) personal expectations of the 31 study observers (as stated in the personal questionnaire).

Though not strictly comparable – since the observations on the scene contained purely personal, very short-range and factually limited expectations – both series of data provide confirmation of a basic pattern of observer expectations. The persons on the scene *anticipated* 'mobs' and 'wild crowds.' They expected some disruption of transportation. Their journey downtown was in search of adventure and excitement. Leaving out such purely personal expectations as 'seeing' and 'greeting,' the second most frequent preconception emphasizes the extraordinary nature of the preparations and the entertaining showmanship connected with the spectacle.

As a result of an unfortunate collapsing of several questions regarding personal data into one, the response did not always focus properly on what the observers 'expected to see.' In some cases no evidence or only an incomplete description of this aspect was obtained. Of those answering, 68 per cent expected excited and wildly enthusiastic crowds. But it is a safe inference from the discussion during the briefing session that this figure tends to underestimate the number who held this type of imagery. The main incentive to volunteer resided, after all, in the opportunity to study crowd behaviour at first hand.

To sum up: most people expected a wild spectacle, in which the large masses of on-lookers would take an active part, and which contained an element of threat in view of the absence of ordinary restraints on behaviour and the power of large numbers.

The role of mass media in the pattern of expectations

A more detailed examination of the data supports the original assumption that the pattern of expectations was shaped by way of the mass media. For it is in that way that the picture of the larger world comes to sophisticated as well as unsophisticated people. The observers of the study were no exception to this dependence on what newspapers, newsreels, and television cameras mediated. They were, perhaps, better able than others to describe the origin of these impressions. Thus Observer 14 wrote in evaluating his report and his subjective feelings:

> I had listened to the accounts of MacArthur's arrival in San Francisco, heard radio reports of his progress through the United States, and had heard the Washington speech as well as the radio accounts of his New York reception. . . . I had therefore expected the crowds to be much more vehement, contagious, and identified with MacArthur. I had expected to hear much political talk, especially anti-Communist and against the Truman administration.
>
> These expectations were completely unfulfilled. I was amazed that not once did I hear Truman criticized, Acheson mentioned, or as much as an allusion to the Communists. . . . I had expected roaring, excited mobs; instead there were quiet, well ordered, dignified people. . . . The air of curiosity and casualness surprised me. Most people seemed to look on the event as simply something that might be interesting to watch.

Other observers made statements of a very similar content.

Conversation in the crowd pointed to a similar awareness. Talk repeatedly turned to television, especially to the comparative merit of 'being there' and 'seeing it over TV.' An effort was consequently made to assess systematically the evidence bearing on the motives for being there in terms of the patterns of expectations previously built up. The procedures of content analysis served as a useful tool, allowing the weighing of all evidence *directly* relevant to this question in terms of confirmatory and contrary evidence. The coding operation involved the selection of two types of indicators: (1) general evaluations and summaries of data; and (2) actual incidents of behaviour which could support or nullify our hypothesis.

Insofar as the observers had been instructed to report concrete behaviour rather than general interpretations, relatively few such generalizations are available for tabulation. Those given were used to formulate the basic headings under which the concrete evidence could be tabulated. The generalizations fall into two types: namely, the crowds had turned out to see a great military figure and a public hero 'in the flesh'; and – its logical supplement – they had turned out not so much 'to see *him*, as I noticed, but to see the spectacle (Observer 5).' Six out of eleven concretely stated propositions were of the second type.

An examination of the media content required the introduction of a third heading, which subdivided the interest in MacArthur into two distinct interpretations: that people had come to find vantage points from which to see the man and his family; or, as the official (media and 'Chicago official') version held, that they had come to welcome, cheer, and honour him. Not one single observer, in any generalized proposition, confirmed the official generalization, but there was infrequent mention of isolated incidents which would justify such an interpretation.

The analysis of actual incidents, behaviour, and statements recorded is more revealing. A gross classification of the anticipations which led people to participate is given (according to categories outlined above) in Table 1.

Table 22.1 Types of Spectator Interest

Form of Motivation	Per Cent
Active hero worship	9.2
Interest in seeing MacArthur	48.1
Passive interest in spectacle	42.7
Total	100.0

A classification of these observations by area in which they were secured gives a clear indication that the Loop throngs thought of the occasion *primarily* as a spectacle. There, the percentage of observations supporting the 'spectacle hypothesis' was 59.7. The percentage in other areas was: Negro district, 40.0; Soldiers Field, 22.9; Airport, 17.6; University district, 0.0. Moreover, of the six generalizations advanced on crowd expectations in the Loop, five interpreted the prevalent motivation as the hope of a wild spectacle.

Thus, a probe into motivation gives a confirmatory clue regarding the pattern of expectations observed. To this body of data, there should be added the constantly over-heard expressions – as the time of waiting increased and excitement failed to materialize – of disillusionment with the particular vantage point. 'We should have stayed home and watched it on TV,' was the almost universal form that the dissatisfaction took. In relation to the spectatorship experience of extended boredom and sore feet, alleviated only by a brief glimpse of the hero of the day, previous and similar experiences over television had been truly exciting ones which promised even greater 'sharing of excitement' *if only one were present.* These expectations were disappointed and favorable allusions to television in this respect were frequent. To present the entire body of evidence bearing on the inadequate release of tension and the widely felt frustration would be to go beyond the scope of this report, in which the primary concern is the study of the television event. But the materials collected present unequivocal proof of the foregoing statements, and this – with one qualified exception – is also the interpretation of each one of the observers.

Moreover, the comparison of the television perspective with that of the participant observers indicates that the video aspects of MacArthur Day in Chicago served to *preserve* rather than disappoint the same pattern of expectations among the viewers. The main difference was that television remained true to form until the very end, interpreting the entire proceedings according to expectations. No hint about the disappointment in the crowd was provided. To cite only one example, taken from what was the high point in the video presentation, the moment when the crowds broke into the parade by surging out into State Street:

> The scene at 2:50 p.m. at State and Jackson was described by the announcer as the 'most enthusiastic crowd *ever* in our city. . . . You can feel the tenseness in the air. . . . You can hear that crowd roar.' The crowd was described as pushing out into the curb with the police trying to keep it in order, while the camera was still focusing on MacArthur and his party. The final picture was of a bobbing mass of heads as the camera took in the entire view of State Street northward. To the monitor, this mass of people appeared to be pushing and going nowhere. And then, with the remark, 'The whole city appears to be marching down State Street behind General MacArthur,' holding the picture just long enough for the impression to sink in, the picture was suddenly blanked out.

Observer 26, who was monitoring this phase of the television transmission, reported her impression:

> . . . the last buildup on TV concerning the 'crowd' (cut off as it was, abruptly at 3:00 p.m.) gave me the impression that the crowd was pressing and straining so hard that it was going to be hard to control. My first thought, 'I'm glad I'm not in that' and 'I hope nobody gets crushed.'

But observers near State and Jackson did not mention the event in an extraordinary context. For example, Observer 24 explained that as MacArthur passed:

> Everybody strained but few could get a really good glimpse of him. A few seconds after he had passed most people merely turned around to shrug and to address their neighbours with such phrases: 'That's all,' 'That was it,' 'Gee, he looks just as he does in the movies,' 'What'll we do now?' Mostly teenagers and others with no specific plans flocked into the street after MacArthur, but very soon got tired of following as there was no place to go and nothing to do.

> Some cars were caught in the crowd, a matter which, to the crowd, seemed amusing.

The structure of the TV presentation

The television perspective was different from that of any spectator in the crowd. Relatively unlimited in its mobility, it could order events in its own way by using close-ups for what was deemed important and leaving the apparently unimportant for the background. There was almost complete freedom to aim cameras in accordance with such judgments. The view, moreover, could be shifted to any significant happening, so that the technical possibilities of the medium itself tended to play up the dramatic. While the spectator, if fortunate, caught a brief glimpse of the General and his family, the television viewer found him the continuous centre of attraction from his first appearance during the parade at 2.21 p.m. until the sudden blackout at 3.00 p.m. For almost 40 minutes, not counting his seven minute appearance earlier in the day at the airport and his longer appearance at Soldiers Field that evening, the video viewer could fasten his eyes on the General and on what could be interpreted as the interplay between a heroic figure and the enthusiastic crowd. The cheering of the crowd seemed not to die down at all, and even as the telecast was concluded, it only seemed to have reached its crest. Moreover, as the camera focused principally on the parade itself, the crowd's applause seemed all the more ominous a tribute from the background.

The shots of the waiting crowd, the interviews with persons within it, and the commentaries, had previously prepared the viewer for this dramatic development. Its resolution was left to the inference of the individual. But a sufficient number of clues had already come over television to leave little doubt about the structure. Out of the three-hour daytime telecast, in addition to the time that MacArthur and party were the visual focus of attention, there were over two hours which had to be filled with visual material and vocal commentary. By far the largest amount of time was spent on anticipatory shots of the crowd. MacArthur himself held the picture for the second longest period; thus the ratio of time spent viewing MacArthur to time spent anticipating his arrival is much greater for the TV observer than for the spectator on the scene.

The descriptive accounts of the commentators (also reflected in the interviews),[5] determined the structure of the TV presentation of the day's events. The idea of the magnitude of the event, in line with preparations announced in the newspapers, was emphasized by constant reference. The most frequently employed theme was that 'no effort has been spared to make this day memorable' (eight references). There were seven direct references to the effect that the announcer had 'never seen the equal to this moment' or that it was the 'greatest ovation this city had ever turned out.' The unique cooperative effort of TV received five mentions and was tied in with the 'dramatic' proportions of the event. It was impossible to categorize and tabulate all references, but they ranged from a description of crowded transportation and numerical estimates of the crowd to the length of the city's lunch hour and the state of 'suspended animation' into which business had fallen. There was repeated mention that nothing was being allowed to interfere with the success of the celebration; even the ball game had been cancelled.[6] In addition to these purely formal aspects of the event, two – and only two – aspects of the spectacle were *stressed*: (1) the unusual nature of the event; (2) the tension which was said to pervade the entire scene. Even the references to the friendly and congenial mood of the waiting crowd portended something about the change that was expected to occur.

Moreover, in view of the selectivity of the coverage with its emphasis on close-ups,[7] it was possible for each viewer to see himself in a *personal* relationship to the General. As the announcer shouted out: 'Look at that chin! Look at those eyes!' – each viewer, regardless of what might have been meant by it, could seek a personal interpretation which best expressed, for him, the real feeling underlying the exterior which appeared on the television screen.[8]

It is against the background of this personal inspection that the significance of the telecast must be interpreted. The cheering crowd, the 'seething mass of humanity,' was fictionally endowed by the commentators with the same capacity for a direct and personal relationship to MacArthur as the one which television momentarily established for the TV viewer through its close-up shots. The net effect of television thus stems from a convergence of these two phenomena; namely, the seemingly extraordinary scope of the event together with the apparent enthusiasm accompanying it and personalizing influence just referred to. In this way the public event was interpreted in a very personal nexus. The total effect of so many people, all shouting, straining, cheering, waving in personal welcome to the General, disseminated the impression of a universal, enthusiastic, overwhelming ovation for the General. The selectivity of the camera and the commentary gave the event a personal dimension, non-existent for the participants in the crowds, thereby presenting a very specific perspective which contrasted with that of direct observation.

Other indices of the discrepancy

In order to provide a further objective check on the discrepancies between observer impressions and the event as it was interpreted by those who witnessed it over television, a number of spot checks on the reported amount of participation were undertaken. Transportation statistics, counts in offices, and the volume of sales reported by vendors provided such indices.

The results substantiate the above finding. The city and suburban lines showed a very slight increase over their normal loads. To some extent the paltry 50,000 increase in inbound traffic on the street cars and elevated trains might even have been due to rerouting. The suburban lines had their evening rush hour moved up into the early afternoon – before the parade had begun.

Checks at luncheonettes, restaurants, and parking areas indicated no unusual crowding. Samplings in offices disclosed only a minor interest in the parade. Hawkers, perhaps the most sensitive judges of enthusiasm, called the parade a 'puzzler' and displayed unsold wares.

Detailed illustration of contrast

The Bridge ceremony provides an illustration of the contrast between the two perspectives. Seven observers witnessed this ceremony from the crowd.

> TV perspective: In the words of the announcer, the Bridge ceremony marked 'one of the high spots, if not the high spot of the occasion this afternoon. . . . The parade is now reaching its climax at this point.'
>
> The announcer, still focusing on MacArthur and the other participating persons,

took the opportunity to review the ceremony about to take place. . . . The camera followed and the announcer described the ceremony in detail. . . . The camera focused directly on the General, showing a close-up. . . . There were no shots of the crowd during this period. But the announcer filled in. 'A great cheer goes up at the Bataan Bridge, where the General has just placed a wreath in honor of the American boys who died at Bataan and Corregidor. You have heard the speech . . . the General is now walking back . . . the General now enters his car. This is the focal point where all the newsreels . . . frankly, in 25 years of covering the news, we have never seen as many newsreels gathered at one spot. One, two, three, four, five, six. At least eight cars with newsreels rigged on top of them, taking a picture that will be carried over the entire world, over the Chicagoland area by the combined network of these TV stations of Chicago, which have combined for this great occasion and for the solemn occasion which you have just witnessed.'

During this scene there were sufficient close-ups for the viewer to gain a definite reaction, positive or negative, to the proceedings. He could see the General's facial expressions and what appeared to be momentary confusion. He could watch the activities of the Gold Star mothers in relation to MacArthur and define this as he wished – as inappropriate for the bereaved moment or as understandable in the light of the occasion. Taking the cue from the announcer, the entire scene could be viewed as rushed. Whether or not, in line with the official interpretation, the TV viewer saw the occasion as *solemn*, it can be assumed that he expected that the participant on the scene was, in fact, experiencing the occasion in the same way as he.

Actually, this is the way what was meant to be a solemn occasion was experienced by those attending, and which constitutes the crowd perspective. The dedication ceremony aroused little of the sentiment it might have elicited under other conditions. According to Observer 31, 'People on our corner could not see the dedication ceremony very well and consequently after he had passed immediately in front of us, there was uncertainty as to what was going on. As soon as word had come down that he had gone down to his car, the crowd dispersed.' Observer 8 could not quite see the ceremony from where he was located on Wacker Drive, slightly east of the bridge. Condensed descriptions of two witnesses illustrate the confusion which surrounded the actual wreath-laying ceremony (three other similar descriptions are omitted here).

It was difficult to see any of them. MacArthur moved swiftly up the steps and immediately shook hands with people on the platform waiting to greet him. There was some cheering when he mounted the platform. He walked north on the platform and did not reappear until some minutes later. In the meantime the crowd was so noisy that it was impossible to understand what was being broadcast from the loud-speakers. Cheering was spotty and intermittent, and there was much talk about Mrs. MacArthur and Arthur . . . (Observer 2).

Those who were not on boxes did not see MacArthur. They did not see Mrs. MacArthur, but only her back. MacArthur went up on the platform, as we were informed by those on boxes, and soon we heard some sound over the loudspeakers. Several cars were standing in the street with their motors running. . . . Some shouted to the cars to shut their motors off, but the people in the cars did not care or did not hear. . . . The people in our area continued to push forward trying to hear. When people from other areas began to come and walk past us to go toward the train, the people in our area shrugged their shoulders. 'Well, I guess it's all over. That noise must have been the speech.' One of the three men who had stood there for an hour or more, because it was such a good spot, complained, 'This turned out to be a

> lousy spot. I should have gone home. I bet my wife saw it much better over television' (Observer 30).

Regardless of good intentions on the part of planners and despite any recognition of the solemn purpose of the occasion by individuals in the crowd, the solemnity of the occasion was destroyed, if for no other reason, because officials in the parade were so intent upon the time-schedule and camera-men so intent upon recording the solemn dedication for the TV audience and for posterity that the witnesses could not see or hear the ceremony, or feel 'solemn' or communicate a mood of solemnity. A crowd of confused spectators, cheated in their hopes of seeing a legendary hero in the flesh, was left unsatisfied.

Reciprocal Effects

There is some direct evidence regarding the way in which television imposed its own peculiar perspective on the event. In one case an observer on the scene could watch both what was going on and what was being televised.

> It was possible for me to view the scene (at Soldiers Field) both naturally and through the lens of the television camera. It was obvious that the camera presented quite a different picture from the one received otherwise. The camera followed the General's car and caught that part of the crowd immediately opposite the car and about 15 rows above it. Thus it caught that part of the crowd that was cheering, giving the impression of a solid mass of wildly cheering people. It did not show the large sections of empty stands, nor did it show that people stopped cheering as soon as the car passed them (Observer 13).

In much the same way, the television viewer received the impression of wildly cheering and enthusiastic crowds before the parade. The camera selected shots of the noisy and waving audience, but in this case, the television camera itself created the incident. The cheering, waving, and shouting was often largely a response to the aiming of the camera. The crowd was thrilled to be on television, and many attempted to make themselves apparent to acquaintances who might be watching. But even beyond that, an event important enough to warrant the most widespread pooling of television facilities in Chicago video history, acquired in its own right some magnitude and significance. Casual conversation continually showed that being on television was among the greatest thrills of the day.

Conclusion

It has been claimed for television that it brings the truth directly into the home: the 'camera does not lie.' Analysis of the above data shows that this assumed reportorial accuracy is far from automatic. Every camera selects, and thereby leaves the unseen part of the subject open to suggestion and inference. The gaps are usually filled in by a commentator. In addition the process directs action and attention to itself.

Examination of a public event by mass-observation and by television revealed considerable discrepancy between these two experiences. The contrast in perspectives points to three items whose relevance in structuring a televised event can be inferred from an analysis of the television content:

1 technological bias, i.e., the necessarily arbitrary sequence of telecasting events and their structure in terms of foreground and background, which at the same time contains the choices on the part of the television personnel as to what is important;
2 structuring of an event by an announcer, whose commentary is needed to tie together the shifts from camera to camera, from vista to close-up, helping the spectator to gain the stable orientation from one particular perspective;
3 reciprocal effects, which modify the event itself by staging it in a way to make it more suitable for telecasting and creating among the actors the consciousness of acting for a larger audience.

General attitudes regarding television and viewing habits must also be taken into account. Since the industry is accustomed to thinking in terms of audience ratings – though not to the exclusion of all other considerations – efforts are made to assure steady interest. The telecast was made to conform to what was interpreted as the pattern of viewers' expectations. The drama of MacArthur Day, in line with that pattern, was nonetheless built around unifying symbols, personalities, and general appeals (rather than issues). But a drama it had to be, even if at the expense of reality.

Unlike other television programmes, news and special events features constitute part of that basic information about 'reality' which we require in order to act in concert with anonymous but like-minded persons in the political process. Action is guided by the possibilities for success, and, as part of this constant assessment, inferences about public opinions as a whole are constantly made. Even though the average citizen does, in fact, see only a small segment of public opinion, few persons refrain from making estimates of the true reading of the public temper. Actions and campaigns are supported by a sense of support from other persons. If not, these others at least constitute an action potential that can be mobilized. The correct evaluation of the public temper is therefore of utmost importance; it enters the total political situation as perhaps one of the weightiest factors.

Where no overt expression of public opinion exists, politicians and citizens find it useful to fabricate it. Against such demonstrations as the MacArthur Day, poll data lack persuasiveness and, of necessity, must always lag, in their publication, behind the development of popular attitudes. For the politician who is retroactively able to counter the errors resulting from an undue regard for what at a given time is considered overwhelming public opinion, there may be little significance in this delay. The imagery of momentary opinion may, however, goad him into action which, though justified in the name of public opinion, may objectively be detrimental. It may prevent critics from speaking out when reasoned criticism is desirable, so that action may be deferred until scientific estimates of public opinion can catch up with the prior emergence of new or submerged opinion.

Above all, a more careful formulation of the relations among public opinion, the mass media, and the political process is vital for the understanding of many problems in the field of politics. The reports and telecasts of what purports to be spontaneous homage paid to a political figure assume added meaning within this context. The most important single media effect coming within the scope of the material relevant to the study of MacArthur Day was the dissemination of an image of overwhelming public sentiment in favour of the General. This effect gathered force as it was incorporated into political strategy, picked up by other media, entered into gossip, and thus came to overshadow immediate reality as it might have been recorded by an observer on the scene. We have labelled this the 'landslide effect' because, in view of the wide-spread dissemination of a particular public welcoming ceremony the imputed unanimity gathered tremendous force.[9] This 'landslide effect' can, in large measure, be attributed to television.

Two characteristics of the video event enhanced this effect (misevaluation of public sentiment). (1) The depiction of the ceremonies in unifying rather than in particularistic symbols (between which a balance was maintained) failed to leave any room for dissent. Because no lines were drawn between the conventional and the partisan aspects of the reception, the traditional welcome assumed political significance in the eyes of the public. (2) A general characteristic of the television presentation was that the field of vision of the viewer was enlarged while, at the same time, the context in which these events could be interpreted was less clear. Whereas a participant was able to make direct inferences about the crowd as whole, being in constant touch with those around him, the television viewer was in the centre of the entire crowd. Yet, unlike the participant, he was completely at the mercy of the instrument of his perceptions. He could not test his impressions – could not shove back the shover, inspect bystanders' views, or attempt in any way to affect the ongoing activity. To the participant, on the other hand, the direction of the crowd activity as a whole, regardless of its final goal, still appeared as the interplay of certain peculiarly personal and human forces. Political sentiment, wherever encountered, could thus be evaluated and discounted. Antagonistic views could be attributed to insufficient personal powers of persuasion rather than seen as subjugation to the impersonal dynamics of mass hysteria. The television viewer had little opportunity to recognize this personal dimension in the crowd. What was mediated over the screen was, above all, the general trend and the direction of the event, which consequently assumed the proportion of an impersonal force, no longer subject to influence.

This view of the 'overwhelming' effect of public moods and the impersonal logic of public events is hypothesized as a characteristic of the perspective resulting from the general structure of the picture and the context of television viewing.

23

Broadcast TV as sound and image

John Ellis

From Ellis, John 1982: *Visible fictions*. London: Routledge and Kegan Paul, 127–37.

TV offers a radically different image from cinema, and a different relation between sound and image. The TV image is of a lower quality than the cinematic image in terms of its resolution of detail. It is far more apparent that the broadcast TV picture is composed of lines than it is that the cinema image is composed of particles of silver compounds. Not only this, but the TV image is virtually always substantially smaller than the cinema image. Characteristically, the size of TV sets ranges from the 12 inch portable to the 24 inch or sometimes 30 inch model: all these measurements refer to the distance across the screen diagonally. The TV image shows things smaller than they are, unless it is a close-up of a small object, or of a person in head and shoulders only, when they appear more or less their real size. Such simple observations have profound effects on the kind of representations and spectator attitudes that broadcast TV creates for itself.

First, it is a characteristic of broadcast TV that the viewer is larger than the image: the opposite of cinema. It seems to be a convention also that the TV image is looked down on, rather than up to as in cinema. TV sets that are produced with stands are about two feet off the floor, which gives the effect of being almost but not quite level with the eyes of an individual lounging in an easy chair (as indeed we are meant to watch TV according to the advertisements it screens for itself.) TV takes place in domestic surroundings, and is usually viewed in normal light conditions, though direct sunlight reflects off the screen to an unacceptable degree. The regime of viewing TV is thus very different from the cinema: TV does not encourage the same degree of spectator concentration. There is no surrounding darkness, no anonymity of the fellow viewers, no large image, no lack of movement amongst the spectators, no rapt attention. TV is not usually the only thing going on, sometimes it is not even the principal thing. TV is treated casually rather than concentratedly. It is something of a last resort ('What's on TV tonight, then?) rather than a special event. It has a lower degree of sustained concentration from its viewers, but a more extended period of watching and more frequent use than cinema.

This has two major effects on the kind of regime of representation that has developed for TV. First, the role that sound plays in TV is extremely important. Second, it engages the look and the glance rather than the gaze, and thus has a different relation to voyeurism from cinema's [. . .].

The role played by sound stems from the fact that it radiates in all directions, whereas the view of the TV image is sometimes restricted. Direct eye contact is needed with the TV screen. Sound can be heard where the screen cannot be seen. So sound is used to ensure a certain level of attention, to drag viewers back to looking at the set. Hence the importance of programme announcements and signature tunes and, to some extent, of music in various kinds of series. Sound holds attention more consistently than image, and provides a continuity that holds across momentary lapses of attention. The result is a slightly different balance between sound and image from that which is characteristic of cinema. Cinema is guaranteed a centred viewer by the physical arrangement of cinema seats and customs of film viewing. Sound therefore follows the image or diverges from it. The image is the central reference in cinema. But for TV, sound has a more centrally defining role. Sound carries the fiction or the documentary; the image has a more illustrative function. The TV image tends to be simple and straightforward, stripped of detail and excess of meanings. Sound tends to carry the details (background noises, music). This is a tendency towards a different sound/image balance than in cinema, rather than a marked and consistent difference. Broadcast TV has areas which tend towards the cinematic, especially the areas of serious drama or of various kinds of TV film. But many of TV's characteristic broadcast forms rely upon sound as the major carrier of information and the major means of ensuring continuity of attention. The news broadcast, the documentary with voice-over commentary, the bulk of TV comedy shows, all display a greater reliance on sound than any form that cinema has developed for itself. The image becomes illustration, and only occasionally provides material that is not covered by the sound-track (e.g. comedy sight-gags, news actuality footage). Sound tends to anchor meaning on TV, where the image tends to anchor it with cinema. In both, these are a matter of emphasis rather than any simple reliance one upon another. Sound and image exist in relation to each other in each medium rather than acting as separate entities. However, the difference of emphasis does exist between the two. It gives rise to two critical attitudes that are fundamental to the way in which newspaper critics and practitioners alike tend to conceive of the two media. Any film that contains a large amount of dialogue is open to the criticism that 'it could have been a radio play', as was Bergman's *From the Life of Marionettes (Aus dem Leben der Marionetten,* 1980). A similar accusation is never hurled at a TV play, however wordy. Instead, there are unwritten rules that govern the image for TV. Especially in British broadcast TV (possibly the most hide-bound in the world), the image is to be kept literal for almost all the time. There are licensed exceptions (science fiction, rock music programmes) where experimentation with the physical composition of the video image can take place, but the rule is that the image must show whatever is before the camera with the minimum of fuss and conscious technique. The image is to be kept in its place. Both these attitudes refer to occasions in which the subservient partner in the sound/image relationship tends to assert itself too much: for cinema, it is sound; for TV it is the image.

TV's lower level of sustained concentration on the image has had another effect upon its characteristic regime of representation. The image on broadcast TV, being a lower grade image than cinema's, has developed in a particular way. Contrasting with cinema's profusion (and sometimes excess) of detail, broadcast TV's image is stripped-down, lacking in detail. The visual effects of this are immediately apparent: the fussy detail of a film shown on TV compared to the visual bareness of a TV cop series, where cars chase each other through endless urban wastes of bare walls and near-deserted streets. The broadcast TV image has to be certain that its meaning is obvious: the streets are almost empty so that the movement of the car is all the more obvious. The walls are bare so that no writing distracts attention from the segmental event. This is not an effect of parsimony of production investment (rather, it enables it to happen). It is a more fundamental aspect of the broadcast TV image coming to terms with itself. Being small, low definition, subject to attention that will not be sustained, the TV image becomes jealous of its meaning. It is unwilling to waste it on details and inessentials. So background and context tend to be sketched rather than brought forward and subject to a certain fetishism of details that often occurs in cinema, especially art cinema. The narratively important detail is stressed by this lack of other detail. Sometimes, it is also stressed by music, producing an emphasis that seems entirely acceptable on TV, yet would seem ludicrously heavy-handed in cinema. This is particularly so with American crime series, where speed of action and transition from one segment to another dictates the concentration of resources on to single meanings. Where detail and background are used in TV programmes, for example the BBC historical serials, action tends to slow down as a result. The screen displays historical detail in and for itself; characters are inserted around it, carrying on their lengthy conversations as best they can. Segments are drawn out, their meanings unfolding gradually. For historical dramas, especially of the Victorian and Edwardian era, this tends to lend greater authenticity to the fiction: these are assumed to have been the decades of leisure and grace.

The stripped-down image that broadcast TV uses is a central feature of TV production. Its most characteristic result is the TV emphasis on close-ups of people, which are finely graded into types. The dramatic close-up is of face virtually filling the screen; the current affairs close-up is more distant, head-and-shoulders are shown, providing a certain distance, even reticence. Close-ups are regularly used in TV, to a much greater extent than in cinema. They even have their own generic name: talking heads. The effect is very different from the cinema close-up. Whereas the cinema close-up accentuates the difference between screen-figure and any attainable human figure by drastically increasing its size, the broadcast TV close-up produces a face that approximates to normal size. Instead of an effect of distance and unattainability, the TV close-up generates an equality and even intimacy.

The broadcast TV image is gestural rather than detailed; variety and interest are provided by the rapid change of images rather than richness within one image. TV compensates for the simplicity of its single images by the techniques of rapid cutting. Again, the organization of studios is designed for this style of work. The use of several cameras and the possibility of alternation between them produces a style of shooting that is specific to TV: the fragmentation of events that keeps strictly to the continuity of their performance. There is much less condensation of events in TV than in cinema. Events are shown in real time from a multiple of different points of camera view (all, normally, from the same side of the action). Cinema events are shot already fragmented and matched together in editing. Still today, video editing is expensive, and the use of the studio set-up (in which much is already invested: capital and skills alike) provides instantaneous editing as the images are being transferred to tape for later transmission. This enables a rapid alternation of images, a practice which also affects the editing of TV programmes made on film. The standard attitude is that an image should be held on screen only until its information value is exhausted. Since the information value of the TV image is deliberately honed-down, it is quickly exhausted. Variation is provided by changing the image shown rather than by introducing a complexity of elements into a single image. Hence the material nature of the broadcast TV image has two profound effects on the regime of representation and working practices that TV has adopted. It produces an emphasis on sound as the carrier of continuity of attention and therefore of meaning; it produces a lack of detail in the individual image that reduces the image to its information value and produces an aesthetic that emphasizes the close-up and fast cutting with strict continuity.

However, this is to compare the broadcast TV image with the cinema image. The TV image has further distinct qualities of its own, no doubt the result of a tenacious ideological operation, that mark it decisively as different from the cinema image with its photo effect. The broadcast TV image has the effect of immediacy. It is as though the TV image is a 'live' image, transmitted and received in the same moment that it is produced. For British broadcast TV, with its tight schedules and fear of controversy, this has not been true for a decade. Only news and sport are routinely live transmissions. However, the notion that broadcast TV is live still haunts the medium; even more so does the sense of immediacy of the image. The immediacy of the broadcast TV image does not just lie in the presumption that it is live, it lies more in the relations that the image sets up for itself. Immediacy is the effect of the directness of the TV image, the way in which it constitutes itself and its viewers as held in a relationship of co-present intimacy. Broadcast TV very often uses forms of direct address from individual in close-up to individuals gathered around the set. This is very different from cinema's historic mode of narration, where events do not betray a knowledge that they are being watched. Broadcast TV is forever buttonholing, addressing its viewers as though holding a conversation with them. Announcers and news-readers speak directly from the screen, simulating the eye-contact of everyday conversation by looking directly out of the screen and occasionally looking down (a learned and constructed technique). Advertisements contain elements of direct address: questions, exhortations, warnings. Sometimes they go further, providing riddles and jokes that assume that their viewers share a common frame of reference with them. Hence advertisements for various staple commodities, beer for example, tend to make oblique and punning references to each other's advertising campaigns. The audience is expected to understand these references. This also is an operation of direct address: an ephemeral and immediate knowledge is assumed in the viewer, who otherwise would have no understanding of the reference or the joke. Hence these advertisements are addressing a viewer as an equal: 'we both know what we are talking about'.

Direct address is recognized as a powerful effect of TV. Its most obvious form, that of an individual speaking directly (saying 'I' and 'you'), is reserved for specific kinds of people. It can be used by those who are designated as politically neutral by TV itself (newsmen and women), or by those who have ultimate political power: heads of state. Otherwise, direct address is denied to individuals who appear on TV. Important personalities are interviewed in three-quarter face. Other strategies of address are open to them, that of recruiting the audience against the interviewer by appealing to a common sense that media persons do not share, for instance. Interviewers in their turn tend to construct themselves as asking questions on behalf of the viewers: 'what the public/the viewers/ordinary people *really* want to know is. . . .' This is also a form of highly motivated direct address. Again, it assumes an audience who is there simultaneously, for whom events are being played out.

Direct address is not the only form of construction of broadcast TV's effect of immediacy. Broadcast TV's own perpetual presence (there every night of the year), and its series formats, breed a sense of the perpetual present. Broadcast TV declares itself as being in the present tense, denying recording as effectively as cinema uses it. TV fictions take place to a very large extent as though they were transmitted directly from the place in which they were really happening. The soap opera is the most obvious example of this, where events in a particular milieu are everlastingly updated. Broadcast TV has a very marked sense of presence to its images and sounds which far outweighs any counterbalancing sense of absence, any sense of recording. The technical operation of the medium in its broadcast form strengthens this feeling. The tight scheduling that is favoured by most large broadcast operations means than an audience wanting to see a particular programme has to be present at a very precise time, or they miss it. This increases the sense that broadcast TV is of the specific present moment. In addition, unlike cinema, the signal comes from elsewhere, and can be sent live. The technical origin of the signal is not immediately apparent to the viewer, but all the apparatus of direct address and of contemporaneity of broadcast TV messages is very present. The broadcast signal is always available during almost all normal waking hours. It is ever-present.

The favoured dramatic forms of broadcast TV work within this framework of presence and immediacy. Besides the obvious forms of soap opera, entirely cast in a continuous present, the series format tends towards the creation of immediacy and presence. The open-ended series format of the situation comedy or the dramatic series tends to produce the sense of immediacy by the fact that it presents itself as having no definite end. Unlike the cinema narrative, the end of the episode and the end of the series alike leave events unresolved. They are presented as on-going, part of the texture of life. This sense even extends to the historical reconstruction dramas beloved of the BBC, through the operation of a further mechanism that produces the effect of immediacy.

Immediacy is also produced by the logical extension of the direct address form: by echoing the presumed form of the TV audience within the material of the TV fiction itself. The institution of broadcast TV assumes its audience to be the family; it massively centres its fictional representations around the question of the family. Hence TV produces its effect of immediacy even within dramas of historically remote periods by reproducing the audience's view of itself within its fictions. Hence TV dramas are concerned with romance and the family, both conceived of within certain basic kinds of definition. Broadcast TV's view of the family is one which is at variance with the domestic practices of the majority of domestic units and of individuals in Britain at the moment. Yet broadcast TV's definition has a strength in that it participates in the construction of an idea of the normal family/domestic unit, from which other forms are experienced and remembered as temporary aberrations [. . .].

The centring of all kinds of broadcast TV drama upon the family (much as direct address assumes a family unit as its audience) produces a sense of intimacy, a bond between the viewers' conception of themselves (or how they ought to be) and the programme's central concerns. So a relationship of humanist sympathy is set up, along the lines of seeing how everyone is normal really, how much they really do desire the norm that our society has created for itself. But the intimacy that broadcast TV sets up is more than just this form of sympathy. It is made qualitatively different by the sense that the TV image carries of being a live event, which is intensified by the habit of shooting events in real time within any one segment, the self-contained nature of each segment, and the use of close-up and sound continuity. All of these factors contribute to an overall impression, that the broadcast TV image is providing an intimacy with events between couples and within families, an intimacy that gives the impression that these events are somehow co-present with the viewer, shared rather than witnessed from outside. The domestic nature of the characteristic use of broadcast TV certainly contributes here, but more important is the particular way in which TV has internalized this in its own representations. Broadcast TV has ingested the domestic and bases its dramas upon it. When it does not address its audience directly, it creates a sense of familiarity between its fictions and its audience, a familiarity based on a notion of the familial which is assumed to be shared by all.

Broadcast TV has a particular regime of representation that stresses the immediacy and co-presence of the TV representation. Its particular physical and social characteristics have created a very particular mode of representation that includes the image centred upon the significant at the cost of detail, and sound as carrier of continuity. It gives its audience a particular sense of intimacy with the events it portrays. All of these features of broadcast TV create and foster a form of looking by the TV viewer that is different from the kind of voyeurism (with fetishistic undertow) that cinema presents for its spectators.

TV's regime of vision is less intense than cinema's: it is a regime of the glance rather than the gaze. The gaze implies a concentration of the spectator's activity into that of looking, the glance implies that no extraordinary effort is being invested in the activity of looking. The very terms we habitually use to designate the person who watches TV or the cinema screen tend to indicate this difference. The cinema-looker is a spectator: caught by the projection yet separate from its illusion. The TV-looker is a viewer, casting a lazy eye over proceedings, keeping an eye on events, or, as the slightly archaic designation had it, 'looking in'. In psychoanalytic terms, when compared to cinema, TV demonstrates a displacement from the invocatory drive of scopophilia (looking) to the closest related of the invocatory drives, that of hearing. Hence the crucial role of sound in ensuring continuity of attention and producing the utterances of direct address ('I' to 'you').

24

Visualizing the news

Richard Ericson, Patricia Baranek and Janet Chan

From Ericson, Richard; Baranek, Patricia and Chan, Janet 1987; *Visualizing deviance: a study of news organization*. Milton Keynes: Open University Press, 270–81.

Note: The research reported here is based on studies of a Toronto television station carried out in 1982–1983.

Visuals for television

Television reporters have special requirements in obtaining visuals. These constitute both constraining and enabling features of reporting peculiar to television.

Reporter-cameraman relations

The cameraman is a highly valued member of the television newsroom. This was symbolized by the fact that while he was formally under the direction of the reporter, he was paid substantially more than the reporter, sometimes double or more. Moreover, the reporter was very dependent on the cameraman to understand what the story was about so it could be visualized appropriately. 'While the presence on the screen of the television reporter gives the story journalistic credibility in the traditional sense, it is heavily reliant on the news judgment of the cameraman' (Schlesinger, 1978, 160–1).

The basis of the reporter-cameraman relationship was tacit understanding of requirements. The cameraman has to become part of the newsroom culture and learn the vocabulary of precedents of reporters, editors and producers. He must learn the approach of each reporter and the subtle differences in the way each works. Over time a reporter develops a preference for working with a particular cameraman and vice versa. However, since reporter and crew assignments were done on the basis of situational availability rather than keeping a team together, it was not possible to expect a regular matching of those who felt that they worked well together.

Beyond tacit understandings and cues in the field, cameramen often offered specific advice and direct assistance. Advice was given freely on what shots to take, what clips or source quotations to use from shots taken, and what reporter voice-over phrasing might be appropriate for particular visuals. The cameraman also helped to encourage sources to co-operate generally, and to phrase particular notions when they stumbled or seemed unable to articulate in the manner expected.

As would be expected where aesthetic preferences are constantly being worked at rather than already worked out, reporter-cameramen relations were often marked with tension and conflict. Reporters carped about cameramen who took poor shots; packed up too early and therefore missed something of relevance; did not follow directions well; and failed to 'hustle' in the face of time limitations. In turn, cameramen were critical of reporters' judgements about the selection of sources; questioning of sources; overshooting, and otherwise obtaining too much information that risked missing the slot; and making stories where there were none in the opinion of the cameraman. Some of these disputes were taken to the level of supervisors. Underlying the reporter's concern was the fact that he was dependent on the cameraman for a good public face, while the cameraman could remain invisible after completing his visualization work.

On locations

One picture may not be equivalent to a thousand words, at least not in television news. However, television does have the ability to show as well as to tell. It can show the location of an event and can also picture what the people involved look like, thereby providing for the audience readings of their moral character. It provides a form of tertiary understanding not available to print except through still photographs: during our observations, efforts ranged from committing a crew for hours to obtain a visual of a suspect leaving court with a bag over his head, to hiring a helicopter so a crew could obtain an overview of a fire.

Television actuality shots are quite rare. Visuals rarely capture an actual event in the world, only talk of an event. Hence, most often the choice of a location is a matter of finding some place or someone that will stand for what is being talked about. In terms of physical locations, these were termed 'generic spots'. The release of a government report on domestic violence was represented by a reporter stand-up in front of the legislative building. A politician whose election campaign included disputing a high-rise building development was interviewed near the site of the development to represent this point of the campaign. When a place of a type could provide the sign, then the most convenient one would do: for a story about the contamination of a pharmaceutical product sold in drugstores, the handiest local drugstore was chosen for the visuals; for a story about lawyers' fees for real estate transactions, a street near the newsroom was selected for shots of houses with 'for sale' signs.

If official reports or other documents were being talked about, these were sometimes also shown to exist. These visuals focused on title pages or bold headings, as well as taking key words or phrases to print across the screen with the document shown above or in the background. As in showing events and people, this treatment did not provide documentation of the matter being addressed but only a sign that the document being talked about actually existed and appeared in a particular form.

Most television stories were built around interview clips with a few sources, 'the talking head'. While the key aspect of these clips was in the person's words rather than the context in which they were said, as shown in the visuals, the context was often deemed by the reporter to be a relevant component of his visualization of how the source was to be represented. An important consideration for the reporter was how to represent the source's authority, in the context of the matter being reported and the other sources involved. The basic choice was whether talking-head visuals should be 'on the street', or in the physical office which represented the source's official office. One executive producer encouraged 'on the street' accounts from ordinary folk, the vox pop. He circulated memos to reporters to this effect, including one that stressed the need generally of 'going to the street' rather than letting 'institutional mouthpieces get air time with no real people'. However, what was done in this respect depended on the particular story and the contexts in which sources were dealt with.

In stories of the basic point/counterpoint format, involving leading representatives of each of two organizations in dispute, the usual approach was to represent each in the authority of their physical office. An instance of this, in which complications arose, is illustrative. A representative of a citizens' organization trying to establish a halfway house for prisoners was also an executive of a large corporation, but did not want visuals to be taken in his corporate office because the matter did not pertain to that corporation. Similarly, the representative of a citizens' organization in opposition to the establishment of this halfway house did not wish to have his unrelated corporate office used for visuals. The reporter, wishing to represent both as men of significance, found alternative physical space to convey the authority of their offices visually. One person agreed to come to the television station for interview. He was interviewed sitting behind the desk of one of the station's senior executives, appearing as if it was his own. The other interview was done in the lobby of a quality hotel across the street from the source's corporate offices, with permission of the hotel. The lobby furniture was rearranged so that the source could be suitably visualized sitting in a stolid wingback chair, indicating that he too was talking from, and within, an authoritative office.

A reporter was assigned to do a story on the reaction of various ethnic groups to the revelation that RCMP security services had an 'ethnic list' of suspicious persons connected to various ethnic types. The reporter planned to do the story by filming the representatives of different ethnic groups 'on the street' to emphasize the 'voice of the people,' against what was portrayed as a possible sign of totalitarian practice of government. The first representative agreed to a street interview. The reporter then proceeded to interview a doctor who said that in his practice he had many patients who complained of abuse by immigration and RCMP policing authorities. However, upon arriving at this doctor's office it was discovered to be in a poor part of town and the doctor himself did not look the part: he was in his late twenties or early thirties, and casually dressed without tie or jacket. The reporter decided that a street interview with this person would undercut his authority as a family physician speaking sympathetically about aggrieved patients. In spite of crew grumblings about having to set up their lights inside, the reporter insisted on representing the source in the authority of his office to add credence to his story. In contrast, the third source, a representative of a Polish group, was wearing a quality business suit. He refused an interview on the street in spite of the reporter's efforts to convince him. Instead he insisted on being shot behind his desk, flanked by flags of Poland, Canada, and Ontario that were placed there by him for the purpose of the interview. In this instance the source's desire to appear authoritative won out, and the reporter ended up, against her initial wish, with two or three sources 'off the street.'

It was a convention to shoot two sources with opposing views from opposite angles. On one occasion a reporter thought that the cameraman tacitly understood this procedure when he asked him to obtain a clip from each of two opposition-party leaders making statements outside the legislature. When he discovered that the cameraman had taken shots of both from the same angle he was most annoyed, stressing that it was simply 'common sense' to shoot from opposite angles.

When sources were not available to be filmed, or a crew was not available to film them, other means were devised. A common practice was to telephone a source and obtain a voice recording that could be used along with a still photograph of the source and/or a shot of the reporter in the newsroom making the call. This was done in particular when a source was some distance from Toronto and the newsroom did not want to go to the expense of obtaining something that they knew could be captured in essence through a voice clip. In these circumstances something had to be shown with the voice clip, and if a still photograph of the source was lacking, that something was usually the reporter shown to be doing his job. While this representation was useless in terms of its relation to the source's statement, it was at least useful in showing the authority of the reporter's office.

Staging

The imperative of television journalists to represent visually people doing things related to the story often poses difficulties. Crews cannot always be on time at the place where the events are occurring, and even if they are they often face technical difficulties in obtaining shots with proper lighting, sound and focus. When such difficulties arise the solution is to have sources stage their activities, or in the words of the vocabulary of precedents, do a 'fake'. Indeed, the resource limitations and technical requirements of television news are so well known to sources that they usually take elaborate measure to script the event in advance and stage everything so that television crews can obtain their clips expeditiously. In this activity Goffman's dramaturgical model of social life is acted out in considerable detail.

When the requirement is talk about events, and 'talking heads' for television visuals, the news conference is a standard format. Well-prepared and well-groomed organizational representatives represent what is organizationally in order. After the basic presentation at a news conference, sources are asked by a television journalist to repeat a part of their performance for the purposes of a television clip. This approach saves film and editing time, since shooting the entire presentation expends a lot of film and requires a lot of sorting out back in the film-editing room.

The two leaders of the Guardian Angels citizen-patrol organization, based in New York City, came to Toronto on several occasions to stage news events that might help them with their cause. They chose significant locations – such as the city-hall square, the main indoor shopping precinct, and a space in a park where a woman had been sexually assaulted – to conduct interviews and give displays of self-defence techniques. On these occasions television journalists cooperated in the staging of each event by helping with locations, props, and suggestions for the enactment of representative action such as self-defence-technique displays.

Similar staging techniques were employed by citizens' groups with other causes. Groups in opposition to nuclear weapons held several marches and demonstrations, engaging in acts guaranteed to result in arrest by police, and equally guaranteed to be visually represented as the dominant frame of television-news items. At one demonstration television reporters from two different outlets discussed the various staging techniques being employed. They were very negative about the sources' representations, but nevertheless felt 'forced' to treat them as a central aspect of how they visualized the deviance. They pointed out that the source organization selected a key location (a plant involved in weapons manufacture); a good day in summer when 'news' was likely to be 'slow' because key source bureaucracies were in recess; and, sent a release, along with promises via telephone calls, that there would be 'violence' of some sort. One television station had three crews on location. The source organization arranged for five people to jump over a fence to throw red paint, signifying blood, on a wall. They were stopped and arrested for trespass or public mischief. Meanwhile two women were able to paint with stencils six green doves on a wall before they were arrested for public mischief. A spokesperson for the organization, wearing a shirt displaying an anti-nuclear statement, was asked by a reporter why these acts were necessary. The source replied that it was a way of educating the public, through the media, about the role of Toronto-based firms in the manufacture of nuclear weapons. The reporter asked what acts of civil disobedience had to do with educating the public and was told that the organization wanted to remind employees of these firms, as well as the general public, that people have strong feelings about disarmament. The source organization had also come equipped with a variety of props – including banners (e.g., 'Ban the Bomb') and a coffin – all of which were focused on by cameramen, along with shots of the paint on the wall and the persons responsible being arrested. The reporters recognized that they were simply following the script of their source, and one remarked while editing that this item was a prime example of creating the news.

Virtually all sources work to create appearances for visuals in some way. Persons are selected who photograph and speak well. Background materials are also attended to, for example, piling papers and documents on the office desk to sustain the dominant cultural impression that good people work hard. If television crews are to be allowed inside private space to visualize what goes on there, it is in the terms of public culture.

Reporters did on occasion object to a source's staging techniques or props and were sometimes able to get rid of them. Ultimately it was for the reporter to decide what was an appropriate sign, and whether it was useful or in good taste. A reporter doing a story on a citizens' group campaigning against anti-Semitism included an interview with a source who had a collection of artifacts that were anti-Jewish, such as bumper-stickers, pamphlets and a Nazi flag. The source offered to hold the Nazi flag as she was being filmed, but the reporter regarded this as out of place and rejected it.

Reporters sometimes moved into the director's seat and used equally elaborate techniques of perfecting the 'fake'. A reporter doing a story on the introduction of a major foot-patrol-beat system for police arranged through the public-relations office of the police to meet two officers on patrol. The location selected was the 'Yonge Street strip,' a segment of a street widely visualized in the city as a centre of vice and disorderliness. One police officer was wired up with a tape recorder, and two were followed down the street by the reporter, cameraman and researcher. Shots were taken of the officers going into stores, talking to people on the street, interviewing people in an automobile accident, and writing a traffic ticket.

Efforts to film events occurring 'naturally' sometimes ran into difficulty so that staging was necessary. Television crews waited in the mayor's office for the arrival of two people who represented an organization wanting to meet the mayor's executive assistant about a social problem. When the two people arrived and entered the office of the secretary to the executive assistant, they and the television crews 'broke up' laughing at the effort to appear 'natural' in face of a pack of reporters and cameramen. After everyone had regained composure, one television reporter asked the two people to go out of the office and enter again. A proper representation of the event was made on this 'retake'.

Reporters sometimes set out to recreate what had occurred previously without always stating to the audience that it was a recreation. The RCMP had sent some illegal drugs to an incinerator, but part of the shipment was lost. A reporter decided to visualize the incineration process on film. At the incineration plant, she had visuals taken of the fire itself, a 'pretend' run of a crane picking up the garbage and dropping it into the incinerator, the temperature gauge, and a conveyor belt carrying indestructable materials left behind after the ashes were washed away. These visuals, visualizing the process, were then used by the reporter in juxtaposition with a description of what was believed to have taken place weeks before, when the drugs went missing. The result was a blend of illusion and accounts of reality, as is evident in the following script of the story.

Anchorperson roll-up: The RCMP had it . . . but it slipped out of their hands. In the process of destroying the largest-ever cache of drugs seized in Canada . . . one million dollars' worth went missing. The mounties had wanted to get rid of the stuff but not in the way it disappeared. It has been recovered . . . but police are still trying to figure out how it got lost. [Names reporter] has more.

Reporter: They tried to burn it away . . . faced with the problem of disposing 6 tonnes of methagaeline . . . heroin substitute . . . [fire burning in furnace]. RCMP took it to the Metro Toronto incinerating plant [Sign: Metro Inc. plant]. On that day in June, an RCMP truck arrived with its precious cargo . . . [truck dumping garbage] it came and went 4 times with 200 million dollars worth of drugs, packed in Sunlight detergent boxes . . . [Sunlight detergent box].

While 3 RCMP officers supervised . . . a crane picked up the drugs with the other garbage and dropped it into the hopper leading to the incinerator. But inside the incinerator . . . [a Metro works official] explains, the drug made it too hot [crane and hopper travelling and dropping]. [Metro works official talking head]

Reporter: When the temperature went up to a critical point they shut down the incinerator . . . One truck load was turned away and the RCMP were informed the drug clogged the incinerator. Somehow one box was neither burned away nor returned to the RCMP [temperature gauge, followed by shot of truck coming down ramp].

The RCMP didn't even know 1 million dollars' worth of drugs was missing . . . that is until Metro Toronto police stopped a car 3 weeks later and found hidden in a suitcase 35 pounds of meth . . . 3 people were subsequently charged for possession of drugs and one of them is an employee of the incinerating plant . . . [stand-up] Inspector [name] and his staff were the ones who announced the huge heist of meth . . . last October . . . [Super: stock film showing RCMP] it was a dramatic seizure of drugs at Collingwood airport . . . [still graphic: police firing guns with an airplane in the background].

Even though the missing box went unnoticed for 3 weeks . . . [the inspector] says he's sure there isn't any more on the streets . . . [box]

RCMP Inspector: I would be very surprised if there's another box missing . . . [etc] [talking head]

Reporter: The RCMP has used this plant three times . . . [Sign: Metro Inc. Plant] workers don't have special security clearances . . . the RCMP assumes full responsibility [truck going down ramp]

RCMP Inspector: Our procedure . . . [etc] completely recalculated to make sure they get into the fire box. [talking head]

Reporter: The remaining drugs will be neutralized by a chemical process [pick up drug in box].

This story used visuals of the incineration process – including 'fakes' – to represent what happens in talking about what happened. The audience was offered the illusion of watching the actual dumping of the drugs, the detergent box used, etc., at the time. There was no mention that this was a reconstruction of what *may* have happened. Similarly, the artist's graphic of the original seizure of drugs ten months before was simply his visualization of what happened at that time. Apart from the talking heads, all the visuals were simply visualizations of how events might have appeared, and of how things might have looked in the improbable event a 'live-eye' television crew had been there at the time. Faced with the impossibility of being 'everywhere', television news is left to visualize things as if they are.

Stocks

Time and space considerations influenced decisions about whether to use stocks (old visuals) from previous stories to represent something. A reporter faced with a distant event, an event that had already occurred, an event occurring in a physical space he could not penetrate, and/or an encroaching deadline, often turned to the stocks for something to add vision to his story. A considerable amount of 'research' in the television newsroom consisted of searches for appropriate stocks, and special 'researchers' were employed to assist reporters in this regard.

The choice of stocks also involved 'fake' elements. A reporter doing a story on an Ontario government report concerning domestic violence, including police response to it, had difficulty finding appropriate shots of police attending a domestic call. He and an editor searched for and eventually found stock footage from a story done five months earlier, involving the domestic-response team for the city of Detroit, Michigan. This footage visualized police officers talking with a woman, and was used with a voice-over stating. 'The report says wife battering must be treated as a crime . . . that the onus for laying charges in such cases should be with the police.' Of course the reporter's intention was not to lie with these shots of police – not to indicate that Detroit police should be seen to be helping Ontario women – but that the police institution should be doing more of what was represented in these pictures.

The only real concern in using this form of representation was that it not be seen through as an obvious fake. Thus, there was concern that repeated use of the same stock clip, especially to represent aspects of different stories rather than the same continuing story, would look bad. Moreover the stock clips usually had to look as if they appeared to have been taken at the time of the shots actually taken on the day. A reporter was preparing a story about an accused person, released after a preliminary hearing, who had decided to sue various officials for malicious prosecution and negligence. The reporter relied on stocks of the attorney general's office building to use with a voice-over stating this office was named in the suit. This story was in August, but the stock shot of the building showed it to be winter. The reporter had a close-up 'freeze' of the building edited to get rid of the snow in the wider shot. The reporter relied on stocks of police officers on motorbikes to use with a voice-over stating the chief of police and investigating officers were also named in the suit. This stock film had a background of trees with no leaves, indicating that it was not in August, but in this case the reporter did not ask for it to be 'frozen'. The regular memos from the executive producer included critiques about the use of stocks which laid bare the fact that they did not represent the time the story was done. One such memo complained that a particular reporter had all day to shoot new footage, but had relied on stock footage with snow on the ground when it was summer. Even though everyone could see that a particular visual clip was a representation, it was not to appear as a 'fake'.

Visuals and Texts

The research literature emphasizes that television journalism is still dominated by the written text, and most of the conventions of print journalism are therefore applied in television journalism. If anything, the visual requirement serves as a limiting factor on television texts, meaning that they have less depth and range of opinion than is available in the quality press. Nevertheless the construction of a coherent text is paramount in most instances. 'The serious business of agenda setting is too important, as far as we can tell, for the visual imperative to be a major consideration' (GUMG, 1980, 248). The early news broadcasts of BBC television included sound only, and even a decade after its establishment in 1957, the BBC television news bulletin was in essence a radio bulletin read to camera, with a few illustrations. Even in the contemporary period television news has been portrayed as 'visual radio' (Gans, 1980) because the spoken word predominates, linking and ultimately binding the visual images (Williams, 1974).

The traditional interpretive role of the print journalist is sustained through the considerable degree to which the journalist himself appears, or voices-over, other visuals, talking about things that happened elsewhere. While clips with sources are termed 'actualities', they too involve the source talking about what has gone on, or is going on, in another time at a different place. The need for talk from sources, and for the reporter to connect their talk in the maintenance of a flow, means the text forecloses on the range of visual options. Visuals only rise to the forefront 'when the material is exclusive, exceptionally dramatic and has unusual immediacy' (GUMG, 1976, 121). A shot of a person leaping from a building that is in flames may speak for itself. However, most visual material consists of people speaking for themselves. In this respect television most often pictures the sources who form part of the text, a capability also available to newspapers through still photographs.

The reporter constantly visualizes what visuals can be obtained in relation to the text he imagines to be appropriate and what can fit with what his sources have said. This relation is dialectical, with amendments to both text and visuals being made at every stage of the reporting process. The extent to which one or the other predominates is related to the nature of the matter being reported on. In a story on a public demonstration several video cameramen were sent to obtain copious material; video was selected because it does not require processing and can be edited with much greater speed. The reporter then returned to the newsroom to screen the videotapes and in that process took notes for an accompanying text. Only after the screening was the text prepared, with the reporter indicating explicitly to the researcher how the script was written in accordance with the visuals obtained. In contrast we witnessed scripts being written in advance and then tinkered with as required by source-interview material obtained and stocks available. When the requirement was simply to update a continuing story, then the updated information – augmented by background information, usually from newspaper clippings – was the core of the matter and written much like a newspaper journalist would write it. However, because something had to be put up on the screen, the journalist scrambled to obtain generic visuals, stocks, graphics, or whatever else would fill the bill.

The importance of visuals was revealed when there were problems in obtaining them. If particular visuals could not be obtained, or if obtained were deemed to be lacking in drama, immediacy, or exclusivity, then the entire story was sometimes dropped. Thus a reporter assigned to a story on Monday regarding a murder on the previous weekend initially screened some film a weekend crew had obtained. The film was of poor quality and there were only two things shown: the apartment building where the incident occurred, showing a third-floor balcony, and two detectives looking around for evidence. With no body and no police-car visuals, the reporter said he did not think he could build a story around it and the item was eventually dropped.

When there were inadequate visuals – often in the face of severe time limitations to search for stocks or to dispatch a crew, or in face of the material limitation of no crew available – one option was to do a script for the anchorperson to read. This script was sometimes accompanied by whatever visuals were available, but usually only included a head-on shot of the anchorperson reading what the reporter had written for him. This situation arose also when physical space was difficult or impossible to penetrate for visuals. For example, reporters doing court stories were prevented from filming in the courtroom or courthouse. They usually relied upon a graphic artist's representation of the key actors during the day in court, supplemented if possible by source interviews outside the courtroom and stocks of the original incident or investigation related to the court hearing. A typical court story we observed entailed the use of five graphics with the reporter's voice-over. We also observed instances in which not even an artist's conception was obtained, and the anchorperson was left to say what the reporter had written.

Visuals were used to make statements beyond the scripted text. A reporter was doing a story on fear among women, in the context of a reported series of incidents of attacks on women. She decided to take shots of a 'secluded' area in a park where a woman had been raped while sunbathing in her swimming suit. On location, there were approximately one hundred men, women, and children in the area. The reporter had the cameraman isolate two women in swimming suits sitting on a bench, with no other people in the background. This clip was later used in the story, suggesting that even after the rape there were women foolhardy enough to be in this 'secluded' area of the park and wearing only scanty clothing.

In the ideological use of visuals there was usually textual material to set the stage and give preferred readings of what the reporter was representing. '[S]uch commentary does not tend iconically to describe the shots, but rather forces the viewer to see shots indexically or even symbolically within a framework of understanding thereby established by the professionals' (GUMG, 1980, 332). A reporter covering a well-policed demonstration by persons opposed to nuclear weapons began his item with a series of shots of police dragging demonstrators off a roadway. Instead of his own words, he used a background audio clip of demonstrators elsewhere on the site singing 'Give Peace a Chance.' A shot of a woman being led by a policeman who had hold of her hair was selected to fit with a textual statement, 'For the most part police used as little violence as possible.' Here the reporter was suggesting visually that this was a possible instance of excessive use of force, even while the text standing by itself was less suggestive of police excesses.

In a story on local residents' opposition to the establishment of a halfway house for prisoners, the reporter wanted to visualize a view that people felt threatened and there was potential for danger. In order to do so she not only obtained a talking-head of a spokesperson for the citizens in opposition, but went to the location of the house proposed for this use to obtain suggestive visuals. On location she had the cameraman picture the house alongside the one next door, which had a for-sale sign on it. This was done in the context of a newspaper story which cited the potential centre's next-door neighbour as being concerned about the safety of her two children. 'Are they sex offenders, drug offenders, or what?' The suggestion to be left was that the house was for sale because the neighbour was concerned about 'criminals' moving into the neighbourhood. Shots were also taken of young children walking through the park across the street from the house, suggestive that with the park so accessible to the house, children might be easy prey. Visualizations of deviance of this type were rendered more easily and subtly with the aid of television visuals. In print, reporters would be quite unlikely to say, 'Since this park is directly across the street from the house, criminals are likely to watch for children playing in the park and calculate whom they can victimize.' They might obtain a quotation from a local resident expressing this fear, but even that is arguably less immediate and dramatic than *showing* the physical possibility with visuals.

25

Documentary meanings and the discourse of interpretation

John Corner and Kay Richardson

From Corner J. (ed.) 1986: *Documentary and the mass media.* London: Edward Arnold, 141–53. [Only the earlier sections of the article are reprinted below.]

Across both its richly various social and technical history and its diverse employment of words and images, 'the language of documentary' has shown an abiding concern with self-effacement. Through whatever devices of naturalization have been available and judged to be stylistically and functionally appropriate, documentaries have regularly sought to present audiences with accounts in which the viewed is to be taken as effectively indistinguishable from the real. Life not only takes precedence over art in such accounts, it frequently replaces it altogether within a discourse and aesthetics of maximum 'transparency'.

Undoubtedly, the particular conventions by which this effect is contrived have been extended and made more subtle by quite recent developments in lightweight television technology. Since documentary television first began to offer viewers a regular 'window on the world' however, these conventions have often been at the heart of discussion and controversy about television's function as a source of popular knowledge and as an 'agenda-setter' on public issues.

But to talk of television's power in this respect – in gaining widespread acceptance of its lucidly 'obvious' interpretations of 'what is going on' – risks begging a number of questions about the *ways* in which programmes are subject to interpretative activity themselves as an integral part of the viewing process. Part of this activity of making sense from what is seen and heard may include not only the attribution of particular motives and intentions to programme-makers, but also the recognition of the different levels at which visual and verbal representations are organized accordingly. In the case of documentary then, whilst the techniques of naturalism may to the film-maker's satisfaction bind together world and 'document' with all the immediacy of revealed truth, the lines of communicative relation with *viewers* may be such as not necessarily to confirm this truth at all but to open up gaps and to render awkward any offered coherence.

In this study we are interested in the way in which a sample group of viewers understood, and responded to, specific sequences in one documentary programme, *A Fair Day's Fiddle,* broadcast on BBC2 in March 1984 but viewed by our respondents on video just before one-to-one discussion sessions.

Our particular concerns thus cut into both sets of general issues outlined above. We wanted to explore questions of documentary *form,* especially the ways in which the speech and behaviour of the participants in the programme were organized within its mode of address to the viewer and its overall version of 'what was going on'. Here, two aspects of the programme attracted our close attention, due both to their prominence in respondent accounts and to their significance for television documentary more generally. First of all, and of primary interest to us in this article, there were those scenes of 'enactment' in which participants supposedly went through daily routines, with the cameras (and viewer) looking on unacknowledged. These scenes sometimes modulated into sequences of interview response and we have focused on one such scene in our discussion below. The second aspect – again, a staple element of documentary production – was what we termed the 'narratorial function', meaning by this the range of possibilities and effects relating to voiced-over commentary in which some authority for the general guidance of understanding appears to be invested. In exploring these formal questions, however, we saw it as a primary aspect of our investigation (not just a continuation of it) to research what the forms actually 'delivered' to the viewers. We wanted to get as close as possible to the *interpretative work* which viewers performed by paying the strictest attention to the language of their descriptions and judgements – what did they see happen on the screen? How did what was said seem to relate to what was shown? What bearing did one sequence have on another? What, if anything, was the programme 'trying to say'?

It might be useful for us to proceed by giving a general description of the programme we chose for the study and to offer a very brief account of previous work on audience interpretations of television insofar as this work relates to our own ideas and methods. The analysis of the sequence selected and the respondents' comments can then follow.

A fair day's fiddle – access documentary?

A Fair Day's Fiddle was a 50-minute programme made for BBC2's *Brass Tacks* series by BBC North West in Manchester. Its ostensible topic was the various 'fiddles' by which recipients of social security payments on one Liverpool estate supplemented their income. Its chief formal characteristic was its relatively loose structuring as a series of episodes in which various inhabitants of the estate recounted their experiences and offered comments to an off-camera reporter without their contributions being linked or voiced-over by any 'official' commentary. Such commentary was used only in the introduction to the programme – a brief 'scene-setting' passage spoken over long-shots of estate housing and finishing with the remark that 'in this programme, local people speak for themselves about a thriving unofficial economy'.

This 'speaking' occurred within a number of different forms. There were scenes in which participants spoke on-camera to an out-of-frame interviewer. Often the interviewer's questioning was edited out, intensifying the quality of personal testimony, at times close to that of the 'confessional'. There were other scenes in which participants were seen going about 'routine' business (e.g. entering derelict premises in search of scrap materials, welcoming home a child from school, assembled in a group at a friend's house to have their fortunes told, consulting a Citizens' Advice Bureau officer about a debt problem). Sometimes these scenes involved spoken exchanges, thus adding speech 'enactment' to 'enactment' of physical behaviour. This mode of presentation is perhaps pushed to its most dramatized level in a scene where a husband and wife are seen to open the morning's mail and to have a lengthy and emotionally charged conversation about their debts and the possibilities for repayment. A number of devices employed here (e.g. use of held facial close-up during pauses in 'dialogue') seemed to us to construct participant behaviour as 'performance' more intensively than elsewhere in the programme. Where scenes of activity involved one person only, they were usually accompanied by voice-over from the person in shot.

At points, the documentary included contributions from those who worked on the estate (the vicar, the staff of the Citizen's Advice Bureau). In one sequence, it shifted into the conventions of television investigative journalism by having its reporter call on a local moneylender and extract a reluctantly offered series of on-camera answers to questions concerning circumstances which we had heard about in an earlier scene.

For most of its length however, the programme attempts to address its topic 'from the inside', as it were. Access to the 'inside' is through an intimate, cumulative rendering of personal experience – the people of Netherley, in their own homes, speaking for themselves about the difficulties of 'getting by'.

As we indicated above, we were interested in aspects of how this articulation was put together and in the positions of understanding and evaluation which our respondents took up towards it. Together, we saw these as constituting the ground for an inquiry into the *pragmatics* of television documentary form.

Researching reception

Detailed empirical research on the interpretative schemes which audiences use to understand and assess television programmes is a relatively recent development in mass communication studies. However, in the last few years a number of projects have followed Morley's seminal study, *The Nationwide Audience*,[1] in attempting to explore, either by group or individualized discussion with sample viewers, the various sense which has been made of particular items of output and the range of interpretative resources which the 'sense-making' activity draws on.

Elsewhere, we have discussed theoretical and methodological aspects of these developments in some detail,[2] but it may be useful for us to itemize here the following points, most of which are taken up practically in the ensuing analysis and commentary.

(a) We wanted to find out as much as we could about the primary levels of interpretation at which visual and verbal elements were *understood* rather than simply being concerned to identify *attitudes* towards the material.

(b) This suggested the use of one-to-one discussion sessions which, moving from more open to more closed questioning, allow for exploratory, follow-up points to be made without the problems this can cause in group work. Given sessions of adequate duration (ours were approximately one hour long) it is possible for the conventions of conversational exchange partly to modify the dominant ones of question and answer. At these points, we did not feel constrained from indicating some of our own speculations as to how various items might be understood or from mentioning the views of other respondents in order perhaps to 'trigger' a more focused response and reaction. What might be considered 'leading questions' or 'prompts' in the context of the fixed range of responses in questionnaire method, seem to us to be a useful late stage in sessions of the type undertaken. We felt that the effects of any 'over-prompting' that might occur could be identified and allowed for in the analysis of the recordings and transcripts.

(c) It also suggested a need to pay close attention to respondent's use of language as an indicator of, for instance, variations in the framing, status and perceived intentions used in the construing of items. One part of this close attention was a respect for the particular speech contexts of utterances under analysis. We have followed this through in our presentation by giving some of our transcribed data as continuous exchange rather than in edited chunks.

With these broad considerations in view, we set about the business of recording the accounts of our sample viewers and analysing as thoroughly as we could their commentaries on what they saw and heard.

'The toy scene'

This episode involves a boy, his mother, and another woman who speaks about her own child. It begins as enactment: the boy returns from school, greets his mother and runs upstairs to play. She then joins him. When they are together we hear their dialogue, when he is playing alone we hear the two women in voice-over. The naturalism of the scene is grounded partly in the home location and partly upon 'spontaneous' human touches – their hug as he gets home (Figure 25.1): a joke, 'Don't break the bed springs, mum', when she sits down on his bed (Figure 25.3). Before he enters his bedroom the camera is focused upon a toy lying on the bed. He bursts in, grabs it, plays with it for a few seconds of screen-time then moves on to another one (Figure 25.2). His mother comes in and lets him turn the TV on. Some of this action is voiced over by the mothers, who say:

Voice 1: I think I try to make up for being one parent instead of two, and I try to get him more to compensate for the lack of a father.

Voice 2: And you feel as if you don't get it you let them down sort of thing.

Voice 1: I find the pressure from the school. I mean it's, every week, it's he needs trainers for running, he needs pumps for gym, he needs a football outfit, he needs shorts for something else. Week by week there are more pressures just from the school, I find.

Thus, whilst the verbal track speaks about sports kit, the visual track shows toys. Toys then become the overt topic, as the enactment gives way to an interview with the mothers using, firstly, shots of toys they have bought (Figure 25.4) and, secondly, shots of the speakers themselves (as Figure 25.6). The sequence begins with a list:

Voice 1: This one cost twenty-five pounds fifty. That one cost me nineteen pounds. There's another one here for twenty-five pounds. There's a thirteen pound fifty one. That cost me about seven pounds for the men.

Against the background of a pile of toys, mostly wrapped in plain brown paper, the camera moves from one parcel/item to another, following the speaker's pointing hand as she costs each item (See Figure 25.5 for example).

Many respondents speculated that these toys might be Christmas or birthday

25.1

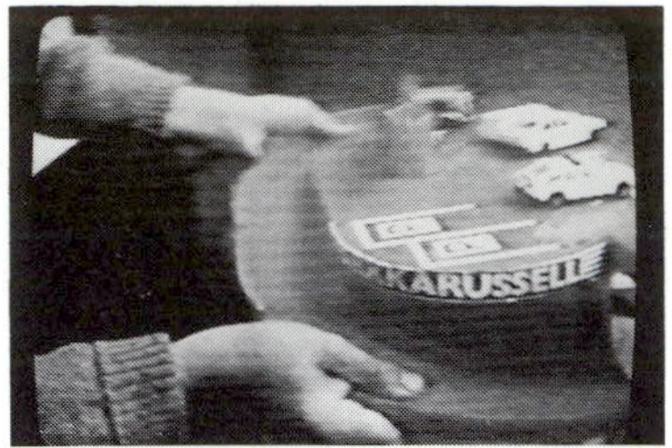

25.2

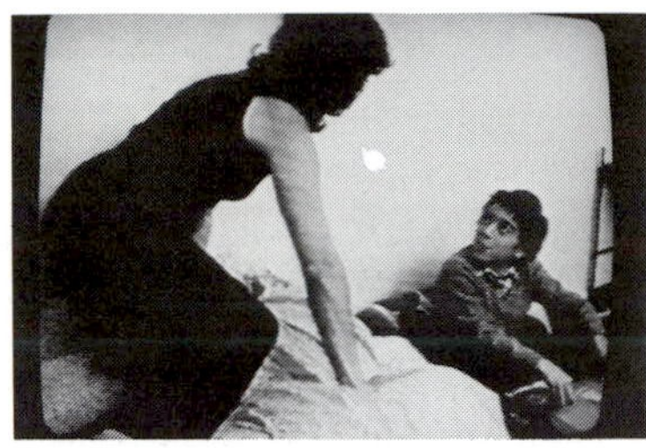

25.3

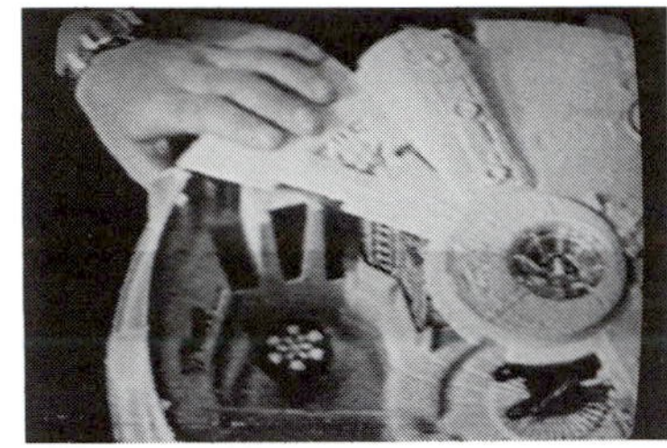

25.4

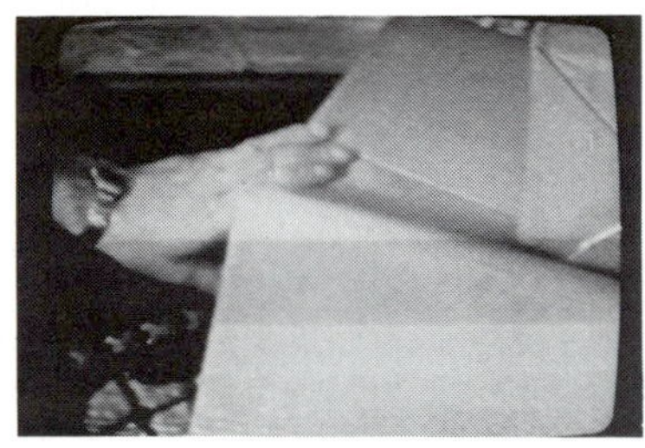

25.5

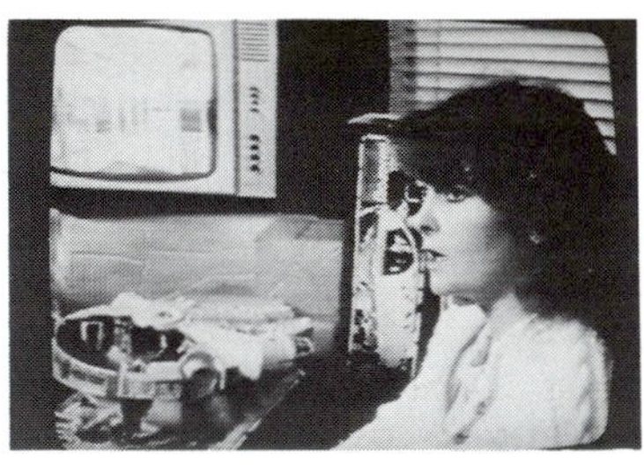

25.6

presents recently arrived by post from the catalogue (one of the mothers says she goes into debt in the catalogue in order to buy things for her son).

This speculation suggests a need to *make* sense of something which lacks a *given* sense in these 'transparent' terms. It is not obvious why there should be toys in such quantity, wrapped and stacked as they are. The possible 'explanations' are all underdetermined textually – including the nonrealist explanation that the arrangement of the toys was at the request of the production team.

Certainly, the toys are put to rhetorical use. Unspecified plurality signifies 'a lot of' toys for one child. Then there is the work done in constituting that plurality out of individual items, which by implication gradually increase both the total number of items and the total bill. The wrapping paper gives them a visual 'sameness' which may play a part in this incremental effect.

If this is the 'rhetoric of excess' – an idea we wished to explore in relation to our respondents' accounts – it is underlined later when the other mother speaks (See Figure 25.6).

> He wanted this electric car which was two hundred pounds, but he also went through the things saying 'I want this' and 'I want that'. I said, 'If you have the car you can't have all these little things as well'. I wouldn't let him go without or go less just because I'm on my own.

The mothers' talk is also important in deciding whether they see their expenditure as excessive or normal and whether they are defensive or matter-of-fact about it. One says: 'Well, you can't deny them, can you?', though most respondents thought that children *could* be denied many/expensive toys. But, as we've suggested, it is not perhaps the women themselves who are emphasizing these factors. They seem very open; they speak quietly and evenly – but we don't know what questions they are answering. By the editing out of the reporter, visually and verbally, whilst the speakers address fairly private matters, a somewhat 'confessional' style of unsolicited testimony is produced whether or not the spirit of confession informed the women's remarks when the interview was recorded.

As the scene ends (Figure 25.6), and the remarks about the two hundred pound car quoted above) a television set is seen to be on in the background. The next words to be heard on the verbal track are 'British television . . .' followed by an argument about the social pressure of advertising: 'It doesn't ask me to purchase the product, it dares me not to purchase the product.' This is in a new scene with a different speaker. Nevertheless the thematic continuity is strong and, because of it, the argument about advertising can be read back into the circumstances of the previous participants.

Respondents' accounts

In this section we discuss our respondents' interpretative uptake on the toy scene. The first two lengthy transcripts represent everything that two of the respondents said on this subject. These accounts are revealingly contrastive and the more selective quotation which follows samples the broader range of possibilities amongst respondents who found the toy scene worthy of extended comment. In discussing the long transcripts we point to shifts and inconsistencies of interpretation, themselves very pertinent to our study. This is not possible with the shorter quotations and it must be remembered that these quotes are extracted from accounts just as complex in this respect as those represented in the longer texts.

First respondent

Q Can I just move you on now to talk about another scene which one or two people have commented on? That's the scene in which two mothers discuss buying toys. I think first of all it will be useful perhaps if . . .

R From the catalogue?

Q Yes that's right, from a catalogue. Can you remember what we see in that scene? It starts off with a little boy, doesn't it?

R Yes. The little boy comes in from school, gives her a big hug, runs off up to his bedroom. There's a toy ready waiting for him on the bed. This has all been, he's been told to come in, hug his mother, run upstairs. I know because I've seen them make a documentary for *Gaskin** as I told you and

*A drama-documentary broadcast by BBC2 in 1983

so I recognize the fact that that's a script from that, which I might not have done if I hadn't seen one made.
Q Mm
R He plays with the toy, he picks it up, he does a face.
Q Ah, yes, that's right.
R Doesn't put the nose on. Then he puts that down immediately that he's done the face. He's done as he's been told to do, he's run upstairs, he's sat on the bed, he's played with the toy and he puts that aside, picks up something else.
Q And then the mothers' . . .
R Very fabricated.
Q Mm. And what about what we see subsequently then, when we . . .
R The mother comes in and sits on the bed. He tells her not to crease the bed or something. She says 'oh that doesn't matter'. They're trying to keep things natural.
Q Mm. So it's another little scene in a way, rather like . . .
R Yes. And then they go and talk about the toys.
Q Toys
R And they show us the price of them.
Q And what do we see about the, how do we see the toys?
R There's some in paper parcels, brown paper, wrapped up in brown paper boxes. If they've come from the catalogue that's not really the way they come, I've got a catalogue and they come in a sort of sack, so, but I presume that's to put across that they have actually come from the catalogue. They're all brand new. If they are his toys he's either keeping them very well or they've got replicas of what he owns.
Q Mm, or some people suggest Christmas on the way or whatever, and they . . .
R Yes, maybe it is. You said the film was made at that time. But there wasn't any sign of Christmas trees or Christmas festivities.
Q No, that's true.
R And why would they be in brown paper, they'd be in wrapping paper?
Q Yes.
R But,
Q And then we hear about the prices of them.
R Mm. And it ranges from seven pounds to twenty five, thirty odd. And then of course there's the big electronic toy that they have to have, that's two hundred pounds, but if they have the two hundred pound one they can't have the twenty five, seven pounds and all the rest of it, out the catalogue. And their children are going to have these toys no matter what. They're in a poverty trap but they have to have these toys.
Q Mm. Do you feel that at that point the programme or the film portrays their toy-buying as really rather extravagant?
R Yes, they're certainly not looking for sympathy in that scene. In fact they've overdone it, for they've chosen to put those toys out and everything, they're letting you see that, that is putting across a different view. OK they're complaining that they're hard up but look here, they're getting all this stuff, they're determined to have it no matter how many bills they run up. So yes, that's giving you the chance to turn against them if you were on their side previously.

Q How do you feel that the mothers sound about it? Do they sound confident or defensive or a bit edgy or embarrassed?
R Defensive with a ring to it, 'Don't dare to try to pass a comment on this, it's nobody's business but ours'.
Q Yes. Do you think that most people would tend to regard that scene as a scene in which a certain extravagance was on display?
R I think, yes, it certainly does show that they wanted to put across that there was extravagance. But I think that the whole film shows how extravagant these people are.
Q Mm. In a number of ways, some of which we've discussed.
R In a number of ways. And the bloke in the taxi says 'they expect us not to have a video or a car or a television', or everything and he mentions that at the end of the film. And in truth I suppose you wouldn't expect them to have all those things.
Q Mm.
R But I think, it's not so much that you wouldn't expect them to have them as you wouldn't expect them to be priorities, and these people use it as a priority.
Q Yes.
R Now the gas bill and the electric bill and food in the cupboard and the rest of it. The children's clothes are not a priority. They go down to the social and they get given, the man with the beard again was saying 'the right of clothing our children is taken away from us'. That's not a priority, clothing a child is not a priority, the toys are priority, the television's a priority, drink's a priority. The taxi's a priority.

The first respondent indicates that the experience of having seen a documentary made informs her perception of the toy scene as 'fabricated'. Her stance allows her to infer production intentions and motivations for programme features – up to a point. 'They're trying to keep things natural' she says about the mother/son dialogue. The talk has been artificially constructed to give the 'natural' effect.

Our general term for interpretations that are intention/motivation-conscious is *mediation reading*; as against *transparency reading* where comments are made about the depicted world as if it had been directly perceived reality. This respondent's account shows some tension between transparency and mediation readings. She takes mediation reading to the point of articulating doubts about the reality-status of programme depictions, saying 'If they are his toys he's either keeping them very well or they've got replicas of what he owns' ('they' quite clearly suggesting the programme makers). The boxes may just be *representations* of his toys. Such doubts do not prevent the respondent from producing transparency readings which go beyond information explicitly encoded in the programme and have judgmental overtones: 'clothing a child is not a priority, the toys are priority'.

This may show that she takes her sense of the real more from verbal testimony than from visual images. But notice also that she says of the scene 'It certainly does show that they wanted to put across that there was extravagance. But I think that the whole film shows how extravagant these people are.' In this mediation reading the idea of the people's extravagance is understood as something 'put across' (by the programme-makers?) and not as an inference derived independently of the programme's authored discourse, as in a transparency reading. The idea-as-shown is *then* taken as truth, without any extra-textual evidence. This suggests either a faith in the veracity of the film's account in this respect or a slide towards transparency reading. The verb 'shows', with impersonal subject 'film', can support either interpretation.

In the passage beginning 'Yes, they're certainly not looking for sympathy in that scene', the pronoun 'they' is used eight times. Sometimes a reference to the participants is indicated, 'they're complaining that they're hard up'; others could be references to the programme makers, as in the opening sentence. Ambiguities of pronominal reference are particularly significant across the initial two sentences, where the first proposes that 'they' were not looking for sympathy whilst the second, in suggesting that 'they' have overdone it seems to imply that 'they' *were* looking for sympathy but have made tactical mistakes. If the pronouns are taken as coreferential, the account is inconsistent – at some level this respondent is just unsure who is aiming for what response. But even if the pronouns are not coreferential the lack of explicitness about this may also be symptomatic of uncertainty.

In this context, it is interesting that the final sentence of the utterance suggests a quite calculated move by the programme-makers to depict the mothers more critically, a move offered as some kind of balance against earlier treatments. Indeed, in the previous sentence, the respondent speaks in the 'voice' of the programme-makers in order to provide a strongly evaluative gloss on what she takes to be their visual undercutting of the mothers 'complaints'.

Second respondent

Q Let's turn to another scene which we've got an interest in. There's a section in the programme where two mothers talk about toy-buying . . .
R Oh yes, the catalogue.
Q Can you just recall what we see, briefly?
R Well, Star Wars, mostly Star Wars figures, you know, there's about four or five boxes,
Q How does it start, can you remember how it starts, that little bit?
R Well, there's a woman, she's a single parent, the kid comes home from school – do you want to go that far back?
Q Well, yes, run through it like that, yes.
R The next, he's up on the bed and she's talking, and then she's with another girl, and they're talking about why their children, why should their children have to go without things. It's all about advertising on the television, and, you know, why should their children have to do without because we're out of work? Then she says, 'I get these out of the catalogue and pay so much a week' which again is true, like.
Q And what about the prices of the toys?
R Yes, twenty seven pounds and fifty, yes. But that again can give you the impression that, you know, some people would say 'well they can't be that bad off, if they can pay you know. She might have got one toy then three months later got another toy, but the way she said it then, she seemed to have bought about one hundred and odd pounds worth of toys and then started paying for them, you know, which, you know, it's probably possible, but,
Q Do you think the programme at that point's generally sympathetic to those the mothers, though?
R Well it gives them the chances to air their views, but it doesn't, I wouldn't say

it was sympathetic, you've got to draw your own conclusions to that. I mean, a lot of people would probably say 'well they can't be that bad off if they've got them models with thirty pound, forty pound', you know. But I agree with what she says, why should her child go short? And she might be paying for them with, she might be working, you know, on the fiddle.

Q There are some things we don't know, yes, about that.

R But, obviously, you can't say that, but she says she gets them on the weekly you know, which a lot of people do.

Q I think that's right. There are some kind of technical aspects of the sequence which you could question. There's the way in which the boxes are all stacked up and the camera goes from box to box, and the way in which they recite the prices.

R Yes – you didn't have to mention any prices, did you? But a lot of them boxes were still in brown paper. Was that made just before Christmas, or something (?)?

Q Well I don't know, what do you reckon?

R Well, it's in the winter isn't it, looks like it was in the winter. But they're giving you the impression there, when she's telling you the prices of each individual toy, then she says 'I'm a single parent benefit family' and all, I think people will probably, a lot of people'll probably say 'Well, you know, she can't be that bad off'. But then people don't know what she's, she might be missing out on something else. She might never go out, she might never have a drink or you know, I don't know. It's just, them scenes, like that, give you the impression that they're getting enough money, you know, to get these things. They aren't really on such a bad amount of money to live on each week.

Q In a sense you think they're probably scenes which are open to interpretation (. . .)

R Yes, debate (?), yes. I mean if someone, if you were talking to a Conservative, typical Conservative like – I'm talking about the people I've seen on television and I've listened to on the radio and that, who've said, 'well there's people in the pubs and there's people . . .,' you know, as if everybody who's on the dole is in the pub everyday. And then if you were talking to someone like that about this documentary they would probably cast something like that up, that you know like 'single parent but she can afford to buy so many toys for him', you know.

Q So the programme gives enough, it gives some evidence,

R of argument, yes.

Q To support that kind of,

R person, yes I think it does, in a few things, you know. Particularly like the pub and the taxi, that would.

The second respondent also shows uncertainty at points in his account, although the general interpretative perspective he employs and his evaluations of what is seen and heard differ from those evident in the first transcript.

For a start, it is interesting how, for this viewer, the scene is perceived to be quite centrally 'about advertising on the television', an interpretation which draws on the comments that bridge across into the subsequent episode (see our discussion above) and uses them retrospectively to organize the significance of the 'enactment and interview' sequence. Use of the comments in this way is clearly part of a sympathetic response to the mothers' circumstances and action ('. . . I agree with what she says'), a sympathy contrasting with the evaluations of our first respondent. But this viewer also regards such sympathy as at least potentially out-of-line with the 'impression' projected by the programme itself, one likely to be uncritically taken up within the interpretations of other viewers (variously, 'some people', 'a lot of people' and, more specifically, a 'typical Conservative').

This feature of interpretative discourse, in which viewers predict the likely interpretations of others, we have called *displaced* reading. It occurred in a variety of forms within our respondents' accounts. Often, as here, it was associated with anxiety about the production of misrepresentations and, in these instances, it was sometimes linked to a perception of the programme-makers' intentions as *manipulative* (i.e. seeking to persuade by covertly strategic management of the programme depictions). Despite his observation that it is 'they' (presumably, the programme-makers) who are behind the 'impression', this respondent does not choose to push the question of production intentions quite so far. Indeed, rather against that line of argument, he remarks that the programme gave participants 'a chance to air their views', suggesting some adherence to a principle of fairness, and cites an aspect of the participants' speech itself as an element in potential misrepresentation ('. . . but the way she said it then she seemed . . .'). Importantly, the question of the programme's own sympathies is regarded as a matter of subjective judgement, of 'drawing your own conclusions' rather than a matter of clear and demonstrable bias or manipulation.

Like many of those viewers we spoke to who, at some point, 'displaced' interpretations on to others, this respondent uses his own social knowledge of what might be 'true' to speculate on what was *not* shown, what we were *not* told and therefore what would not be likely to figure within the understandings and assessments of viewers lacking extra-textual access to the topic of this kind (e.g. 'She might be missing out on something else'; 'she might have got one toy then three months later got another toy'; 'she might be working, you know, on the fiddle').

Comparing the two accounts, we found that the first respondent, despite her declared knowledge of the extent to which the programme is 'fabricated', is untroubled by any sense of possible *mis*representation on this account. In contrast, the second respondent comes to question the *motivation* of programme features (mostly verbal ones) as a result of, so to speak, working 'backwards' from what experience tells him may be misrepresentation or at least ambiguity. He speculates less than the first respondent about the reality-status of programme depictions but both of them, in their different ways, address the programme's status as discourse in relation to the verbal and visual presentation of the toys in the interview phase of the scene.

26

Dallas between reality and fiction

Ien Ang

From Ang, Ien 1985: *Watching Dallas: soap opera and the melodramatic imagination*. London: Methuen, 24–38, 40–2, 44–5.

Note: In the section before the ones extracted, Ang poses some general questions about *Dallas* as a product of the commercial culture industry and about the 'use-values' and pleasures which its viewers actively take from it. She then moves to a more specific level of commentary, using material from her respondents' letters.

DALLAS as text

In reading the letters we encounter an avalanche of self-given 'reasons' why lovers of *Dallas* like watching the programme. The letter-writers extensively describe their viewing experiences and state what does and does not appeal to them.

> I find *Dallas* a super TV programme. For me it means relaxation twice a week, out of the daily rut. You may wonder why twice a week – well, that's because I watch it on Belgian TV too. You have to switch over, but you quickly pick it up again. I'm interested in the clothes, make-up and hair-dos too. Sometimes it's quite gripping too, for example in Miss Ellie's case. [. . .] And I think Ray Krebbs is wonderful. But I think J.R. is a monster, a hypocrite, etc. (Letter 1)

> The reason I like watching it is that you can easily get really involved in their problems. Yet all the time you know it will all turn out all right again. In fact it's a flight from reality. (Letter 5)

> Why do I watch *Dallas* every Tuesday? Mainly because of Pamela and that wonderful love between her and Bobby. When I see those two I feel warmth radiating from them. I am happily married myself too and perhaps I see myself in Pamela. I find her very beautiful too (which I myself am not). (Letter 8)

> First of all it's entertainment for me, part show, expensive clothes, beautiful horses, something I can just do with by the evening. (Letter 11)

> I think it's marvellous to project myself into *Dallas* and in my mind to give J.R. a good hiding when he's just pulled off yet another dirty trick, or admire Miss Ellie because she always tries to see the best in everyone or to bring it out in them. (Letter 13)

> I find *Dallas* marvellous, though it isn't an absolute 'must' for me. Reasons: Everyone is so kind to one another (leaving aside J.R.) and they form a real family, being sociable, having their meals together, for example.
> Witty dialogue.
> Fast, characteristic of an American product. (Letter 17)

> My absorption in *Dallas* has to do with the fact that I follow everything coming from America. I have been there once – last year – and I started watching *Dallas* just to see the American city scene: those beautiful apartment blocks (especially the really

beautiful one you see during the titles) and the cars. (Letter 21)

I don't find everything entertaining. The farm doesn't interest me much. Now and then you get a whole episode with nothing but cowboys and cattle. I find that boring, I'm not keen on Westerns. Too macho. Like the episode when the Ewing men went hunting and were chased. After that it got better again, fortunately. [. . .] I like the pictures of the city too a lot. The office buildings in Dallas. The talks about oil. I really enjoy that. (Letter 23; this letter is from a man)

I find the situations always so well chosen and excellently fitting together and everything runs so well from one thing into another. Then I find the milieu (a rich oil family, etc.) very well chosen. (Letter 40, also from a man)

It is clear that there is not just one 'reason' for the pleasure of *Dallas*, which applies for everyone; each has his or her own more or less unique relationship to the programme. What appeals to us in such a television serial is connected with our individual life histories, with the social situation we are in, with the aesthetic and cultural preferences we have developed, and so on.

But though the ideas of each of the letter-writers are of course personal, they cannot be regarded as a direct expression of their 'motives' or 'reasons' for watching *Dallas*. They can at most be regarded as indications or symptoms of deeper psychological incentives and orientations. Furthermore, although these ideas can *appear* to be strictly personal for the letter-writers themselves, ultimately all these ideas are structured in a specific socio-cultural manner. And so we must take a look behind these ideas; we must subject them to a 'sympathetic reading' to be able to say something about the pleasure of *Dallas* that rises above the merely individual level.

It would be going too far to say that viewers are completely free to handle *Dallas* as they want, as the possibilities of experiencing pleasure in it are not infinite. *Dallas* itself, as an object of pleasure, sets its own limits on those possibilities. From the letter excerpts I have just quoted it emerges that the ideas expressed by these viewers contain many elements referring to what is to be seen in the programme – to its textual characteristics. This fact makes it necessary to go into the specific way in which *Dallas*, as a cultural object, is structured.

Dallas is a weekly television programme. A television programme consists of a series of electronic images and sounds which emerge from a television set. These images and sounds represent something: people talking, walking, drinking, high-rise apartment blocks, moving cars, and so on. From this standpoint a television programme can be looked on as a *text*: as a system of representation consisting of a specific combination of (visual and audible) signs.[1] The problem here, however, is that *Dallas* is a discontinuous text: it is a television serial consisting of a large number of episodes, each more or less forming a separate whole. Each episode can then in its turn be called a textual unit. For the sake of clarity I shall view the television serial *Dallas* as a whole as an incomplete, 'infinite' text.[2]

A text functions only if it is 'read'. Only in and through the practice of reading does the text have meaning (or several meanings) for the reader. In the confrontation between *Dallas* and its viewers the reading activity of the latter is therefore the connecting principle. And this reading does not occur just anyhow. As David Morley says: 'The activity of "getting meaning" from [a] message is . . . a problematic practice, however transparent and "natural" it may seem.'[3] A reader has to know specific codes and conventions in order to be able to have any grasp of what a text is about. So it is not by any means a matter of course for viewers to know directly that in *Dallas* they are dealing with a fictional text and not, for example, with a documentary. A great deal of cultural knowledge is necessary to be able to recognize a text as fiction. In *Dallas* – as is the custom in all television serials – certain hints are given for this, such as the titles, presenting the actors one after another, the music, etc.

Any text employs certain rhetorical strategies to arouse the interest of the viewers, and obviously *Dallas* succeeds in attracting the attention of millions of people with very varied social, cultural and psychological backgrounds and maintaining their involvement in the programme. Very general and widespread structural characteristics of television programmes such as *Dallas* contribute to this.

The function of characters

How do viewers get involved in a television serial like *Dallas*, and what does this involvement consist of? The Belgian media theoretician Jean-Marie Piemme, in his book on the television serial genre,[4] asserts that this involvement occurs because viewers are enabled to participate in the 'world' of the serial. This participation does not come of its own accord, but must be *produced*:

> If, in the serial [. . .] participation can be brought about, this is certainly because this activity has psychological foundations, but it is also because these psychological foundations are confronted by a type of discourse allowing them to be activated. In other words, the structure of the discourse which sustains the serial produces the participation as well as the psychological attitude.[5]

The structure of the text itself therefore plays an essential role in stimulating the involvement of viewers. More importantly still, according to Piemme, it is impossible to watch a television serial without some degree of personal involvement. 'To watch a serial', he states, 'is much more than seeing it: it is also involving oneself in it, letting oneself be held in suspense, sharing the feelings of the characters, discussing their psychological motivations and their conduct, deciding whether they are right or wrong, in other words living "their world".'[6] But what is there so particular about the textual structure of television serials that makes them able to effect such profound involvement?

In commonsense explanations of the attraction of television serials, textual structure and its effects are generally ignored. Often single elements of the story are held responsible for the popularity of a serial. Commentary in the press about *Dallas*, for example, shows a special preference for the striking role of the 'baddie' J.R. One of the letter-writers, however, mentions her preference for another *Dallas* character: 'Sue Ellen is definitely my favourite. She has a psychologically believable character. As she is, I am too to a lesser degree ("knocking one's head against a wall once too often") and I want to be (attractive). Identification, then' (Letter 17). But such identification with one character does not take place in a vacuum. One does not just recognize oneself in the ascribed characteristics of an isolated fictional character. That character occupies a specific position within the context of the narrative as a whole: only in relation to other characters in the narrative is her or his 'personality' brought out. In other words, identification with a character only becomes possible within the framework of the whole structure of the narrative.

Moreover, the involvement of viewers cannot be described exclusively in terms of an imaginary identification with one or more characters. Several other aspects of the text contribute to this, such as the way in which the story is told, or the staging. This does not mean, however, that the characters play a subordinate role in the realization of participation. According to Piemme, in a television serial the characters even function as the pre-eminent narrative element which provides the point of impact for the involvement of viewers. But it is not so much the personalities ascribed to the characters in the story, as their formal narrative status that matters. In a fictional text like the television serial the characters are central. Through the characters the various elements of the text (situations, actions, locations, indications of time and so on) obtain a place and function in the plot. Because the viewer imagines the characters as active subjects, those elements are stripped of their arbitrariness and obtain meaning in the narrative. Furthermore, the 'lifelike' acting style ensures that the distance between actor and character is minimalized, so that the illusion is created that we are dealing with a 'real person'. The character therefore appears for the viewer as a person existing independently of the narrative situations shown in the serial. The character becomes a person appearing to lead an autonomous life outside the fiction of the serial; she or he becomes a person of flesh and blood, one of us. The popular press regularly plays on this illusion: the names of actors and actresses and those of the characters are often used interchangeably or merged – Larry 'J.R.' Hagman.

Being able to imagine the characters as 'real people' thus forms a necessary precondition for the involvement of viewers and is an anchor for the pleasure of *Dallas*. This theoretical assertion is reflected in the letters. When the letter-writers comment on the characters, it is almost always in the same way as we talk about people in daily life: in terms of character traits. The characters are not so much judged for their position in the *Dallas* narrative, as for *how they are*.

That at least is the case for the letter-writers who like *Dallas*. Those who dislike *Dallas* appear to keep a little more distant from the characters. Some of them even criticize their 'unreal' nature.

> One of them (his name escapes me) is always the bastard with his sneaky ideas and tricks, the other son is the goody together with his wife, J.R.'s wife (found the name now) is always 'sloshed' and going off alone to her room. (Letter 32)

> When they can't think up any more problems they send Digger after Miss Ellie and change Sue Ellen around a bit again, while J.R. (over the top) is well away with Sue Ellen's sister. (Letter 36)

> I find the characters appearing in the serial very caricatured [. . .] J.R. with his crazy ideas: always the same teeth-gritting. He is also a very caricatured figure, that is obvious. Oh, how bad he is. It's really laid on thick. I find his wife the most lifelike figure in the serial. I think because she was in such a difficult position the writers had most chances with her. What I really can't stand though is the facial expression she has on. Has on, I can't call it anything else. It looks as though her head is cast in plastic. (Letter 41)

What is striking in these reactions is not only a rejection of the 'personalities' of the *Dallas* people, but also an indignation over their constructedness. Those who like *Dallas*, on the other hand, write much more sympathetically about them. In their descriptions a much greater emotional involvement emerges in the characters as people, even when they find them unsympathetic. As one fan of *Dallas* writes:

> Actually they are all a bit stupid. And oversensational. Affected and genuinely American. [. . .] And yet [. . .] the Ewings go through a lot more than I do. They seem to have a richer emotional life. Everyone knows them in Dallas. Sometimes they run into trouble, but they have a beautiful house and anything else they might want. (Letter 21)

The personalities of the characters are for some fans apparently so important that they have spontaneously included a whole list of characterizations and criticisms in their letters. They make clear to us how central the characters are in their viewing experience.

> I don't think whether it's what you want but I'll write what I myself think of the characters too.
> Miss Ellie: a nice woman.
> Jock: mean, doesn't know himself exactly what he wants, I think.
> Bobby: someone who has respect everywhere and for everyone (except for J.R. but that's understandable).
> J.R.: Just a bastard. I personally can't stand him but I must say he plays his role well.
>
> Pamela: a nice girl (I find her a woman of character; she can be nice, but nasty too).
> Sue Ellen: has had bad luck with J.R., but she makes up for it by being a flirt. I don't like her much. And she's too sharp-tongued.
> Lucy: she has rather too high an opinion of herself, otherwise she's quite nice (she's made up too old).
> I don't know so much about the rest who take part in *Dallas* so I won't write about them. If you need what I've said here about these characters then I hope you can use it. If not tear it up. (Letter 3)

> Now I'll describe the main characters a little, perhaps that might be useful for you too. Here we go then.
> Jock: a well-meaning duffer, rather surly and hard-headed, a very haughty man.
> Miss Ellie: very nice, sensitive, understanding, courageous, in other words a real mother.
> J.R.: very egoistic, hard as nails, keen on power, but a man with very little heart.
> Sue Ellen: just *fantastic*, tremendous how that woman acts, the movements of her mouth, hands, etc. That woman really enters into her role, looking for love, snobbish, in short a real woman.
> Pamela: a Barbie doll with no feelings, comes over as false and unsympathetic (a waxen robot).
> Bobby: ditto.
> Lucy: likeable, naïve, a real adolescent. (Letter 12)

> On the characters: Sue Ellen is definitely my favourite. She has a psychologically believable character. [. . .] (Her friend, Dusty, really loves her and for that reason, although the cowboy business in the serial irritates me and so he does too a bit, I do like him as far as I can judge.)
> Miss Ellie is all right too. She looks good, always knows the right thing to do (conciliatory and firm) within the family and her breast cancer gave her some depth. Lucy has guts, but is a wicked little sod too.
> The others don't offer much as characters, I believe. Pamela pouts, and is too sweet. I have absolutely nothing to say about Jock and Bobby; J.R. is really incredible, so mean. (Letter 17)

What is interesting in these extracts is not so much the content of the character descriptions (although the difference in sympathies in itself is worth some attention), but the fact that 'genuineness' forms the basis for evaluation. The more 'genuine' a character appears to be, the more he or she is valued. But what is even more remarkable is that even for the severe critics 'genuineness' is the criterion by which they judge the characters. The only difference is that the severe critics tend to see them as 'unreal', whereas among the fans the opposite is the case. Characters who are 'caricatures' or 'improbable' are not esteemed, characters who are 'lifelike' or 'psychologically believable' are. Also, casually dropped remarks from fans quoted above ('I must say he plays his role well', 'she's made up too old' and 'tremendous how that woman acts, the movements of her mouth, hands, etc.') make clear that these letter-writers are very well aware that they are only dealing with fictional 'real people'. Such remarks indicate that these viewers would like that fictional element eliminated as far as possible. In their eyes actor and character should merge:

> . . . then I find that all the actors and actresses act very well. So well even that, for example, I really find J.R. a bastard, or Sue Ellen a frustrated lady. (Letter 18)

> . . . Because in my opinion they have chosen awfully good actors. I mean suitable for the role they are playing. The whole Ewing family is played so well that they are really human. Sometimes you get a film or a play and you think: God, if I really had to do that, I'd react quite differently. Then it seems so unreal. But usually *Dallas* could really happen, and the actors and actresses make it credible. (Letter 20)

> The people taking part in it act terribly well. (Letter 4)

The effect of 'genuineness' is then the most important thing these viewers expect. Only when they experience the fiction of the serial as 'genuine' can they feel involved in it. They have to be able to believe that the characters constructed in the text are 'real people' whom they can find pleasant or unpleasant, with whom they can feel affinity or otherwise, and so on. It could be said that such involvement is a necessary condition for the pleasure of *Dallas*.

The (un)realistic quality of DALLAS

But genuine-seeming *people* alone are not enough. According to Piemme, the fictional *world* in which the characters live must seem equally real. But how 'real' or 'realistic' is this world? This rather vague concept of 'realism' also seems to play an important part in the letters. 'Realism' seems to be a favourite criterion among viewers for passing judgement on *Dallas*. And here 'realistic' is always associated with 'good' and 'unrealistic' with 'bad'. So it is not surprising that many haters of *Dallas* express their dislike by referring to its – in their opinion – 'unrealistic' content. Here are some letter extracts:

> In *Dallas* no attention at all is paid to any realistic problems in this world. The problems of ordinary people. (Letter 31) . . . in my eyes the characters appearing in it are totally unreal. (Letter 38)

> It is a programme situated pretty far outside reality. The mere fact that a whole family is living in one house comes over as rather unreal. What happens in this serial you would never run into in the street or in your circle of acquaintance: very unreal events.

> The family relations that are so weirdly involved: this one's married to the sister of the enemy of his brother, etc., etc. (Letter 41)

> 1. It is an improbable story because:
>   1.1 Such a rich family would scarcely live as three families in one house (at least in a Western society), so that privacy for each family is minimal:
>   1.1.1 They breakfast together, etc.
>   1.1.2 Other than the common rooms each family only has one bedroom (no separate sitting room or study, etc.).
>   1.1.3 The whole family concerns itself with everything.
>   1.2 Too much happens in the short time and then it's all dramatic situations, not only for the main characters, but for the minor characters as well. This latter makes things confusing.
>   1.3 The actors are rather clichéd types, i.e. they keep up a certain role or attitude. Normal people are more complex.
> 2. Thanks to the constant drama there is a certain tension in the story, but this is exaggerated. Even in a more realistic story there can be tension and I actually find that nicer. (Letter 42)

In these extracts a number of things are striking. In the first place these letter-writers find *Dallas* 'unrealistic' because in their opinion the world and the events in the story do not coincide with the world and events outside *Dallas*: reality 'as it is'. A text is called 'realistic' here if the 'reality' standing outside and independent of the text is reflected in an 'adequate' way. But – and this is the second striking point – the letter-writers each invest the notion of 'reality' with a different content. For some the represented reality must coincide with the social reality of 'ordinary people' (i.e. 'real' problems such as unemployment and housing shortages and not the 'mock' problems of the rich); for others that reality must be 'recognizable', i.e. comparable to one's own environment; and for others again the world presented must be 'probable', i.e. cohere, be 'normal'. Finally, a text is also occasionally called 'unrealistic' if people find that it simplifies the 'real' reality (whatever that may be), exaggerates it or reflects it in clichés.

As we can see, the significance of the notion '(un)realistic' can assume different forms. Clearly, there is no unambiguous definition of what 'realism' contains. But in the way in which the term is used by the letter-writers quoted, at least an essential community of ideas can be discerned: they all call *Dallas* 'unrealistic' because in their opinion it gives a 'distorted image of reality'. This definition of realism, in which a comparison of the realities 'in' and 'outside' a text is central, we can call 'empiricist realism'.[7] This empiricist concept of realism often fulfils an ideological function in television criticism in so far as its standards are used to furnish arguments for criticizing programmes and to strengthen the concept itself. From this point of view, a text which can be seen as an 'unrealistic' rendering of social reality (however that is defined) is 'bad'. And as we have seen, *Dallas* is often subjected to this judgement.

But having said this, we are immediately confronted by an apparently baffling contradiction. Contrary to the critics and those who dislike *Dallas*, who regard it as particularly 'unrealistic', many fans do find it 'realistic'. Some letter-writers even see the – in their eyes – realistic content of *Dallas* as a reason for the pleasure they experience.

> I find *Dallas* super and for this reason: they reflect the daily life of a family (I find). (Letter 3)

> It is realistic (for me anyway), other people think I'm mad, things happen in it we might well find happening to us later (or have had). (Letter 12)

How should we interpret this contradiction? Should we ascribe to these letter-writers a 'false consciousness' because their judgement on the 'realistic' content of *Dallas* is totally wrong? Or is there more to it? Reasoning from an empiricist-realistic standpoint, we can simply say these letter-writers are misled. In *Dallas*, according to this reasoning, the 'daily life of a family' is certainly not being reflected – for, as one letter-writer suggests: 'I wonder why these people in Heaven's name carry on living in the same house!' (Letter 36). Furthermore, it could be said that the things that happen in it are certainly not things 'we might well find happening to us', for it is clear that in *Dallas* there is an improbable accumulation of sensational events, such as airplane accidents, weird diseases, kidnappings, etc. In short, if *Dallas* is regarded as a mirror of 'the' reality, then we should recognize that it is a big distorting mirror, or more seriously, 'a twisted image of reality'.

But this empiricist conception of realism presents problems for a number of reasons. I shall cite two difficulties here. First, it is wrongly based on the assumption – and this is inherent in empiricism – that a text *can* be a direct, immediate reproduction or reflection of an 'outside world'. This is to ignore the fact that everything that is processed in a text is the result of selection and adaptation: elements of the 'real world' function only as raw material for the production process of texts. The empiricist conception denies the fact that each text is a cultural product realized under specific ideological and social conditions of production. And so there can never be any question of an unproblematic mirror relation between text and social reality: at most it can be said that a text constructs its own version of 'the real'. As Raymond Williams says: 'The most damaging consequence of any theory of art as reflection is that [. . .] it succeeds in suppressing the actual work on material [. . .] which is the making of any art work.'[8]

The second difficulty is connected with this. The empiricist conception of realism cannot do justice to the fact that a large number of *Dallas* fans do seem to *experience* it as 'realistic'. Must we regard this experience merely as the result of incorrect reading and must we, consequently, accuse the letter-writers who read 'wrongly' of a lack of knowledge of reality? It is, to say the least, unsatisfactory just to dismiss this very prevalent way of responding to the programme. A more structural explanation must be possible.

DALLAS and the realistic illusion

In the empiricist conception of realism the thematic content of the narrative becomes the guideline for the assessment of the 'realistic' nature of the text. Some literary and film theoreticians on the other hand make the way in which the story is told responsible for what is called the 'realistic illusion': the illusion that a text is a faithful reflection of an actually existing world emerges as a result of the fact that the constructedness of the text is suppressed. Piemme states that it is this suppression which fosters the involvement of viewers in the serial:

'Participation can only function by denying itself as product of discourse. What produces it must suppress the marks of its production in order that the illusion of the natural, the spontaneous, the inevitable, may function.[9] In other words, the realistic illusion is not something to be blamed on the ignorance or lack of knowledge of the viewers, but is generated by the formal structure of the text itself; the thematic content plays only a subordinate role here.

[The author here offers a brief summary of some of the codes of depiction and of continuity which produce the 'realistic illusion' in film and television fiction – see the item by Ellis in this section. She refers to the work of film theorist Colin MacCabe on the self-concealing nature of 'classical realism'.]

According to MacCabe and others[10] it is precisely this constructed illusion of reality which is the basis for pleasure. It is pleasurable to be able to deny the textuality and the fictional nature of the film and forget it: it gives the viewers a comfortable and cosy feeling because they can 'let the narrative flow over them' without any effort. The apparent 'transparency' of the narrative produces a feeling of direct involvement, because it ensures that the viewer can act exactly *as though* the story really happened. In other words, according to this theory pleasure in *Dallas* could be regarded as a pleasure in the obvious, apparently natural meaningfulness of the ups and downs of the Ewing family and the people around them. It is the *form* of the narrative which produces pleasure, not its content.

Yet this explanation of pleasure is not totally satisfactory, precisely because it abstracts from the concrete narrative-content.[11] Transparent narrativity alone is not enough to get pleasure out of a text; not all transparent narrative texts are experienced as equally pleasurable. On the contrary, the thematic differences between such texts are of interest, as one of the letter-writers states: 'For me *Dallas* is comparable to *Dynasty*. Other American series (*Magnum, Hulk, Charlie's Angels, Starsky and Hutch,* in short, violence) I can appreciate less' (Letter 17). Thus the pleasure of *Dallas* is not only to do with the illusion of reality which is produced by its transparent narrativity – although it might be said that this illusion is a general condition of pleasure as it is experienced by a lot of viewers. *What* is told in the narrative must also play a part in the production of pleasure.

DALLAS and 'emotional realism'

Why then do so many fans call *Dallas* 'realistic'? What do they recognize as 'real' in its fictional world?

A text can be read at various levels. The first level is the literal, denotative level. This concerns the literal, manifest content of the *Dallas* narrative: the discussions between the characters, their actions, their reactions to one another, and so on. Is this literal content of the *Dallas* story experienced as realistic by viewers? It does not look like it. Indeed, we can say that the above-quoted letter-writers who dislike *Dallas* are talking on this literal narrative-level when they dismiss the programme as *un*realistic. Let us repeat a letter extract:

> It is a programme situated pretty far outside reality. The mere fact that a whole family is living in one house comes over as rather unreal. What happens in this serial you

would never run into in the street or in your circle of acquaintance: very unreal events. The family relationships that are so weirdly involved: this one's married to the sister of the enemy of his brother, etc., etc. (Letter 41)

This indicates that the *Dallas* narrative at the level of denotation is not exactly regarded as realistic; literal resemblances are scarcely seen between the fictional world as it is constructed in *Dallas* and the 'real' world. Again the inadequacy of the empiricist-realistic approach becomes clear here. It is only sensitive to the denotative level of the narrative. Therefore it can only see the fact that so many *Dallas* fans obviously do experience the programme as realistic as a paradox.

A text can, however, also be read at another level, namely at the connotative level.[12] This level relates to the associative meanings which can be attributed to elements of the text. The same letter-writer we have just quoted also wrote the following: 'The nice thing about the serial is that it has a semblance of humanity, it is not so unreal that you can't relate to it any more. There are recognizable things, recognizable people, recognizable relations and situations in it'. (Letter 41)

It is striking; the same things, people, relations and situations which are regarded at the denotative level as unrealistic, and unreal, are at connotative level apparently not seen at all as unreal, but in fact as 'recognizable'. Clearly, in the connotative reading process the denotative level of the text is put in brackets.

[Ang here quotes further from her letters to indicate in what ways viewers find *Dallas* 'recognizable' and 'taken from life' at this level of reading. She notes that it is primarily at this level that the programme is judged to be realistic.]

In naming the 'true to life' elements of *Dallas* the concrete living circumstances in which the characters are depicted (and their wealth in particular springs to mind here) are, it is true, striking but not of significance as regards content; the concrete situations and complications are rather regarded as symbolic representations of more general living experiences: rows, intrigues, problems, happiness and misery. And it is precisely in this sense that these letter-writers find *Dallas* 'realistic'. In other words, at a connotative level they ascribe mainly emotional meanings to *Dallas*. In this sense the realism of *Dallas* can be called an 'emotional realism'.

And now it begins to become clear why the two previous conceptions of realism discussed above, empiricist realism and classical realism, are so unsatisfactory when we want to understand the experience of realism of *Dallas* fans. For however much the two approaches are opposed to one another – for the former realism is a token for a 'good' text, and for the latter for a 'bad' text – in both a cognitive-rationalistic idea dominates: both are based on the assumption that a realistic text offers *knowledge* of the 'objective' social reality. According to the empiricist-realists a text is realistic (and therefore good) if it supplies 'adequate knowledge' of reality, while in the second conception a classic-realistic text is bad because it only creates an illusion of knowledge. But the realism experience of the *Dallas* fans quoted bears no relation to this cognitive level – it is situated at the emotional level: what is recognized as real is not knowledge of the world, but a subjective experience of the world: a 'structure of feeling'.[13]

27

Captured on videotape: camcorders and the personalization of television

Lawrence J. Vale

From Dobrow, J. (ed.) 1990: *Social and cultural aspects of VCR use.* New Jersey: Hove: and London: Lawrence Erlbaum, 195–209.

On video

The title of Roy Armes' (1988) book, *On Video,* gives it a place in the 'On *X*' genre of analytical essays that ranges from Clausewitz's *On War* to Sontag's *On Photography* (two books whose contents share much more in common than their titles might at first suggest). Armes' title, however, has the added presumed advantage of an implied double meaning; to wit, a pun. Not only does this title offer the suggestion that it will be 'on' video in the sense of being an extended essay-of-definition *about* video, it also opens the possibility that it will delve into the subject more experientially to ask what it means to *appear* 'on' video. This, to the great disappointment of all of us who believe that the title of any profound book should present at least two simultaneous ways to interpret its contents, proves not to be the case. Armes wrote convincingly about the history of video, the social context of video from the point of the professional producer, and the aesthetics of video sound and video image, but did not examine the question of how video is received by those who find themselves captured 'on' it. This next task is the one I take up here.

The camcorder, first launched in 1985 as the one-piece descendent of the two-piece video recorder (Marbach, 1985), would at first seem no more than an accessory to the increasingly ubiquitous VCR.[1] Yet, as I argue here, it is a curious hybrid of many technologies, and is a product that itself seems likely to introduce some new strains. Certainly, in its domestic applications,[2] the 'use value' of a camcorder is close in many ways to that of the home movie camera, the technological dinosaur that it is fast replacing (Chalfen, 1987; Hughey, 1982; Kealy, 1981). Still, as an electromagnetic alternative to the photochemical techniques used in film processing, the camcorder and its videocassettes are technologically quite distinct from the principles that undergird the earlier medium. In this sense, the camcorder is evolutionarily more closely related to the audiocassette recorder and certain other forms of sound recording than it is to film (Armes, p. 9). There are other important linkages, technologically distinct yet socially and culturally interwoven. As a social tool, the camcorder is certainly close to the still camera and the world of home snapshot photography, just as the videocassette – when used as a mode for the interpretation and home storage of recorded family history – bears some relation to the home photo album, as well as to earlier traditions of the family Bible and the painted ancestral portrait (Beloff, 1985; Chalfen, 1983, 1987; Hirsch, 1981; Jacobs, 1981; Lesy, 1980). As a means to produce near-instantaneous reproduction of images, the camcorder is cousin to the Polaroid camera, and the camcorder/VCR/TV ménage-à-trois is the thriving successor to Polaroid's 'Polavision', their abortive attempt to market a system of 'instant home movies' displayed on a non-television monitor (Hughey, 1982; Olshaker, 1978). This last media misconnection suggests also that to begin to assemble the meanings that are associated with the camcorder and the home videos that its users produce, one must consider the importance of the novel intersection that has been created between what Chalfen (1987) called 'home mode communication' and the external mode programming that is more usually associated with television sets. How important is it – and in which ways – that home videos are viewed on *television*? Thus, although the camcorder is closely related to many other media, it is not an exact substitute for anything, even home movie equipment. As a new media technology, it shares many things with the introduction of other new technologies but, nevertheless, it differs – technically, aesthetically, socially, and culturally – from all that has preceded it.

All of the recent commentators on the various modes available for using media to record family history have stressed the ways that this history is edited – both consciously and unconsciously – to reflect positively on lives being lived. A column of helpful hints in *Popular Photography*, entitled 'Family Photography as a Sacrement' (Hattersley, 1971) makes clear that this selective editing should be an explicit ingredient in good photography:

> Use good judgement in picking the time and place for pictures. Children don't worry much about the when and where, and are usually raring to be photographed. With adults, however, it's a different thing, for their worries and concerns leave them feeling pretty bedraggled much of the time. It isn't good to photograph Dad right after a hard day's work, Mom when she's been hassled by a bill collector, and Junior after his girl has given him the brushoff.
>
> Be on the lookout for those golden hours when everything seems to be going right for everyone in the family. It may be the right time on a lazy Saturday morning after a late breakfast, or for the half hour just after the finish of the family's favourite TV show. After finding these sweet times, make sure your approach to photography makes them seem even sweeter. (p. 108).

This set of saccharine suggestions would seem to fit well with what most scholars of 'home made' photography have observed. According to Jacobs (1981), photo albums 'are constructs that propose positive histories . . . With snapshots we become our own historians, and through them we proclaim and affirm our existence' (p. 104). Chalfen (1987) noted wryly that:

> Future anthropologists, if they studied our culture from home photo albums alone, would probably conclude that this breed of man lived mostly at Christmas, indulged in a ritual with coloured eggs at Easter, graduated from institutions frequently, celebrated birthdays mostly while young and had lots of small animals. Further, they would conclude, children were usually fresh-scrubbed, and spent a great deal of time standing around squinting into the sun. (pp. 170–71)

Halla Beloff (1985) commented that 'We are obliged to show what happiness we have experienced, what friends deserved, what ambiences and positions achieved' (p. 190) and concluded that 'What is shown is the past that we can be sentimental about' (p. 196).

'What kinds of information' Chalfen (1987) asked, 'are being transferred from generation to generation between the covers of a family album, in cans of home movies, or in videotape cassettes?' (p. 2) Chalfen's answers, and those conclusions proffered by other studies of various media, confirm a certain consistency across these three media both in terms of subject matter and in terms of attitudes toward it: there are more images of young children than older ones; more of first born than of later births, more emphasis on achievements than defeats, more depictions of vacations than vocations. Armes (1988) continues in his criticism of the superficiality and artificiality of home video:

> [V]ideo is a technology symptomatic of the public role given to images in a capitalist society; it records aspects of the surface of life, but it embellishes, prettifies, as it records. . . . [The video camera] is openly, transparently, both an instrument for celebrating what *is*, rather than what could be achieved by social change, and, at the same time, a machine for making life seem more pleasurable than it is. (p. 197)

Although Armes is in clear agreement with those who suggest that home video (or, more precisely, those persons who make them), like earlier modes of domestic history gathering, tends to put a positive gloss on the events of life, it does not necessarily follow that the camcorder really does act as 'a machine for making life seem more pleasurable than it is.' Even these edited records of life may yield emotions that are far from pleasurable, due both to the process of videomaking and the resultant product.

To test this hypothesis about attitudes toward the process of videomaking with home camcorders, this chapter employs data from a set of nearly 200 interviews with VCR owners conducted in early 1989 by graduate students from Boston University's College of Communication. In the course of each interview, each person who reported having watched him or herself on video was asked to comment on the experience. In addition, those in this sample who said that they owned, rented, or had access to a camcorder or videocamera were asked not only to identify the kinds of things they chose to videotape, but also to state whether there were certain things that they would *never* want to portray on videotape. The responses to these questions suggest that, even if the finished video products are sanitized selected views of atypical situations, these products may have little to do with enhancing life's pleasures.

Video do's and video taboos

More than two thirds of this sample of VCR owners reported having watched themselves on video, a reminder that the social impact of camcorders goes far beyond the statistics about camcorder ownership penetration. Although camcorder ownership in the United States has grown rapidly and at this writing stands at about 7 million – equal to about 7 per cent or 8 per cent of all television households (*Videomaker*; Electronics Industry Association, December 1988, cited in Collins, 1989) – the prevalence of VCRs – approximately seven times the penetration rate of camcorders – has already greatly extended the accessibility of the video products created by home camcorder technology.

Before turning to a discussion of the questions about how respondents said they reacted to viewing themselves on television videos and what camcorder users would never want to videotape, some comment seems useful about how camcorder use may be related to other patterns of media use. Of 49 respondents who owned, rented, or had access to a camcorder or video camera, nearly three quarters said they had at some point owned a still camera and made photo albums.[3] Of the remainder who did not make photo albums, nearly half said that they had made photo albums before they began using a camcorder. Although the sample size for these interviews is too small to draw any firm conclusions, it does seem likely that the home-produced videotapes may, in many cases, supersede the family photo album as the primary repository for domestic documentary history.

About 40 per cent of those using camcorders reported having owned a Polaroid camera at some time, a proportion consistent with the range of instant camera penetration figures over the last decade (Wolfman, 1980; Photo Marketing Association, 1985, cited in Wolfman, 1987). Approximately the same proportion stated that they had, at some time, owned an 8 or 16mm movie camera. Not surprisingly, the correlation between camcorder ownership and previous movie camera ownership is high. The percentage of camcorder users reporting previous movie camera ownership is about one third higher than the maximum penetration rate of movie cameras in the United States (achieved in 1977). Indicative of its increasing obsolescence of home moviemaking equipment – 80 per cent of which still relied on soundless systems as late as 1980 (Wolfman, 1981), it is estimated that only 3 per cent of American households were still using movie cameras in 1988, down from 12 per cent in 1980 (Wolfman, 1980, 1988). In other words, current camcorder use is estimated to be nearly three times greater than current movie camera use.

Of those who said they had watched themselves on video, most reported that the videos showed them at social events with family and friends – primarily weddings, parties, and birthdays. The next most commonly mentioned subject matter focused on the workplace. Other subject matter frequently mentioned included performances such as recitals and skits (10 per cent), school-related activities (8 per cent), and sporting events (8 per cent). What is perhaps surprising is that only 5 per cent of those who said they had watched themselves on video reported that this video was taken during a vacation. This surprisingly low proportion may be explained, in part, by the fact that only 37 per cent of those who said they had watched themselves on video personally owned, rented, or had access to a camcorder. Of those who themselves *owned* a camcorder, 42 per cent reported using it on vacations.

When asked to identify anything that they would *never* want to film with a camcorder, only 17 per cent of the camcorder users said that there was nothing that they would not want to film. An additional 17 per cent gave noncommittal or uncertain responses. By way of contrast, a full two thirds of the sample, without any prompting from the interviewers, had a clear idea about situations that they would not want to film (cf. Chalfen, 1984). Not surprisingly, perhaps, the responses of camcorder users centred on the Freudian duo of sex and death. Repeatedly, both male and female respondents stated their refusal to let the camcorder into the bedroom. In addition to prohibitions on filming sexual relations and home-produced pornography, several respondents (mostly East Asians) said that they would never wish to film someone sleeping or just awakened. A few respondents thought to extend the camcorder ban to the bathroom as well. A substantial number of respondents stated that they would not want to film situations connected with death: no 'funerals'; 'no wakes, no wills'; 'nothing bad – like the report of a death.' In at least one case, the ban on funeral filming was based on actual intergenerational experience, directly prompted by a woman's distress at her mother's cinematic treatment of her own mother's funeral. Another respondent, an Hispanic obstetrician, said that although he regularly helps film childbirth, he would never want to film the moment when 'the baby learns [about] death.'

Whatever its potential to document individual and family experience, most camcorder users wished to impose clear spatial and ethical boundaries. Their responses suggest a powerful underlying fear of both the intrusiveness of this technology on closely guarded realms of privacy and a further fear of the camcorder's potential to perpetuate or revive unpleasant memories. Even more frequently than mentions of sex or death, respondents stressed the more general concern that the camcorder not be used to expose undesirable parts of private life. Many camcorder users insisted that they did not want to film what they variously called 'intimate life,' 'very private life,' 'anything private or personal,' 'private things,' 'anything private or concerning the private life of someone,' 'personal things,' and 'anything personal.' Some respondents, perhaps out of this same kind of discomfort, preferred to give a humorous response. One man said he would never film his 'stepmother in a swimsuit'; another proposed a special ban on depicting his mother's cooking. Humour, as always, remains intimately tied to anxieties. Other camcorder users dispensed with qualifying words like 'swimsuit' or 'cooking' and made clear their desire to exclude completely certain people from appearing in their videos. One woman singled out her stepfather for exclusion, whereas another, more categorically, said she would not wish to film 'people who I don't like.' A college student preferred to edit out any activity associated with her relatives, 'like family things, you know.'

In contrast to this pervasive and conscious desire to edit out unpleasantries and limit visual access to personal and private affairs, some members of one category of respondents seemed more concerned with using their camcorders to document something closer to a more balanced interpretation of their positive and negative experiences. This category of camcorder users is comprised of recent immigrants to the United States, who use their videos as a means to communicate their American experiences to those still living in the 'old country' (an ancestral home presumably well-endowed with the proper VCR receptor). Although the number of individuals in this sample who are engaged in First World/Third World home video transfer is too small to derive anything beyond anecdotal evidence, alternative patterns of camcorder use and significance seem quite possible (see Dobrow, 1989). One Haitian man, for example, reported that his primary use for a camcorder was less to film 'family or social events' than to depict 'political events or to record some aspect of American society such as the homeless.' He wished to do this, he said, 'because some fellow countrymen [in Haiti] still believe everything is gold in the US.' A Cambodian refugee, less concerned perhaps with documenting the downside of life in America, also differs from the mainstream of home video tendencies when he says that he would never wish to film 'things that are not important to my family.' Although it is not clear exactly what would constitute an 'important thing' to him, he too seeks self-consciously to document his family's accommodation and adaption to a new culture. For him, the camcorder is used 'to record our life in the US to send to our family abroad to see how we progress.' Whether these documents of progress are accompanied by snippets of setbacks is uncertain. What seems clear is that the exported home videos are intended to be used as evidence. Each of these camcorder users describes ways that they edit their lives for television, but video lives may be edited to serve many different purposes.

In accord with previous studies that have examined the content of home movies and home snapshots, this investigation of home video suggests that image content is selectively cultivated to favour special events and gatherings not part of the normal routines of daily life. There are clear – if often unspoken – boundaries of acceptability, and there is a general agreement among camcorder users that the machine must not be allowed to intrude too closely on the private or personal life of those who are being filmed. How, however, are these restraints received by those who see and hear themselves depicted on video? Does the camcorder user's proclaimed desire to keep the camera away from the most private realms of experience permit those 'captured" on video to be seen only as they would like to be seen?

The televideo self

To get at this issue, 122 VCR users who reported having seen themselves on video were asked 'What was it like to see yourself on TV?' Interviewers were asked to try to record each respondent's exact words. These responses were then coded in several ways. First, key words from the responses were coded and divided according to type of reaction denoted: positive, negative, and neutral/mixed feelings. Next, for further analysis, responses were divided into those persons who had seen themselves on video and who themselves used a camcorder and those who reported having seen themselves on video but did not personally use a camcorder. Finally, the responses were subdivided according to sex.

The first striking observation was that remarkably few people – only 15 per cent – expressed opinions that could be categorized as 'neutral' or 'mixed feelings': This was clearly a subject where people reported clear views, whether positive or negative. Moreover, of this 15 per cent who could not be easily categorized at one extreme, 39 per cent responded by expressing marked ambivalence, characterized by strong but conflicting feelings. Thus, out of the whole sample, only 9 per cent described the experience as being 'no big deal' or were unable to articulate any particular response.

Of the remaining 91 per cent who reported more clearly resolved sentiments, nearly two thirds expressed a negative reaction to seeing and hearing themselves videotaped on television. The negative reactions ranged from the rather mild – 35 per cent used the words 'weird,' 'strange,' 'funny,' 'odd,' or 'peculiar' to describe their experience – to the thoroughly traumatized – 22 per cent used words such as 'horrible,' 'awful,' 'scary,' depressing,' and 'grotesque.' The rest of the negative responses could be grouped somewhere in the more middle regions of tolerated unpleasantries. Many respondents used the words 'uncomfortable' or 'embarrassed' to describe their reaction in more general terms, whereas others zeroed in on more specific aspects of their own behaviour that served as the source of the discomfort and embarrassment. Some described their distress at finding their body movements appeared 'awkward,' 'goofy,' or 'foolish,' whereas others bemoaned certain more static aspects of telegenic inadequacy – they saw themselves as 'ugly,' 'fat,' or 'old.' One man said, 'It was shocking – I realized I'm really getting bald.' Another noted that 'It shows every wrinkle and fat.' A third concluded that 'ever since the first day I saw myself on video I've always considered that I have a face for radio.' A fourth respondent said: 'I have never been satisfied with what I see. Although I knew I was an ugly guy, on TV I am 10 times uglier.' Another summed up: 'Sad and depressing.'

Even more than men, women who were asked to comment on the experience of watching themselves on television reported seeing evidence of their own physical inadequacy. Statements included such things as 'I looked fatter than I had expected,' 'I didn't like myself in a bathing suit,' 'I looked 10 pounds heavier,' 'It was sad and disappointing – I was too pale,' 'Embarrassing, and I thought I looked older than my age,' and 'Horrible – I didn't look good. You look fatter on TV.' These comments about visual self-perception, however, constitute only one part of the way that discussion about the experience of video was structured.

Although many respondents answered in a general way that did not identify which component of their televideo selves was the greatest source of pleasure or displeasure, responses ranged across three aspects of the video experience – the perceptions of audio, visuals, and kinesics. A man, one of many who reported the experience as 'embarrassing,' identified the source of discomfort in terms of audio rather than visual experience: 'I never thought that my voice was that high-pitched.' A woman remarked, 'It was funny to see me and *hear* me. Finally, I realized that I had a false image of myself.' Other respondents were most disturbed by kinesic aspects: A woman related her embarrassment at realizing she 'put [her] pelvis up front too much.' Another, describing the experience as 'weird,' said that 'It's funny to see yourself on TV because you do not think you act that way.' And one man reported experiencing the visual, audio, and kinesic aspects of video in terms of a triple threat of unpleasantries. 'It was really strange. It wasn't so much seeing the video, but my mannerisms and the way I spoke. And I wasn't thrilled about the shirt I was wearing, although it was one of my favourites.'

As to the minority of respondents who felt more positively about the experience of seeing themselves appear in a video, there were three basic tiers of response. The least enthusiastic yet still positive group of respondents (24 per cent of the total number of positive respondents) were coded according to their consistent use of two words: 'interesting' and 'educational.' A second, somewhat higher tier of positive response, also could be coded according to the repeated use of only two key words: 'amusing' and 'fun.' This group, clearly entertained but reporting no great thrills, constituted the largest proportion of the positively inclined part of the sample – 38 per cent. Only 35 per cent of the positively inclined respondents used words such as 'great' or 'exciting' although a few were totally enraptured. One man commented: 'I find it a real thrill to be able to see myself on say a TV set which is so alien from us; we are attached to TV but at the same time apart. It's real – you become a part of TV life,' whereas a woman said: 'I suppose orgasmic isn't something you want to write on your survey – it was . . . very special.' Such comments were clearly exceptional, however. Out of this whole sample of VCR users who reported experiencing themselves on video, only 11 per cent of them regarded the experience as overwhelmingly positive.

The perception of the televideo self as an unpleasant experience is even more marked if one looks only at the portion of the sample who reported seeing themselves on television but who did not themselves ever use a camcorder. Although it may not be surprising that those who are presumably in closer contact with camcorders (and who – in many cases – invested $1,000 or more in purchasing one) would be more comfortable with this technology and feel more positively about its use, it is nonetheless striking to observe the extent of displeasure that this experience seems to cause most non-camcorder users. Although only 30 per cent of the total sample reported viewing themselves on TV as any sort of positive experience, this approval rating drops even lower if one simply looks at the responses of those who do not own, rent or have access to a camcorder. Among this group, the approval rating is only 25 per cent. Among women, the reported approval rating was even more miniscule – less than 7 per cent. Taken overall, only 5 per cent of the non-camcorder-using subset reported the experience in overwhelmingly positive terms, not one of them a woman.

Conclusions

Although most accounts of home media use have stressed the ways that it is used to produce highly selective positive interpretations of life, this data on camcorder use and misuse and the reactions to viewing home video suggests a considerably less rosy picture. Even though most camcorder users claim to proceed only with extreme caution as they prepare to videotape the private lives of their subjects, this does not seem to have prevented a largely negative response to the experience of being videotaped. It would seem that the perception of excessive intrusion into personal realms is not limited to the boundaries of the bedroom or the bathroom, but may be found even in some of the most casual exchanges. Even where videotaping is tolerated or actively advocated, a certain ambivalence may remain. In a 1985 interview, one man described the act of videotaping his family reunion as 'a cleverly disguised method of crowd control' (Waters, 1985, p. 53). In the 1989 survey of VCR users, one woman, herself a camcorder user, said that the experience of seeing herself on video was 'really awful' because 'most people see you from different angles' which made her 'very self-conscious.' Moreover, she added, her 'whole family feels that way.' Nonetheless, she observed, 'they still want to see themselves.' Another woman – who had not used a camcorder herself – also described the experience of seeing herself on video as 'awful' but noted that, since 'everybody else' looked just as terrible, she 'didn't feel so bad.' At present, any overarching social or psychological assessments of the reasons for this ambivalence, this tolerated displeasure, must remain largely speculative. Yet, certain perceptual aspects of the televideo self seem logical outgrowths of camcorder technology and the social ritual of video replay on television.

Surely there is some appeal to the interface between home video and the television set, because the act of watching home video on television extends the personal and group flexibility made possible by the VCR.[4] Whereas the VCR enables individuals to shift broadcast timetables to fit their own schedules and allows them to choose from an increasingly diverse range of non-network content, the camcorder/TV interface gives them the ability to create their own programming in terms of content as well as genre. In so doing, the phenomenon of 'time-shifting' is applied to personal and family history: Like the VCR itself, the camcorder helps advance the illusion that past events can be brought forward and rendered convincingly in the present.

Although time-shifting was, perhaps, always inherent in the act of looking at family photo albums, this flexibility and personalization is now paired with the authority of the medium of television. Although home movie viewing required the assembly of often awkward systems of projectors, sound equipment and free-standing screens, with home video the new technology becomes experientially consolidated into the TV/VCR. No longer projected onto a screen on a shaft of light – with sound behind and picture out front – the televideo family image seems to emerge outwards from within the television box, emanating from the same mysterious source as the nightly news or the latest soap opera episode. As Armes observed (1988), 'When tapes are viewed, the hierarchical distinction between a multi-million dollar feature film and domestic recording of a family wedding is erased' (p. 83). A single screen is now shared by multiple media. The seeming self-sufficiency of the video cassette, hidden deep within the interstices of the TV/VCR media complex, would seem to yield at least some degree of conflation – or at least confusion – in a culture that does not now tend to differentiate between 'watching TV' and 'watching the VCR.' In those cultures where the village VCR actually preceded the reach of broadcast television the effects of hierarchical uncertainties might be even greater.

The infiltration of home-produced video into the television world of network-produced broadcasts and studio-produced films seems likely to affect home video's meaning. On the one hand, there is a significant sense of empowerment gained through association with television; on the other hand, the family subjects are implicitly placed up for comparison with professional actors who are even more self-consciously coiffed, made-up, and scripted. Does this, perhaps, account for some part of the widely reported distress at seeing our televideo selves?

Most forms of broadcast television and professional film involve significant amounts of rehearsal and editing before the material is released for distribution. Home video, by contrast, tends to be 'live' and largely undigested. Unlike still photographs, where it is the usual practice to throw out bad snapshots and reprint only the most flattering ones for wider distribution, with home video this selection process is more difficult, both technically and conceptually. Given the temptation to review the results of a videotaping session immediately – 80 per cent of respondents reported viewing the work 'immediately' or 'as soon as possible' – and the further tendency of the vast majority to show the video in a group setting, there is little opportunity for editing, even in the unusual cases where the person in charge of the videotape's contents might actually be interested in taking the time to do this. Although the most significant forms of 'editing' happen in the process of deciding what to videotape in the first place, after-the-fact editing is not a simple task. Although it may not be technically difficult to erase undesirable parts of the tape, the nature of the moving image makes diagnosing and ameliorating undesirability an extremely complex activity. In the multifaceted context of captured kinesics, it is not easy to edit-out all ill-considered remarks and uncharacteristic grimaces. Video gets uncomfortably beyond the pose and undiplomatically preserves the casual comment, made, perhaps, by someone still clinging to the illusion that cameras photograph but do not 'listen.' This last aspect – the lack of an audio self-consciousness – may indicate, in part, a sort of 'media lag,' wishful or otherwise. Nevertheless, even more than in the case of still photography, video may reveal 'too much.'

At the same time that viewing a videotape may produce discomfort because of its unwanted intrusion into private realms, there is also the related danger that the video rendition of life may be accepted as an alternative to memory, rather than as its supplement. It would seem plausible that the meaning of events is altered by the immediateness of the experience of video replay; reflection occurs in response to the video, rather than to the totality of the event. Participants captured on video are asked to accept the sights and sounds of an event shot from a point of view that may differ significantly from that of their own perceptual experience, an alternative perspective that is then given the added legitimacy of portrayal on television. Also subordinated, or at least disrupted, is the cultivation into memory of other sensory experiences – the tastes, smells, and touches that are even less directly recordable by video technology. Still, at least with these three kinds of sensory data that the videotape does not attempt to replicate or represent, there remains a greater latitude for letting each individual's memory intervene; the visual cues to remembered tastes, smells, and touches may be quite valuable.

Nonetheless, taken overall, the seeming 'real'-ness of the videotape medium can encourage a dangerous substitution. 'Is it live or is it Memorex?' becomes 'Is it life or is it Memovid?' and the consequences of being unable to distinguish become ever greater. In some cases, the temptation toward instantaneous video playback may cause video to not only serve as an important way to memorialize an event but may actually become the event itself: the second half of a party can be the video view of the first half. As one man put it, 'If we record a kid's birthday party everyone runs into the other room to see themselves and it breaks up the party. I don't like it because the taping becomes more a part of the event than the event itself' (Stocker, 1988, pp. 54–55).[5] Video's ability to transform 'the event itself' thus may occur not only because of the intrusions of the videomaking process, but also because of the intrusion of the product. At all times, social situations are altered and cultural meanings transmuted. As Armes (1988) commented, 'Electro-magnetic recording in particular can take our most intimate situations and, while apparently preserving them, simply turn them into mere information flow, the very opposite of lived experience' (p. 5).

Many of the most interesting questions about the meanings of camcorders and home video will require close observation and analysis over a long period of time. What will it mean for a new generation to hear and see the movements of unmet ancestors? How are camcorders being used differently in different cultures? As photography continues to lose some of its authority in the face of digital retouching processes (Lasica, 1989), will we continue to regard video as a superior form of cultural authenticity? Benjamin (1936/1969), in his classic discussion of the loss of aura in art in an age of mechanical reproduction regards this loss, in part, as a positive shift, a welcome and democratizing liberation from the tyranny of upper class high culture. What about this new form of reproduction, even more mechanical than film? Does the camcorder, currently far more expensive to own than any other media with which it is compared, represent a return to mediated elitism, a privileged but illusory personalization of television? Is it possible that the camcorder brings us only the worst of all worlds: a new form of reproduction that is both aura-less and anti-democratic?

28

Music, text and image in commercials for Coca-Cola*

Hroar Klempe

Some assertions

According to some commentators, much advertising in industrialised societies during recent decades has undergone a significant change, one which from a rhetorical point of view can be formulated as a change in argumentative technique. Whereas such argumentation was formerly more overt and explicit, often taking the form of a syllogism, in more recent years it has become far more implicit and indirect and has often involved what can be termed understatement.

In the present paper I will try to demonstrate that any explanation of this change in what we can call the language of advertising has to have recourse to the concept of *repetition*. As a point of departure I will consider the assertion that the sort of change I have mentioned involves a kind of 'telling' which goes beyond traditional views of narrative meaning. I will illustrate my argument by reference to the commercial for Coca-Cola entitled 'First time'. Following this, I will consider the rôle played by repetition in music, text and image in advertisements for Coca-Cola spanning a number of years, and by means of these examples I will attempt to show how central the musical way of using repetition is to an understanding of these commercials.

Very many commercials for cleaning agents argue in an explicit way. The argumentation in such commercials may follow a kind of syllogistic structure as follows:

Premise 1:	Washing the bath tub with scouring powder (as evidence shows us) creates scratches.
Premise 2:	This cleaner is not an ordinary scouring powder, it is a liquid.
Conclusion:	This cleaner is better than any ordinary cleaning agent (as the evidence confirms).

The argumentation makes a claim which is explicit, clear and persuasive.

If, on the other hand, we turn our attention toward the Coca-Cola commercial 'First Time', we will find that there is no explicit argumentation, either in the images or in the lyrics.

*Coca-Cola is a registered trademark of The Coca-Cola Company. Permission to use granted by the Company.

*First Time**
First time, first love, oh what feeling is this,
electricity flows like the very first kiss,
like a break in the clouds and the first ray of sun,
I can feel it again, something new has begun.
You can tell all the pieces are starting to fit,
like the first time you said share my Coke
and I knew that was it.

It's an uncharted sea, an unopened door,
ya got to reach out, oh, ya gotta explore.
I can do anything that life will permit.
You can't beat how it feels, share my Coke,
Coca-Cola is it.
Coca-Cola is it.

However, although overt argumentation is totally absent from this text, it may still contain a lot of implicit arguments. This is because some of the sentences may serve as premises and conclusions in imaginary constructed syllogisms. Syllogisms based on the story told in the text might be as follows:

Premise 1:	The first kiss is something to share
Premise 2:	Coca-Cola is something to share
Conclusion	Coca-Cola is the first kiss

Premise 1:	'It', is the experience of my first kiss
Premise 2:	Coca-Cola is a part of experiencing the first kiss
Conclusion:	Coca-Cola allows me to re-experience the first kiss.

Despite the fact that none of these syllogisms is literally true, they are no more untrue than the argumentation in commercials for cleaning agents.

The *images* also produce a similar effect. So far as the persons and places depicted are concerned, the different shots do not have much in common:

1 A boy and a girl at a party to which he has taken her, probably to dance.
2 The boy and the girl in the next shot are much younger, probably Asian. He gives his bottle of Coke to her.
3 A couple running to a big boat who are impossible to identify.
4 A new setting with people kissing.
5 A new pair embracing.
6 A new pair, in love and playing.
7 Three girls laughing, one of them is prevented from grasping a bottle of Coke.
8 Two boys terminating a phone call. They have probably made a date with some girls.
9 A girl and a boy, they are both very young, sharing a Coke. These persons are also new to us.
10 A new loving pair who are playing the piano together.
11 The first of three in a short story. The two are taking their seats in a subway station, and he gives her his bottle of Coke.
12 Then we get a close up while she is drinking the Coke with a sentimental expression on her face.
13 She is about to give him a kiss.

*Lyrics by Tom Anthony and Lyle Greenfield.

14 The fourteenth shot is also the first of three in a short story. He is probably at school, staring at someone in the classroom.
15 She casts her gaze upon him.
16 He becomes embarrassed by her response.
17 An isolated situation where two girls, one of them holding a bottle of Coke, are passing by some boys who probably make comments.
18 Two children who are about two years old are kissing each other.
19 A girl dancing in a summer field.
20 A pair kissing each other by the sea. The waves are making them wet.
21 A pair embracing each other on the street.
22 A pair sharing a Coke on the bus.
23 A new pair by a lake in sunset.
24 A boy and a girl walking around.
25 A girl drinking Coke next to the dynamic ribbon logo.
26 A boy and a girl kissing each other with the written text 'Coca-Cola is it' and the logo and the rings of the Olympic Games.

The persons depicted change from shot to shot. The exceptions are shots eleven to sixteen. These tell two separate short stories, each of three shots duration. Nevertheless, it is difficult to isolate a clear story throughout this commercial. From a narrative perspective the story is cut up into separate and very short sequences that do not form any clear narrative continuity. Some researchers would say that this commercial is an example of how advertising has changed from telling a story containing a more or less explicit argument to not telling any story at all (Johansen, 1990).

Of course, there are many metaphorical and metonymical connections in this commercial. First of all, there is a metaphorical relation between the first kiss and the taste of Coca-Cola. They are both something to share, and therefore they might be rhetorically replaced by each other. On a more sophisticated level, a similar metaphorical relation exists between the title 'First time' and Coca-Cola. They both seem to share the attribute of chastity. This is most directly expressed in the eighteenth shot.

At the same time, a metonymical relation is also present. As a matter of fact, Coca-Cola is 'always there'. In the successive images it is a part of everything, and that includes the first love and the first kiss. A causal relation between Coca-Cola and the first kiss could also be pointed out. That might be the content of shots eleven to thirteen. He gives his Coke to her, then she turns her face as if she is going to give him a kiss. We do not see them kissing each other, but we suppose they do so.

Based on these concepts of opposition, it might be claimed that there is a continuity between the different shots. It is not the same form of continuity as we will meet in a story, but a lot of stories are told in a metonymical way. We see parts of different stories based on the same subject. But this unity of subject is only a metaphorical unity. Basically, there are two different subjects: a boy and a girl sharing a Coke, and a boy and a girl sharing their love for each other. From this point of view, there is a very tight connection between the different shots.

However, it is not that easy to explain the connection between these shots. The problem is how the category 'metaphor' is used. Two different expressions that are metaphorically linked together can not *explain* the fact that they are linked together. It is rather the opposite: they are linked together, and this relation is of a metaphorical nature. This means that categorizing two different expressions as metaphorically connected presupposes that there ***is*** a connection. In this case, it is not obvious that Coca-Cola is something to share in the same way as people share their reciprocal love. This is an attribute that is *added* to Coca-Cola. The fact that this attribution is a metaphorical one is only a description of a result and not any explanation of *why* Coca-Cola has these different kinds of attribute.

Basically, Coca-Cola is a carbonated soft drink with a certain kind of flavour. Its objective attributes are directly generated from its ingredients. In short: because of its flavour, it may taste good, because of the water, it may quench people's thirst, and because of the phosphoric acid, it may produce a feeling of refreshment. This is the reason why the American film researcher John Paul Freeman (1986), in his theses about the Coca-Cola television commercials, has made a distinction between three different categories of advertising (Freeman, 1986, p. 11):

1 Product attribution
2 Cultural appeals
3 Lifestyle appeals.

As Freeman says: 'In a product attribution advertisement, the consumer is offered a concrete "reason why" the product is worth buying' (Freeman, 1986, p. 12). In other words, in product-attribution advertisement, there is a rational connection between the product and the information the advertisement gives. This is not the situation in advertisements with cultural or lifestyle appeals. Freeman defines cultural appeals as '. . . those which create favourable associations through use of American symbols and representations of American life', and lifestyle appeals such as '. . . those which create favourable impressions toward Coca-Cola through the manners, behaviour, apparel and appearance of desirable role models drinking Coke in the ads' (Freeman, 1986, p. 12f).

Freeman is investigating 'The Real Thing' and 'Look up America' campaigns from the period 1969 to 1976. These campaigns are very deep-rooted in American life and history. Moreover, he points out that there had been a change in the way that Coca-Cola was marketed in the campaigns just before these ones. In the campaign 'Refreshing new feeling' (1962) there was a tendency to use lifestyle appeals in the commercials, but the 'Things go better with Coca-Cola' – campaign (1963–68) was what Freeman calls a '. . . fully developed lifestyle campaign' (Freeman, 1986, p. 117). Freeman's aim is to show how 'The Real Thing' campaign differs from the lifestyle campaigns in using broader cultural appeals. With his introduction of the concept 'cultural appeals', he opens the possibility of looking at the commercials as documents for understanding cultural history.

This is a very important conclusion, but there are some problems connected with the concept of 'lifestyle appeals'. Lifestyle appeals are easy to define, but the problem is that they are omnipresent in advertisements. It is almost impossible to find any advertisements that do not have lifestyle appeals. This is also one of Freeman's conclusions. 'Even "product attribution" advertisements for Coca-Cola, which lend themselves to "rational" approaches, tend to use some emotional appeals.' (Freeman, 1986, p. 405).

Emotional appeals are fundamental to the communication of advertising. Therefore it is an illusion to believe in pure product attribution advertisements. This does not mean that the concept 'product attribution' cannot be fruitful. As a point of departure, it is. What happens in the communication of an advertisement, however, is that the attributes of the product become expanded or replaced.

The commercial 'First Time' is an example of this. Coca-Cola is supplied with new attributes. It is not only a pleasant-tasting soft drink. It has a taste of the first kiss too. And because of the fact that the first time such things happen in one's life is a unique experience, they represent a kind of crisis or climax. The commercial 'First Time' is an example of how such life-climaxes may be the only content of advertising language. And because these climaxes are shown in short, disconnected shots, Jørgen Dines Johansen has referred to 'the effect of epiphany' (Johansen, 1990, pp. 12f). The different shots represent a kind of 'revelation'.

A basic conflict

The effect of epiphany is a kind of telling of a nonstory. 'First Time' shows only isolated shots from different and unrelated stories. But here we meet a fundamental contradiction. On the one hand, the different shots are disconnected, and because of this they represent a discontinuity in the narrative sequence. On the other hand, these disconnections are not experienced as any discontinuity, rather the reverse. Very many of the shots are experienced as coherent because it looks as if they are circling around the same subject.

As pointed out earlier, this commercial deals with more than one subject. There are at least two: people sharing their love for each other, and people sharing a bottle of Coke. The connection between these topics is of a formal character: both instances involve sharing something. And because of this, they may be perceived as a unity.

This, in my opinion, is the key for understanding what kind of repetitions dominate the language of advertising. *The repetitions are not concrete, but only formal.*

This description of advertising language is very similar to a general description of musical language. Development in music is a kind of continual transformation of structures which do not have any precise meanings. In this sense it is a very *formal* language. The similarity between the different manifestations of a musical theme do not have to be on either a denotative or a connotative level, but only on a structural one.

One of the most influential musicologists of this century, Heinrich Schenker used this point as a motto for his work: '*Semper idem sed non eodem modo*' (Schenker, 1979) ('Always the same, but not in the same way'). In this sentence, the principle of his philosophy of music is formulated. He suggests that music is fundamentally based on some few deep structures which follow the same principle all the way through, namely a development from tension to relaxation. He divides these structures into three different levels. The musical structure with the explicit movement from tension to relaxation is what he calls the 'background'. The other two, 'middleground' and 'foreground' represent different kinds of elaboration of the background.

Despite the fact that the different compositions each sound unique, their similarities will suggest that they constitute a unity. This is manifested on the background level. But because of this principle of unity in the musical deep-structures, there must be a certain degree of unity in the surface structures too. The French structuralist and amateur musician Claude Lévi-Strauss has pointed this out in his discussion of music. In his analysis of Maurice Ravel's *Bolero* (1928), he attempts to show how the music goes through simple transformations in one direction without coming back to the point of departure ((Lévi-Strauss 1981, p. 660). That is, the composition is built on repetitions, but none of these repetitions is identical. Disregarding the musical theme and the rhythmical pattern, everything changes in orchestration and texture. The musical theme is recognized primarily because of its structure and not by the associations it creates in the mind. Thus, the similarity between the different parts in the musical piece is a kind of structural similarity.

Lévi-Strauss uses music as an example of the structure of the mythical. And he builds on the German philosopher Ernst Cassirer's attempt to explain the scientific way of thinking by finding out how myths are structured. 'Thus, taken abstractly both the mythical and the scientific explanations of the world are dominated by the same kind of relations: unity and multiplicity, coexistence, continuity and succession.' (Cassirer 1955, p. 60) Cassirer's point is that the mythical way of thinking is an attempt to create continuity between different phenomena in the world. From an objective point of view, it is impossible to prove that there is a continuity, but the mythical way of thinking makes us believe that there must be.

On the other hand, Lévi-Strauss asserts that myths lost their popular force in the seventeenth and eighteenth centuries. Thereafter, music took over the role of the mythical. (Lévi-Strauss, 1981, p. 653f.) From now on, the mythical way of creating continuity in one's life is primarily kept alive through music.

This might sound strange to modern ears, but there are a lot of tendencies that confirm these assertions. The debate about the different arts in the rationalistic movement of the eighteenth century is a debate about their ability to impart knowledge. In this debate, literature is the winner. That is, literature is the art form that is able to impart relatively precise knowledge. The loser in this discussion is, among others, music. Music is not able to carry any unambiguous meanings, it is an open kind of expression.

Music is the most mythical kind of expression because of this capacity, and I will claim that this is the reason why it has become so central to advertising strategies. As Roland Barthes has pointed out, the language of advertising is not only rhetorical, it is also mythical. If we take a look at the liquid cleaner commercial, it is not only a rhetorical argumentation, it is a kind of mythical explanation. But it is easy to see through this explanation. As educated people, we do not believe in it. On the other hand, it is not that easy to see through 'First Time'.

If we believe in 'First Time', it is neither because of the lyrics nor the images. The separate repetitions of structures are not sufficient to constitute a myth. It has to be perceived as a continuity. In 'First Time', the continuity is present because of the music.

Repetitions and developments in advertising music

'First Time' is a part of the 'Coca-Cola is it' campaign. This is a campaign that dominated the eighties. It started in 1982 with a tune composed by Ginny Redington, one of the most central composers of American advertising music in the seventies and eighties. She also formulated the slogan 'Coke is it'. But this was a slogan for the US. In international marketing, the name 'Coca-Cola' had to be used. This had to be added to the original text, and this is the reason why the lyrics often have so tautological an ending: '. . . Coke is it, Coca-Cola is it!'

The eighties were troublesome years for The Coca-Cola Company. Pepsi-Cola presented an increasing challenge and destabilized the market. Robert Woodruff, the Chairman of the Board of Directors since 1923, died in 1985, and the same year, the company made a decision that has been characterized as one of the greatest blunders ever made in the business world. They changed the taste of Coke!

After this crisis, the only one in the company's history, its reputation had to be reshaped. But they did not change the campaign. Instead they renewed it. They kept the slogan 'Coca-Cola is it' but changed the melody and the lyrics. It is impossible to say why it was done in this way, but the effect was a renewal of the campaign without disturbing its continuity. In other words, it was an attempt to combine repetitions with a certain degree of change.

This tendency is more obvious if we look at the music. But it was not because of the music that the new composition was chosen, it was because of the lyrics. The advertising agency McCann-Erickson, Coca-Cola's representative for thirty years, was searching for new ways to revive the slogan 'Coca-Cola' is it! After listening to Art Garfunkel's 1975 version of the 'Waters of March', they decided to use this song because of its lyrics.

This composition was created by the Brazilian bossa-nova composer Antonio Carlos Jobim in 1973, the Portuguese title being 'Aguas de março'. There are many different recorded versions of the composition, and Garfunkel's interpretation is not very different from the other versions. It is a typical bossa-nova arrangement. However, the original bossa-nova style is not present in Coca-Cola's version of 'Aguas de março'. The beat is more hard-rock.

It is a quite different melody, but it is still Antonio Carlos Jobim's composition. The main difference between the version arranged for Coca-Cola and the original composition is the way each uses the motifs that constitute the melody. In the original composition, the motifs are repeated very many times. In the arrangement for Coca-Cola the same melodic motifs are used, but they are not repeated as often. The motif:

Figure 28.1

is repeated seven times in the original composition. In the Coca-Cola arrangement, it is repeated twice. Another motif:

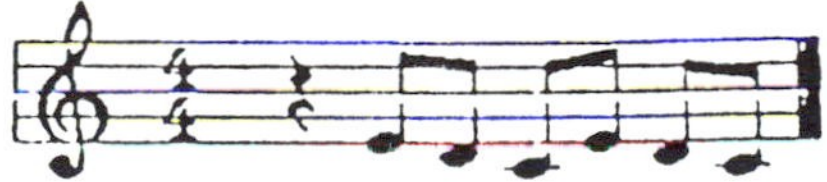

Figure 28.2

is repeated thirty-two times in the original, but only three times in the arrangement for Coca-Cola.

This different way of using repetitions creates quite different kinds of narrativity in the music. 'Aguas de março' is organized in the same way as minimal music. That is, a few elements are repeated several times. The effect of this is to create a kind of standstill. There is no movement towards any climax. The highest notes do not form any culmination of a development, they inject a kind of variation in the monotony. In the commercial for Coca-Cola, the music is fashioned in the opposite way. There are so few repetitions because the different motifs form different stages on the way to a climax.

This illustrates various points. First of all, the music has to be organized in a special way when it has to function as advertising music. It has to follow a certain kind of ascending development in which the climax is coincidental with the conclusion and termination. In this way, 'narrativity' in commercial music is expressed through a certain pattern that is in part very well known from the history of music and in part quite unique.

The differences between the original version of 'Aguas de março' and the arrangements for Coca-Cola also illustrate the manner in which music is or is not mythical. The repetitions in commercial music go through certain transformations, and these create development and continuity in the music. In 'Aguas de março' the motifs are not transformed, and because of this, the music is perceived as more discontinuous. 'Aguas de março' is therefore mythical to a lesser degree than are the arrangements for Coca-Cola.

Repetitions and developments in advertising slogans

This formal structural way of creating continuity is not only a characteristic of the music, it is also one of the most central aspects of how the different slogans are created. If we look at the successor to the slogan 'Coca-Cola is it', it may be perceived as a fundamentally new creation. 'You can't beat the feeling' and 'Coca-Cola is it!' do not have much in common. Seen from a perspective of marketing, the current slogan is something totally new and different.

If we investigate other central parts of the lyrics from the 'Coke is it!' promotion. We will find that the current slogan evolved from previous campaigns. Some central parts of the lyrics had to be changed, because of the new tune composed by Jobim, and the sentences of the ending, constituting a kind of refrain, run: 'It's just how you feel when you know it's for real. You can't beat how it feels, it's a Coke. Coca-Cola is it.' In other words, 'You can't beat the feeling' and 'You can't beat how it feels' have very much in common.

The 'Coke is it!'– campaign differed markedly from its predecessors. The two previous campaigns, 'Coke Adds Life' and 'Have a Coke and a Smile' have very little in common with the 'Coca-Cola is it!' campaign and were relatively anonymous. But before these promotions, there was a campaign that was much more successful: the 'Real thing' campaign whose formulation goes back to the forties. In 1942 slogans such as 'Everybody welcomes Coca-Cola. It's the real thing.' (Freeman op.cit. 1986, p. 601), 'With a taste all of its own. It's the real thing.' (op.cit.) and 'The only thing like Coca-Cola is Coca-Cola itself. It's the real thing.' (op.cit.) had been used. And in the 'Real Thing' campaign from the beginning of the seventies, the formulation 'It's the real thing' is *the* most important formulation of the slogan. This 'it' establishes connections with the 'Coke is it' – campaign. Such an assertion is strengthened by the fact that the singers in the 'It's the Real Thing' song, sometimes sing: '. . . What the world wants today is the real thing (Coke is) it's the real thing.'* This is a kind of overlapping that is quite well known in music, but not so frequent in literature. The musical phrase goes like this:

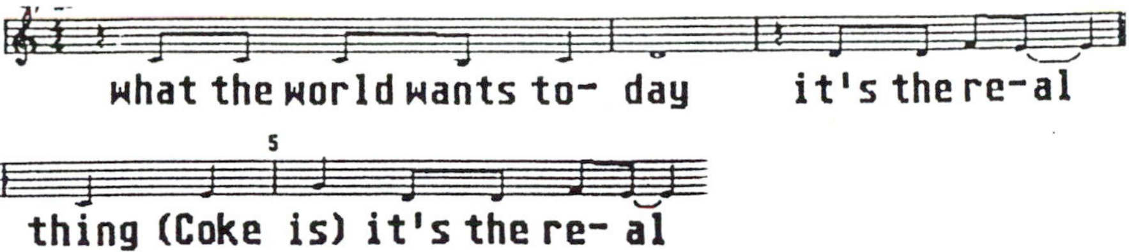

Figure 28.3

The last two bars can be divided into two separate groups:

Figure 28.4

If we add lyrics to these groups, then we get:

Figure 28.5

I would say that the overlap in the lyrics is not natural from a literary point of view, but is more the result of the music. This kind of overlap in the music sounds quite natural, but the sentence 'Coke is it's the real thing' is not grammatically correct. In other words, if this sentence does not follow the grammar of the language, it follows the musical grammar.

On the basis of these examples, I would argue that the slogans represent a continuity in that they draw identifiable themes from previous campaigns. It is always possible to find the formulation of a slogan in earlier campaigns, not as a dominating slogan, but as a subordinate one. I will also draw the conclusion that this continuity is neither on the level of denotations nor of connotations, but on a structural level which is primarily given by the rules that govern the music. The repetition of slogans, therefore, serves the function of creating a kind of musical continuity.

Repetitions and developments in advertising images

I have mentioned earlier that the discontinuity of the images in the commercials has the function of telling many different stories. In this way, they are a kind of quotation. The audience has to create the beginnings and the endings of these stories by themselves in most cases. However sometimes the images can be quotations from other commercials. This, to a lesser degree, is similar

*Lead sheet 'It's the real thing', words and music by William Backer

to the repetitions of slogans. One shot might be taken from one commercial and integrated into another commercial.

The most brilliant example of this effect of quotation is the commercial 'We are'. This is a very special commercial because it was probably made for the big Jubilee Exhibition in 1986. Most of the shots are quotations from other successful commercials. Except for the fact that we are familiar with the different stories from which they are quoted, the commercial as such is constructed in the same pattern as 'First Time'. By means of all these quotations, it is able to show us a new story. Most of the shots present American national symbols, victorious situations, friendship and leisure. We are able to construe a coherence between the shots and then subsume them under a limited number of subjects because of the likeness in structure between many of the images. In this way this commercial tells us a new story by means of quotations. Here again, the music is very important. By means of repetitions and development it serves the function of keeping coherence and continuity in the story. The structure of the music is very similar to that of Ravel's Bolero.

'We are' is not a representative commercial because the music is original and unique and it serves the function of summing up the television commercials of the previous thirty years. However, the principle by which this summary is constructed is not unique.

The repetitions in the images are on a structural level in advertising language, because this will ensure a higher degree of identification on the part of the viewer. And this is what it is all about. The commercials have to lead to positive associations in the pre-release tests. People accept what they can identify with. That is, *audiences have to be a part of the commercial,* which cannot engage the viewer without telling everyone's story – a mythical project. Everyone can put their own concrete experiences of life into a structure that invites such participation. Music does so automatically. The lyrics and the images are structured in a way that results in the same kind of invitation. The bottom line of advertising language, however, is letting this kind of invitation be a part of the product itself. That is, the product has to be structured in a way that everyone's experiences of life can be a part of it. The product itself has to become mythical. This is what I have called replacement of attribution, and the commercials for Coca-Cola show that the music is the most important factor in achieving this effect.

Notes and references

I 1: What is communication? (Colin Cherry)

Notes

1 But such reflexes do not form part of true human language; like the cries of animals they cannot be said to be *right* or *wrong* though, as signs, they can be interpreted by our fellows into the emotions they express.
2 John Donne, the Sixteenth Devotion.
3 See Reusch and Kees (1964) for many illustrations and examples of pictures, icons, motifs, gestures, manners, etc.
4 With kind permission of the Clarendon Press, Oxford.
5 With kind permission of the *Journal of the Acoustical Society of America.*

References

Gallie, W.B. 1952: *Peirce and pragmatism.* Harmondsworth: Pelican Books. An outline of Peirce's scattered work.

Locke, John 1689: *An essay concerning human understanding.* Numerous editions: e.g. London: Ward, Locke & Bowden.

McDougall, W. 1927: *The group mind.* London: The Cambridge University Press.

Morris, C.W. 1938: Foundations of the theory of signs. In *International encyclopedia of unified science series,* Vol. 1.2. Chicago: University of Chicago Press. A brief introduction.

Morris, C.W. 1946: *Signs, language and behaviour.* New York: Prentice Hall Inc.

Reusch, J. and Kees, W. 1964; *Non-verbal communication.* Berkeley and Los Angeles: University of California Press.

Stevens, S.S. 1950: Introduction: a definition of communication. In Proceedings of the speech communication conference at MIT, *Journal of the Acoustical Society of America* 22, 689–90.

I 2: A generalized graphic model of communication (George Gerbner)

Reference

Halloran, James 1969: The communicator in mass communications research. In Halmos, P. (ed.), *The Sociological Review Monograph* 13. University of Keele, January 1969.

I 3: Defining language (Jean Aitchison)

References

Evans, W.E. and Bastian, J. 1969: Marine mammal communication: social and ecological factors. In H.T. Anderson (ed.), *The biology of marine mammals*. New York: Academic Press.

Hockett, C.F. 1963: The problem of universals in language. In J.H. Greenberg (ed.), *Universals of language*. Cambridge, Mass.: MIT Press.

Marshall, J.C. 1970: The biology of communication in man and animals. In Lyons, J. (ed.), 1970: *New horizons in linguistics*. Harmondsworth: Penguin.

McNeill, D. 1966: Developmental psycholinguistics. In Smith, F. and Miller, G.A. 1966: *The genesis of language*. Cambridge, Mass: MIT Press.

Morton, J. 1971: What could possibly be innate? In Morton, J (ed.), *Biological and social factors in psycholinguistics*. London: Logos Press.

Robins, L 1971: *General linguistics: an introductory survey*. Second edition, London: Longman.

Struhsaker, T.T. 1967: Auditory communication among vervet monkeys (*Cercopithecus aethiops*). In Altman, S.A. (ed.), *Social communication among primates*. Chicago: Chicago University Press.

Thorpe, W.H. 1961: *Bird song: the biology of vocal communication and expression in birds*. Cambridge University Press.

Thorpe, W.H. 1963: *Learning and instinct in animals*. Second edition. London: Methuen.

Von Frisch, K. 1950: *Bees: their vision, chemical sense and language*. Ithaca: Cornell University Press.

Von Frisch, K. 1954: *The dancing bees*. London: Methuen.

Von Frisch, K. 1967: *The dance and orientation of bees*. Translated by L.E. Chadwick, Cambridge, Mass.: Harvard University Press.

I 4: Deceit and misrepresentation (W.T. Scott)

Reference

MacKay, D.M. 1972: Formal analysis of communicative processes. In Hinde, R.A. (ed.), *Non-verbal communication*. Cambridge: Cambridge University Press, 24.

I 5: Verbal and non-verbal communication (Michael Argyle)

References

Abercrombie, K. 1968: Paralanguage: *British Journal of Disorders in Communication* 3, 55–9.

Argyle, M. 1972: Non-verbal communication in human social interaction. In R. Hinde (ed.), *Non-verbal communication*. London: Royal Society and Cambridge University Press.

Argyle, M., Salter, V., Nicholson, H., Williams, M. and Burgess, P. 1970: The communication of inferior and superior attitudes by verbal and non-verbal signals. *British Journal of Social and Clinical Psychology* 9, 221–31.

Brown, R. 1965: *Social psychology*. New York: Collier-Macmillan.

Burns, T. 1964: Non-verbal communication. *Discovery* 25(10), 30–7.

Chapple, E.D. 1956: *The interaction chronograph manual*. Moroton, Connecticut: E.D. Chapple Inc.

Cook, M. 1969: Anxiety, speech disturbances, and speech rate. *British Journal of Social and Clinical Psychology* 8, 13–21.

Crystal, D. 1969: *Prosodic systems and intonation in English*. London: Cambridge University Press.

Davitz, J.R. 1964: *The communication of emotional meaning*. New York: McGraw-Hill.

Ekman, P. and Friesan, W.V. 1969: The repertoire of non-verbal behaviour: categories, origin, usage, and coding. *Semiotica* 1, 49–98.

Hall, E.T. 1963: A system for the notation of proxemic behavior. *American Anthropologist* 65, 1003–26.

Jourard, S.M. 1966: An exploratory study of body-accessibility. *British Journal of Social and Clinical Psychology* 5, 221–31.

Kendon, A. 1972: Some relationships between body motion and speech: an analysis of an example. In A. Siegman and B. Pope (eds.), *Studies in dyadic communication*. Elmsford, New York: Pergamon.
Lott, R.E., Clark, W. and Altman, I. 1969: *A propositional inventory of research on interpersonal space*. Washington: Naval Medical Research Institute.
Melly, G. 1965: Gesture goes classless. *New Society*, 17 June, 26–7.
Sarbin, T.R. and Hardyk, C.D. 1953: Contributions to role-taking theory: role-perception on the basis of postural cues. Unpublished, cited by T.R. Sarbin 1954: Role theory. In G. Lindzey (ed.) *Handbook of social psychology*. Cambridge, Mass.: Addison-Wesley.
Scheflen, A.E. 1965: *Stream and structure of communication behavior*. Eastern Pennsylvania Psychiatric Institute.
Schlosberg, A. 1952: The description of facial expressions in terms of two dimensions. *Journal of Experimental Psychology* 44, 229–37.
Sommer, R. 1965: Further studies of small group ecology. *Sociometry* 28, 337–48.

I 6: The analysis of representational images (Bill Nichols)

Note

1 Excerpts from the transcript of Godard-Gorin's *Letter to Jane, Women and Film* 1 3/4 (1973), 48.

II: The socio-cultural relations of language

Notes

1 Levinson, Stephen C., 1983: *Pragmatics*. Cambridge: Cambridge University Press, 35, 38.
2 Leech, Geoffrey 1983: *Principles of pragmatics*. London: Longman, 21.
3 Potter, Jonathan, and Wetherall, Margaret 1987: *Discourse and social psychology: beyond attitudes and behaviour*. London: Sage, 6.

II 7: Social class, language and socialization (Basil Bernstein)

References

Bernstein, B. 1962: Family role systems, socialization and communication. Manuscript, Sociological Research Unit, University of London Institute of Education. Also in: A socio-linguistic approach to socialization. In J.J. Gumpertz and D. Hymes (eds.), *Directions in sociolinguistics*. New York: Holt, Rinehart & Winston.
Bernstein, B. 1970: Education cannot compensate for society. *New Society* 387, February.
Bernstein, B. and Cook, J. 1965: Coding grid for maternal control. Available from Department of Sociology, University of London Institute of Education.
Bernstein, B. and Henderson, D. 1969: Social class differences in the relevance of language to socialization. *Sociology* 3(1).
Bright, N. (ed.) 1966: *Sociolinguistics*. The Hague and Paris: Mouton.
Carroll, J.B. (ed.) 1956: *Language, thought and reality: selected writings of Benjamin Lee Whorf*. New York: Wiley.
Cazden, C.B. 1969: Sub-cultural differences in child language: an interdisciplinary review. *Merrill-Palmer Quarterly* 12.
Chomsky, N. 1965: *Aspects of linguistic theory*. Cambridge, Mass.: MIT Press.
Cook, J. 1971: An enquiry into patterns of communication and control between mothers and their children in different social classes. PhD thesis, University of London.
Coulthard, M. 1969: A discussion of restricted and elaborated codes. *Educational Review* 22(1).
Douglas, M. 1970: *Natural symbols*. London: Barrie & Rockliff, The Cresset Press.
Fishman, J.A. 1960: A systematization of the Whorfian hypothesis. *Behavioral Science* 5.
Gumpertz, J.J. and Hymes, D. (eds.), 1971: *Directions in sociolinguistics*. New York: Holt, Rinehart & Winston. [Volume published 1972.]

Halliday, M.A.K. 1969: Relevant models of language. *Educational Review* 22(1).
Hawkins, P.R. 1969: Social class, the nominal group and reference. *Language and Speech* 12(2).
Henderson, D. 1970: Contextual specificity, discretion and cognitive socialization: with special reference to language. *Sociology* 4(3).
Hoijer, H. (ed.) 1954: Language in culture. *American Anthropological Association Memoir* 79. Also published by University of Chicago Press.
Hymes, D. 1966: On communicative competence. Research Planning Conference on Language Development among Disadvantaged Children. Ferkauf Graduate School: Yeshiva University.
Hymes, D. 1967: Models of the interaction of language and social setting. *Journal of Social Issues* 23.
Labov, W. 1965: Stages in the acquisition of standard English. In W. Shuy (ed.), *Social dialects and language learning*. Champaign, Illinois: National Council of Teachers of English.
Labov, W. 1966: The social stratification of English in New York City. Washington DC centre for Applied Linguistics.
Mandelbaum, D. (ed.) 1949: *Selected writings of Edward Sapir*. Berkeley and London: University of California Press.
Parsons, T. and Shils, E.A. (eds.) 1962: *Toward a general theory of action*. New York: Harper Torchbooks. [Chapter 1 especially.]
Schatzman, L. and Strauss, A.L. 1955: Social class and modes of communication. *American Journal of Sociology* 60.
Turner, G. and Pickvance, R.E. 1971: Social class differences in the expression of uncertainty in five-year-old children. *Language and Speech* 14(4).
Williams, F. and Naremore, R.C. 1969: On the functional analysis of social class differences in modes of speech. *Speech Monographs* 36(2).

II 8: Impossible discourse (Trevor Pateman)

Notes

1 This passage was published in 1956. But in his seminar of 1971–2, Barthes was still speaking of the 'social contract' of language.

2 If 'flower' and 'rose' are not hierarchically ordered, then there is no contradiction for the speaker in such statements as 'It's a rose, not a flower' or 'If it's not a flower, is it a rose?' Note that the speaker is not contradicting himself; there is no contradiction for him, only for the hearer.

3 Not all trees are alike. Whorf (Carroll, 1956, 136: see also the whole of the essay *The relation of habitual thought and behaviour to language*, 134–59) suggests that the word 'stone' logico-linguistically implies 'non-combustibility'. However, this implication seems to exist only as the result of belief in a particular theory, namely that stone is non-combustible, and even to this theory there are exceptions (e.g. brimstone). The link between 'monarchy' and 'government' is different in nature. Here, the link is logico-linguistic. One could say that it is a synthetic truth (or falsehood) that 'stone' and 'non-combustibility' have the link they do or are supposed to have, whereas it is an analytic truth that 'government' and 'monarchy' have the relationship which they do. Whorf fails to come to terms with the fact that different theories can be developed within the same language (by speakers of the same language) and this I presume, because of his identification of thought and language.

4 If I am correct in thinking that the conceptual understanding of a word requires reference to a superordinate term, then the most abstract words – those at the tops of trees – with no superordinate words of their own, would be doomed to remain the words for complexes; they could not be concepts. This seems to me paradoxical, for it is my conventional view that the most abstract words ought to be the most rigorously conceptually definable.

5 To a larger degree, this was recognized by the students of schizophrenic thought being criticized. Thus, for instance, Kasanin writes that abstract or categorical thinking (the absence of which was regarded as a trait of schizophrenia) 'is a property of the educated adult person' (1944, 42). And Hanfmann and Kasanin write that 'the difference in [test] scores of this [group, largely composed of attendants at a state mental hospital] and of the college-educated group is sufficiently striking to warrant the conclusion that the highest performance level in the concept formation test is the prerogative of subjects who have had the benefit of college education' (1942, 59–60).

6 The question then arises, What are the specific features of schizophrenia? Some have concluded that there are none, that schizophrenia as a disease-entity does not exist.

7 Many of Wason and Johnson Laird's undergraduate samples do appallingly on reasoning tests. This cannot just be put down to the nature of the tests, for some do well on them. It could be put down to differential reactions to the test situation. Hudson has explored this (1970).
8 The opposition abstract/concrete helped me a great deal when I started in 1970 on the lines of thought developed here, though it has fallen into a secondary place in this chapter. I would still recommend the books which originally helped me, namely Cassirer (1944) and Goldstein (1963).
9 Compare from a different context the following criticism of Cameron: 'In our opinion all explanations [of failure on the Vygotsky blocks test] in terms of evasion or projection of blame on the task etc., misrepresent the situation of those patients who perform to the best of their ability but are unable even to conceive of the performance required from them' (1944, 96).
10 Based on a discussion in an evening class I once taught.
11 Vygotsky's child has an adequate understanding of both 'cow' and 'dog', otherwise he would not be able to make the particular mistake that he does.
12 This point was brought home to me by Chris Arthur.
13 Some people seek to solve political problems by spatial displacement that is, by *emigration*. It would be interesting to study the political ideology of emigration, perhaps making use of Gabel's theories about spatialization (1969).
14 In contrast, Reich construes apathy as a defence mechanism against recognition of one's class position and interests (1970, 201).
15 'The masses' class consciousness does not consist of knowing the historical and economic laws which rule the existence of man, but: (1) knowing one's own needs in all spheres; (2) knowing the means and possibilities of satisfying them; (3) knowing the hindrances which the social order deriving from private enterprise puts in their way; (4) knowing which inhibitions and fears stand in the way of clearly recognizing one's vital needs and the factors preventing their fulfilment . . .; (5) knowing that the masses' strength would be invincible in relation to the power of the oppressors if only it were coordinated.' (Reich, 1971, 68–9.)
16 'If one wants to lead the mass of the population into the field against capitalism, develop their class consciousness and bring them to revolt, one recognizes the principle of renunciation as harmful, stupid and reactionary. Socialism, on the other hand, asserts that the productive forces of society are sufficiently well developed to assure the broadest masses of all lands a life corresponding to the cultural level attained by society.' (Reich, 1971, 23–4.)
17 Reich, in contrast, has a manipulative concept of song, dance and theatre: 'we must secure the emotional attachment of the masses. Emotional attachment signifies trust, such as the child has in its mother's protection and guidance, and confidence in being understood in its innermost worries and desires including the most secret ones, those relating to sex'. (1971, 58–9). Apart from the last nine words, this reads more like Stalin than Reich.
18 The Maoist criticism of Stalin, that he made mistakes, is no criticism at all. For it does not challenge his right to have held the kind of power which *permitted him* to make such mistakes.

References

Bachelard, G. 1970: *La formation de l'esprit scientifique*. Paris: Bibliothèque des Textes Philosophiques.

Barthes, R. 1967: *Elements of semiology*. London: Cape.

Barthes, R. 1972: *Mythologies*. London: Cape.

Bernstein, B. 1971: *Class, codes and control: volume one: theoretical studies towards a sociology of language*. London; Routledge.

Carroll, J.B. (ed.) 1956: *Language, thought and reality: selected writings of Benjamin Lee Whorf*. Cambridge, Mass.: MIT Press.

Cassirer, E. 1944: *An essay on man*. New Haven, Conn.: Yale University Press.

Gabel, J. 1969: *La fausse conscience, essai sur la réification*. Third edition, Paris: Editions de Minuit.

Goldstein, K. 1963: *Human nature in the light of psychopathology*. New York; Schocken Books.

Halliday, M.A.K. 1969: Relevant models of language. *Educational Review* 22.

Hanfmann, E. and Kasanin, J. 1942: *Conceptual thinking in schizophrenia*. Nervous and Mental Diseases Monograph Series number 67.

Hudson, L. 1970: *Frames of mind*. Harmondsworth: Penguin.

Kasanin, J. (ed.) 1944: *Language and thought in schizophrenia*. Stanford, Calif.: University of California Press.

Laing, R.D. and Esterson, A. 1970: *Sanity, madness and the family*. Harmondsworth: Penguin.

Lyons, J. 1968: *An introduction to theoretical linguistics*. Cambridge: Cambridge University Press.
McKenzie, R. and Silver, A. 1969: *Angels in marble*. London: Heinemann.
Marcuse, H. 1964: *One-dimensional man*. London: Routledge.
Pateman, T. 1973a: Review of R. Barthes, *Mythologies, Human Context* 5.
Pateman, T. 1973b: The experience of politics. *Philosophy and Phenomenological Research* (USA) 33.
Reich, W. 1970: *The mass psychology of fascism*. Trans. V.R. Carfagno. New York; Farrar, Straus & Giroux.
Reich, W. 1971: *What is class consciousness?* London: Socialist Reproduction.
Saussure, F de 1966: *Course in general linguistics*. Trans. W. Baskin. New York: McGraw-Hill.
Vygotsky, L. 1962: *Thought and language*. Cambridge, Mass.: MIT Press.
Wason, P. and Laird, P. Johnson 1972: *Psychology of reasoning: structure and content*. London: Harvard University Press.
Wertheimer, M. 1961: *Productive thinking*. London: Tavistock.
Wittgenstein, L. 1958: *Philosophical investigations*. Second edition, Oxford: Blackwell.
Wood, J. (ed.) 1970: *Powell and the 1970 election*. Surrey: Elliot Right Way Books.

II 9: The structures of speech and writing (Gunther Kress)

References

Brazil, D.C. 1975: *Discourse intonation*. Discourse analysis monographs no. 1. Birmingham: English Language Research.
Hymes, Dell 1972: On communicative competence. In Pride, J.B. and Holmes, J. (eds.) 1972: *Sociolinguistics*. Harmondsworth: Penguin.

II 10: Beyond alienation: an integrational approach to women and language (Deborah Cameron)

Notes

1 Simone de Beauvoir, *Memoirs of a Dutiful Daughter*, Trans. Kirkup (Penguin 1963) p. 17.
2 Camilla Gugenheim, 'Man Made Language?' *Amazon*, no. 4, 1981.
3 Audre Lorde, *The Cancer Journals*, (Spinsters Ink, 1980) p. 19.
4 Genesis, 11: 6–9.
5 For Ferdinand de Saussure see the editors' introduction to this section. Benjamin Lee Whorf is representative of a 'strong' version of linguistic determinism. Basing his views on a study of the language of the North American Hopi Indians, he claimed that because their language structured reality differently from, say, English, the Hopi perceived the world differently from native English speakers. [Editors' note].
6 Trevor Patement, *Language, Truth and Politics*, 2nd edn (Jean Stroud, 1980) p. 129.
7 Cora Kaplan, 'Language and Gender', *Papers on Patriarchy* (WPC/PDC, 1976).
8 *Ibid.*, p. 21.
9 J. Gumperz, *Discourse Strategies* (CUP, 1982) and J. Gumperz (ed.), *Language and Social Identity* (CUP, 1982).
10 Gumperz, *op. cit.*, pp. 4–5.
11 D. Malz and R. Borker, 'A Cultural Perspective on Male/Female Miscommunication', in Gumperz, *op. cit.*
12 Colin MacCabe, 'The Discursive and the Ideological in Film', *Screen*, 19/4.
13 M. Black and R. Coward, 'Linguistic, Social and Sexual Relations', *Screen Education*, 39, p. 78.
14 Dalston Study Group, 'Was the patriarchy conference "patriarchal"?', in *Papers on Patriarchy*, p. 76.
15 M. Jenkins and C. Kramarae 'A Thief in the House', in *Men's Studies Modified*, ed. Spender (Pergamon, 1981).
16 Kaplan, *op. cit.*, p. 21.
17 Shirley Ardener, *Perceiving Women* (John Wiley, 1975); Philip Smith, 'Sex Markers in Speech', in *Social Markers in Speech*, ed. K. Scherer and H. Giles (CUP, 1979).
18 Marielouise Janssen-Jurreit, *Sexism* (Pluto Press, 1982) p. 284.

19 O. Jespersen, *Language: Its Nature, Development and Origin* (Allen & Unwin, 1922), p. 246.
20 Elino Keenan, 'Norm Markers, Norm Breakers' in *Explorations in the Ethnography of Speaking*, ed. R. Bauman and J. Sherzer, (CUP, 1974).
21 E. Marks and I. de Courtivron (eds), *New French Feminisms*, p. 5.
22 J.J. Rousseau, *Emile*, quoted in J. O'Faolain and L. Martines, *Not in God's Image*, (Fontana, 1974) p. 259.
23 Dale Spender, *Man Made Language* (Routledge & Kegan Paul, 1980) p. 107.
24 Cf. Jenkins and Kramarae, *op. cit.*, pp. 16–17.
25 See M. Jenkins and C. Kramer, 'Small group process: learning from women', *WSIQ*, 3, 1980', D. Jones, 'Gossip: notes on women's oral culture', *WSIQ*, 3, 1980.
26 Basil Bernstein, *Class, Codes and Control*, vol. 1, (Routledge & Kegan Paul, 1970).
27 W. Labov, *The Logic of Non-Standard English*, repr. in Giglioli, P. P. (ed.), *Language and Social Context* (Penguin, 1972).
28 Jespersen, *Language*, ch. 24.
29 Dalston Study Group, 'Was the patriarchy conference "patriarchal"?', *Papers on Patriarchy*, p. 77.

II 11: Involvement strategies in a speech by the Reverend Jesse Jackson (Deborah Tannen)

Notes

1 Yet another level of metaphoric play is what Lakoff and Johnson (1980) identify as an 'up is good' metaphor that underlies many of our figures of speech: The ships' success would stem from 'a higher reality' and 'noble instincts' which emerge when we are 'At our highest.' In contrast, the ships will 'drift' if we 'satisfy our baser [i.e. lower] instincts.'

2 So compelling was the rhythm of these repetitions reinforcing the quilt metaphor that it mattered not at all when Jackson omitted a word ('for'). When he told lesbians and gays, 'when you fight against discrimination and a cure for AIDS,' he did not mean that they fight 'against' a cure for AIDS but rather that they fight *for* one. Furthermore, it is gay men, as a group, and not lesbians, who are especially concerned with fighting for a cure for AIDS. But no matter. The message got through in the metamessage established by the list: that all the groups' and individuals' demands would be more likely met if they joined together.

3 *The New York Times* transcription of Jackson's 1984 speech omitted the word 'end' in the phrase 'in the end,' yielding:

Suffering breeds character.
Character breeds faith,
and in the faith will not disappoint.

My conjecture that this is a mistranscription is based in part on the form taken by the same construction in 1988.

References

Davis, Gerald L. 1985: *I got the word in me and I can sing it, you know: a study of the performed African-American sermon*. Philadelphia: University of Pennsylvania Press.

Friedrich, Paul 1986: *The language parallex: linguistic relativism and poetic indeterminacy*. Austin: University of Texas Press.

Lakoff, George and Johnson, Mark 1980: *Metaphors we live by*. Chicago: University of Chicago Press.

Levin, Samuel R. 1982: Are figures of thought figures of speech? In Byrnes, Heidi (ed.) 1982: *Contemporary perceptions of language: interdisciplinary dimensions. Georgetown University round table on languages and linguistics*. Washington, DC: Georgetown University Press, 112–23.

Sapir, J. David 1977: The anatomy of metaphor. In Sapir, J. David and Crocker, J. Christopher (eds) 1977: *The social use of metaphor: essays on the anthropology of rhetoric*. 3–32. Philadelphia: The University of Pennsylvania Press.

II 12: Talk, identity and performance (Graham Brand and Paddy Scannell)

Notes

1 This article is an extensively edited and rewritten version of a Polytechnic of Central London Media Studies undergraduate dissertation (Brand, 1987).
2 For a vivid account of a Blackburn Soul Night Out and its audience, see Brand (1987), pp. 70–3.

References

Atkinson, M. 1984: *Our masters' voices*. London: Methuen.
Blackburn, T. 1985: *Tony Blackburn, 'The living legend'. An autobiography.* (*As told to Cheryl Garnsey*). London: W.H. Allen.
Brand, G. 1987: *Tony Blackburn. The construction and maintenance of a broadcast identity and a broadcast universe.* Polytechnic of Central London: Media Studies dissertation.
Brundson, C. and Morley, D. 1978: *Everyday television: 'Nationwide'*. London: British Film Institute, TV monograph 10.
Garfinkel, H. 1984: *Studies in ethnomethodology*. Cambridge: Polity Press.
Giddens, A. 1984: *The constitution of society*. Cambridge: Polity Press.
Goffman, E. 1969: *The presentation of self in everyday life*. Harmondsworth: Penguin Books.
Goffman, E. 1970: *Asylums*. Harmondsworth: Penguin Books.
Goffman, E. 1974: *Frame analysis*. Harmondsworth: Penguin Books.
Goffman, E. 1981: *Forms of talk*. Oxford: Basil Blackwell.
Lord, A. 1960: *The singer of tales*. New York: Atheneum.
Parry, A. (ed.) 1971: *The making of Homeric verse. The collected papers of Milman Parry*. Oxford: Oxford University Press.
Scannell, P. 1988: *Radio Times*. The temporal arrangement of broadcasting in the modern world'. In Drummond, P. and Paterson, R. (eds): *Television and its audience*. 15–31. London: British Film Institute.
Scannell, P. and Cardiff, D. 1991: *A social history of British broadcasting*. Vol I: Serving the nation, 1922–1939. Oxford: Basil Blackwell.

III 13: The perceptual process (Albert H. Hastorf, David J. Schneider and Judith Polefka)

Reference

Leeper, R. 1935: The role of motivation in learning: a study of the phenomenon of differential motivational control of the utilization of habits. *Journal of Genetic Psychology* 46, 3–40.

III 14: Social communication (Stuart Sigman)

Notes

1 The careful reader will have noted that I alternate between describing communication as a 'process' and as a 'structure.; Communication is an *organized activity*, and can thus be studied from the two interrelated vantage points.
2 'Rule' is used in a most general sense to refer to statements about socially patterned procedures for behaving and interpreting (see Sigman [1980]; Shimanoff [1980] for various uses of the term). Rules are analytic devices to account for behaviour units; relations among behaviour units; contextuality; and normative force (Sigman, 1985). The degree of constraint that rules place on persons, and the definitions of the relevant interacting body or bodies, for example, persons, groups, and so on, to which rules apply are cross-culturally variable and are subject to empirical observation. Thus, social communication theory leaves open the possibility the normative and interpretive views of the social order and communication rules will be appropriate to varying degrees to different cultures (cf. Donohue, Cushman, and Nofsinger, 1980).

3 This is an axiom that guides certain structuralist methodologies (see Birdwhistell, 1970; Scheflen, 1973). For a limited criticism of this approach, see Condon (1980), who presents some data that question the total hierarchical organization of behaviour. Note also that the guiding principle of hierarchy does not directly address the issue of whether a top-down or bottom-up view is most appropriate (cf, Scheflen, 1973; Kendon, 1982), that is, whether the hierarchy is seen as an organization of frame levels that constrain the appearance of behaviour units on lower levels or as a structure built up from lower units.

4 The mechanical model often seems to assume a non-equivalence of the interactants as well, at least in terms of their information states. However, this model usually sees communication as bringing the participants' information states more in line with each other, as in the case of the sender sending information to an unknowledgeable receiver.

References

Aberle, D.F., Cohen A.K., Davis, A.K., Levy, M.J. Jr. and Sutton, F.X. 1950: The functional prerequisites of a society, *Ethics* 6, 100–11.

Aldrich, H. 1972: Sociability in Mensa; characteristics of interaction among strangers. *Urban Life and Culture* 1, 167–86.

Annandale, E. 1985: Work roles and definitions of patient health. *Qualitative Sociology* 8, 124–48.

Berger, P.L. 1963: *Invitation to sociology: a humanistic perspective*. Garden City, NY: Doubleday Anchor.

Berger, P.L. and Luckmann, T. 1967: *The social construction of reality*. Garden City, NY: Doubleday Anchor.

Bernstein, B. 1975: *Class, codes and control: theoretical studies towards a sociology of language*. New York: Schocken.

Birdwhistell, R.L. 1977: Some discussions of ethnography, theory, and method. In Brockham, J. (ed.) 1977: *About Bateson*. New York: Dutton, 103–41.

— 1970: *Kinesics and context*. Philadelphia: U. of Philadelphia Press.

— 1952: *Introduction to kinesics*. Washington, D.C.: Foreign Service Institute.

Blum, A.F. and McHugh, P. 1971: The social ascription of motives. *American Sociological Review* 36, 98–109.

Carey, J.W. 1975: A cultural approach to communication. *Communication* 2, 1–22.

Condon, W.C. 1980. The relation of interactional synchrony to cognitive and interactional process. In Key, M.R. (ed.) 1980: *The relationship of verbal and nonverbal communication*. The Hague: Mouton, 49–65.

Cushman, D.P. 1980: A functional approach to rules research. Paper presented to the Eastern Communication Association, Ocean City, Maryland.

Denzin, N.K. 1984: *On understanding emotion*. San Francisco: Jossey-Bass.

Donohue, W.A., Cushman, D.P. and Nofsinger, R.E. 1980: Creating and confronting social order: a comparison of rules perspectives. *Western Journal of Speech Communication* 44, 5–19.

Durkheim, E. 1938: *The rules of the sociological method*. Chicago: U. of Chicago Press.

— 1933: *The diversion of labor in society*. New York: Free Press.

Erickson, F. and Shultz, J. 1982: *The counselor as gatekeeper: social interaction in interviews*. New York: Academic Press.

Fisher, B.A. 1978: *Perspectives on human communication*. New York: Macmillan.

Frentz, T.S. and Farrell, T.B. 1976: Language-action: a paradigm for communication, *Quarterly Journal of Speech* 62, 333–49.

Gergen, K.J. 1982: *Toward transformation in social knowledge*. New York: Springer-Verlag.

Gerth, H. and Mills, C.W. 1953: *Character and social structure*. New York: Harcourt, Brace.

Giddens, A. 1984: *The constitution of society*. Berkeley: U. of California Press.

Goffman, E. 1967: *Interaction ritual*. Chicago: Aldine.

— 1963: *Behavior in public places*. New York: Free Press.

Golding, P. and Murdoch, G. 1978: Theories of communication and theories of society. *Communication Research* 5, 339–56.

Halliday, M.A.K. 1978: *Language as social semiotic: the social interpretation of language and meaning*. London: Edward Arnold.

Harré, R., Clarke, D. and DeCarlo, N. 1985: *Motives and mechanisms: an introduction to the psychology of action*. London: Methuen.

Heritage, J.C. and Watson, D.R. 1979: Formulations as conversational objects. In Psathas, G. (ed.) 1979: *Everyday language: studies in ethnomethodology*. New York: Irvington, 123–62.

Hothschild, A.R. 1983: *The managed heart*. Berkeley: U. of California Press.

Hymes, D. 1974: *Foundations in sociolinguistics*. Philadelphia: U. of Pennsylvania Press.
Jakobson, R. 1960: Closing statement: linguistics and poetics. In Sebeok, T. (ed.) 1960: *Style in language*. Cambridge, Mass.: MIT Press.
Joos, M. 1967: *The five clocks*. New York: Harcourt, Brace and World.
Kemper, T.D. 1972: The division of labor: a post-Durkheimian analytical view. *American sociological review* 37, 739–53.
Kendon, A. 1982: The organization of behavior in face-to-face interaction: observations on the development of a methodology. In Scherer, K.R. and Ekman, P. (eds.) 1982: *Handbook of methods in nonverbal behaviour research*. Cambridge: CUP, 440–505.
Kockelmans, J.J. 1975: Toward an interpretative or hermeneutic social science. *Graduate Faculty Philosophy Journal* 5, 73–96.
Kroeber, A.L. 1963: *Anthropology: culture patterns and processes*. New York: Harcourt Brace Jovanovich.
LaBarre, W. 1954: *The human animal*. Chicago: U. of Chicago Press.
Lasswell, H.D. 1971: The structure and function of communication in society. In Schramm, W. and Roberts, D.F.W. (eds.) 1971: *The process and effects of mass communication*. Urbana, Ill.: U. of Illinois Press, 84–99.
Leach, E. 1976: *Culture and communication*. Cambridge: Cambridge University Press.
Linton, R. 1940: A neglected aspect of social organization. *American Journal of Sociology* 45, 870–86.
— 1936: *The study of man*. New York: Appleton-Century-Crofts.
Lundberg, G. 1939: *Foundations of sociology*. New York: Macmillan.
Mandelbaum, M. 1973: Societal facts. In O'Neill, J. (ed.) 1973: *Modes of individualism and collectivism*. London: Heinemann, 221–36.
McCall, G.J. and Simmons, J.L. 1978: *Identities and interactions*. Rev. ed. New York: Free Press.
Mead, G.H. 1934: *Mind, self, and society*. Chicago: U. of Chicago Press.
Merton, R.K. 1968: *Social theory and social structure*. Enlarged ed. New York: Free Press.
Nwoye, G. 1985: Eloquent silence among the Igbo of Nigeria. In Tannen, D. and Saville-Troike, M. (eds.) 1985: *Perspectives on silence*. Norwood, N.J.: Ablex, 185–91.
O'Neill, J. (ed.) 1973: *Modes of individualism and collectivism*. London: Heinemann.
Pearce, W.B. and Cronen, V.E. 1980: *Communications, action, and meaning: the creation of social realities*. New York: Praeger.
Pike, K.L. 1967: *Language in relation to a unified theory of the structure of human behavior*. The Hague: Mouton.
Pittenger, R.E., Hockett, C.F. and Danehy, J.J. 1960: *The first five minutes: a sample of microscopic interview analysis*. Ithaca, NY: Paul Martineau.
Poole, M.S. and McPhee, R.D. 1983: A structurational analysis of organizational climate. In Putnam, L.L. and Pacanowsky, M.E. (eds.) 1983: *Communication and organizations: an interpretive approach*. Beverly Hills: Sage, 195–219.
Poole, M.S., Siebold, D.R., and McPhee, R.D. 1985: Group decision-making as a structurational process. *Quarterly Journal of Speech* 71, 74–102.
Radcliffe-Brown, A.R. 1965: *Structure and function in primitive society*. New York: Free Press.
Rapoport, A. 1982: *The meaning of the built environment*. Beverly Hills: Sage.
Rosen, L. 1984: *Bargaining for reality: the construction of social relations in a Muslim community*. Chicago: U. of Chicago Press.
Scheflen, A.E. 1973: *Communicational structure: analysis of a psychotherapy transaction*. Bloomington, Ind.: Indiana University Press.
— 1968: Human communication: behavioral programs and their integration in interaction. *Behavioral Science* 13, 44–55.
— 1965: Systems in human communication. Paper presented to the Society for General Systems Research, Berkeley, California.
Scott, R.L. 1977: Communication as an intentional social system. *Human Communication Research* 3, 258–68.
Shimanoff, S. 1980: *Communication rules*. Beverly Hills: Sage.
Sigman, S.J. 1985: Some common mistakes students make when learning discourse analysis. *Communication Education* 34, 119–27.
— 1983: Some multiple constraints placed on conversational topics. In Craig, R.T. and Tracy, K. (eds) 1983: *Conversational coherence*. Beverly Hills: Sage, 174–95.
— 1982: Some communicational aspects of patient placement and careers in two nursing homes. Unpublished PhD. dissertation, University of Pennsylvania.

— 1980: On communication rules from a social perspective. *Human Communication Research* 7, 37–51.
Sorokin, P.A. 1947: *Society, culture, and personality*. New York: Harper and Brothers.
Thomas, S. 1980: Some problems of the paradigm in communication theory. *Philosophy of Social Sciences* 10, 427–44.
Watzlawick, P., Beavin, J.H. and Jackson, D.D. 1967: *Pragmatics of human communication*. New York: Norton.
Wilden, A. 1979: Changing frames of order: cybernetics and the *Machina Mundi*. In Krippendorff, K. (ed.) 1979: *Communication and control in society*. New York: Gordon and Breach Science Publishers, 9–29.

III 15: Stereotypes (Walter Lippmann)

Notes

[Footnotes in the original text include no publication details for texts cited. Publication details for the first edition of all texts cited have been inserted in square brackets.]

1 Edmond Locard, *L'Enquête Criminelle et les Méthodes Scientifiques* [Paris, 1920]. A great deal of interesting material has been gathered in late years on the credibility of the witness, which shows, as an able reviewer of Dr. Locard's book says in *The Times* (London) Literary Supplement (August 18, 1921), that credibility varies as to classes of witnesses and classes of events, and also as to type of perception. Thus, perceptions of touch, odour, and taste have low evidential value. Our hearing is defective and arbitrary when it judges the source and direction of sound, and in listening to the talk of other people 'words which are not heard will be supplied by the witness in all good faith. He will have a theory of the purport of the conversation, and will arrange the sounds he heard to fit it.' Even visual perceptions are liable to great error, as in identification, recognition, judgement of distance, estimates of numbers, for example, the size of a crowd. In the untrained observer, the sense of time is highly variable. All these original weaknesses are complicated by tricks of memory, and the incessant creative quality of the imagination. *Cf.* also Sherrington, [Sir Charles Scott], *The integrative action of the nervous system*, [1904: New Haven, Yale University Press], 318–27.

The late Professor Hugo Münsterberg wrote a popular book on this subject called [1890]: *On the Witness Stand [: Essays on Psychology and Crime*. New York: Holt.]

2 James, William [1890]: *Principles of Psychology*. Vol. 1. [New York: Holt], 488.
3 Dewey, John [1910]: *How We Think*. [Cambridge, Mass.: D.C. Heath], 121.
4 *Op. cit.*, 133.
5 von Gennep, A: *La Formation des Légendes*, 158–9. Cited van Langenhove, F. [1916]: *The Growth of a Legend*. [New York and London: G.P. Putnam's Sons], 120–22.
6 Berenson, Bernard [1897]: *The Central Italian Painters of the Renaissance*, 60 [New York: G.P. Putnam's Sons] *et seq.*
7 *Cf.* also his comment on *Dante's visual images*, and *his early illustrators* in *The study and criticism of Italian art* (First Series), 13. [Third series 1901–16. London: G. Bell and Sons]. 'We cannot help dressing Virgil as a Roman, and giving him a "classical profile" and "statuesque carriage", but Dante's visual image of Virgil was probably no less mediæval, no more based on a critical conception of the Roman poet. Fourteenth Century illustrators make Virgil look like a mediæval scholar, dressed in cap and gown, and there is no reason why Dante's visual image of him should have been other than this.'
8 *The central Italian painters*, 66–7.
9 Cited by Mr. Edward Hale Bierstadt, *New Republic*, June 1 1921, 21.
10 Wallas, Graham [1921]: *Our social heritage*. [London: G. Allen and Unwin], 17.

III 16: They saw a game (Albert H. Hastorf and Hadley Cantril)

Notes

1 We are not concerned here with the problem of guilt or responsibility for infractions, and nothing here implies any judgement as to who was to blame.
2 The film was kindly loaned for the purpose of the experiment by the Dartmouth College Athletic

Council. It should be pointed out that a movie of a football game follows the ball, is thus selective, and omits a good deal of the total action on the field. Also, of course, in viewing only a film of a game, the possibilities of participation as spectator are greatly limited.

3 We gratefully acknowledge the assistance of Virginia Zerega, Office of Public Opinion Research, and J.L. McCandless, Princeton University, and E.S. Horton, Dartmouth College, in the gathering and collation of the data.

4 The interpretation of the nature of a social event sketched here is in part based on discussions with Adelbert Ames, Jr, and is being elaborated in more detail elsewhere.

References

Cantril, H. 1950: *The 'why' of man's experience*. New York: Macmillan.

Kilpatric, F.P. (ed.) 1952: *Human behavior from the transactional point of view*. Hanover, N.H.: Institute for Associated Research.

III 17: Introduction to the presentation of self in everyday life (Erving Goffman)

Notes

1 Gustav Ichheiser, 'Misunderstandings in Human Relations', Supplement to *The American Journal of Sociology* LV (September 1949), pp. 6–7.

2 Quoted in E.H. Volkart (ed.), *Social behaviour and personality*, contributions of W.I. Thomas to Theory and Social Research (New York: Social Science Research Council, 1951), p. 5.

3 Here I owe much to an unpublished paper by Tom Burns of the University of Edinburgh. He presents the argument that in all interaction a basic underlying theme is the desire of each participant to guide and control the responses made by the others present. A similar argument has been advanced by Jay Haley in a recent unpublished paper, but in regard to a special kind of control, that having to do with defining the nature of the relationship of those involved in the interaction.

4 Willard Waller, 'The Rating and Dating Complex', *American Sociology Review* II, p. 730.

5 William Sansom, *A contest of ladies* (London: Hogarth, 1956), pp. 230–2.

6 The widely read and rather sound writings of Stephen Potter are concerned in part with signs that can be engineered to give a shrewd observer the apparently incidental cues he needs to discover concealed virtues the gamesman does not in fact possess.

7 An interaction can be purposely set up as a time and place for voicing differences in opinion, but in such cases participants must be careful to agree not to disagree on the proper tone of voice, vocabulary, and degree of seriousness in which all arguments are to be phrased, and upon the mutual respect which disagreeing participants must carefully continue to express towards one another. This debaters' or academic definition of the situation may also be invoked suddenly and judiciously as a way of translating a serious conflict of views into one that can be handled within a framework acceptable to all present.

8 W.F. Whyte, 'When Workers and Customers Meet', chap. vii, *Industry and society*, edited by W.F. Whyte (New York: McGraw-Hill, 1946), pp. 132–3.

9 Teacher interview quoted by Howard S. Becker, 'Social Class Variations in the Teacher-Pupil Relationship', *Journal of Educational Sociology* XXV, p. 459.

10 Harold Taxel, 'Authority Structure in a Mental Hospital Ward' (unpublished Master's thesis, Department of Sociology, University of Chicago, 1953).

11 This role of the witness in limiting what it is the individual can be has been stressed by the Existentialists, who see it as a basic threat to individual freedom. See Jean-Paul Sartre, *Being and nothingness* (London: Methuen, 1957).

12 Goffman, 'Communication Conduct in an Island Community', pp. 319–27.

13 Peter Blau, 'Dynamics of Bureaucracy' (PhD dissertation, Department of Sociology, Columbia University, University of Chicago Press, 1955), pp. 127–9.

14 Walter M. Beattie, Jr, 'The Merchant Seaman' (unpublished MA Report, Department of Sociology, University of Chicago, 1950), p. 35.

15 Sir Frederick Ponsonby, *Recollections of three reigns* (London: Eyre & Spottiswoode, 1951).

III 18: Mass communication and para-social interaction (Donald Horton and R. Richard Wohl)

Notes

1 They may move out into positions of leadership in the world at large as they become famous and influential. Frank Sinatra, for example, has become known as a 'youth leader.'

Conversely, figures from the political world, to choose another example, may become media 'personalities' when they appear regularly. Fiorello LaGuardia, the late Mayor of New York, is one such case.

2 Merton's discussion of the attitude toward Kate Smith of her adherents exemplifies, with much circumstantial detail, what we have said above. See Robert K. Merton, Marjorie Fiske and Alberta Curtis, *Mass persuasion: the social psychology of a war bond drive.* (New York: Harper, 1946), Ch. 6.

3 There does remain the possibility that over the course of his professional life the persona, responding to influences from his audience, may develop new conceptions of himself and his role.

4 Dave Garroway as told to Joe Alex Morris 1956: I lead a goofy life. *The Saturday Evening Post,* February 11, 62.

5 Reference note 4, 64.

6 See, for instance: George H. Mead, *Mind, self and society.* (Chicago: University of Chicago Press, 1934) Walter Coutu, *Emergent human nature* (New York: Knopf, 1949.) Rosalind Dymond 1950: Personality and empathy. *Journal of Consulting Psychiatry* 14, 343–50.

7 Burke uses this expression to describe an attitude evoked by formal rhetorical devices, but it seems equally appropriate here. See Kenneth Burke, *A rhetoric of motives* (New York: Prentice-Hall, 1950), 58.

8 This is in contrast to the closed system of the drama, in which all the roles are predetermined in their mutual relations.

9 Kenneth Burke, *Attitudes towards history.* Vol. 1 (New York: New Republic Publishing Co., 1937), see, for instance, 104.

10 See Merton's acute analysis of the audience's demand for 'sincerity' as a reassurance against manipulation. Reference note 2., 142–6.

11 These attributes have been strikingly discussed by Mervyn LeRoy, a Hollywood director, in a recent book. Although he refers specifically to the motion-picture star, similar notions are common in other branches of show business. 'What draws you to certain people?' he asks. 'I have said before that you can't be a really fine actress or actor without heart. You also have to possess the ability to project that heart, that feeling and emotion. The sympathy in your eyes will show. The audience has to feel sorry for the person on the screen. If there aren't moments when, rightly or wrongly, he moves the audience to sympathy, there's an actor who will never be big box-office.' Mervyn LeRoy, and Alice Canfield, *It takes more than talent* (New York: Knopf, 1953), 114.

12 Once an actor has succeeded in establishing a good relationship with his audience in a particular kind of dramatic role, he may be 'typed' in that role. Stereotyping in the motion-picture industry is often rooted in the belief that sustained rapport with the audience can be achieved by repeating past success. (This principle is usually criticized as detrimental to the talent of the actor, but it is a *sine qua non* for the persona whose professional success depends upon creating and sustaining a plausible and unchanging identity.) Sometimes, indeed, the Hollywood performer will actually take his name from a successful role; this is one of the principles on which Warner Brothers Studios selects the names of some of its actors. For instance, Donna Lee Hickey was renamed Mae Wynn after a character she portrayed, with great distinction, in *The Caine Mutiny.* See 1955: Names of Hollywood actors. *Names* 3, 116.

13 The 'loyalty' which is demanded of the audience is not necessarily passive or confined only to patronizing the persona's performance. Its active demonstration is called for in charity appeals, 'marathons' and 'telethons'; and, of course, it is expected to be freely transferable to the products advertised by the performer. Its most active form is represented by the organization of fan clubs with programmes of activities and membership obligations, which give a continuing testimony of loyalty.

14 Comedians on radio and television frequently chide their audience if they do not laugh at the appropriate places, or if their response is held to be inadequate. The comedian tells the audience that if they don't respond promptly, he won't wait, whereupon the audience usually provides the demanded laugh. Sometimes the chiding is more oblique, as when the comedian interrupts

his performance to announce that he will fire the writer of the unsuccessful joke. Again, the admonition to respond correctly is itself treated as a joke and is followed by a laugh.

15 Coutu, reference note 6, 294.

16 See, for example, W. Lloyd Warner and William E. Henry, 'The radio day time serial: a symbolic analysis'. *Genetic Psychology Monographs* (1948) 37, 3–71, the study of a daytime radio serial programme in which it is shown that upper-middle-class women tend to reject identification with lower-middle-class women represented in the drama. Yet some people are willing to take unfamiliar roles. This appears to be especially characteristic of the intellectual whose distinction is not so much that he has cosmopolitan tastes and knowledge, but that he has the capacity to transcend the limits of his own culture in his identifications. Remarkably little is known about how this ability is developed.

17 Most students of the mass media occupy a cultural level somewhat above that of the most popular programmes and personalities of the media, and necessarily look down upon them. But it should not be forgotten that for many millions indulgence in these media is a matter of looking up. Is it not also possible that some of the media permit a welcome regression, for some, from the higher cultural standards of their present status? This may be one explanation of the vogue of detective stories and science fiction among intellectuals, and might also explain the escape downward from middle-class standards in the literature of 'low life' generally.

18 It is frequently charged that the media's description of this side of life is partial, shallow, and often false. It would be easier and more profitable to evaluate these criticisms if they were formulated in terms of role-theory.

From the viewpoint of any given role it would be interesting to know how well the media take account of the values and expectations of the role-reciprocators. What range of legitimate variations in role performance is acknowledged? How much attention is given to the problems arising from changing roles, and how creatively are these problems handled? These are only a few of the many similar questions which at once come to mind.

19 There is a close analogy here with one type of newspaper human-interest story which records extreme instances of role-achievement and their rewards. Such stories detail cases of extreme longevity, marriages of especially long duration, large numbers of children; deeds of heroism – role performance under 'impossible' conditions; extraordinary luck, prizes, and so on.

III 20: The flower dream (Sigmund Freud)

References

Ferenczi, S. 1917: Träume der Ahnungslosen. *Int. Z. ärztl. Psychoanal.* 4(208), 498.

[Translator's note: Dreams of the unsuspecting. *Further contributions to the theory and technique of psycho-analysis*. London, 1926, chapter LVI].

Freud, S. 1905a: *Jokes and their relation to the unconscious. Standard edition of the complete psychological works of Sigmund Freud* vol. 8. London: Hogarth Press, 1953–74.

Freud, S. 1905b: Fragment of an analysis of a case of hysteria. *Standard edition* vol. 7.

Freud, S. 1909: Notes upon a case of obsessional neurosis. *Standard edition* vol. 10.

IV 21: The contours of high modernity (Anthony Giddens)

Notes

1 A fuller exposition of the major points of the next few sections can be found in Anthony Giddens, *The consequences of modernity*. (Cambridge: Polity, 1990).

2 Anthony Giddens, *The nation-state and violence* (Cambridge: Polity, 1985).

3 See Giddens, *Consequences of modernity*.

4 Georg Simmel, *The philosophy of money* (London: Routledge, 1978), p. 179.

5 Alan Fox, *Beyond contract* (London: Faber, 1974). For one of the few generalised discussions of trust in systems, see Susan P. Schapiro, 'The social control of impersonal trust', *American Journal of Sociology*, 93, 1987.

6 *Cf.* Paul Connerton, *How societies remember* (Cambridge: Cambridge University Press, 1989).

7 Giddens, *Central problems in social theory* (London: Macmillan, 1979).

8 Claude Levi-Strauss, *Structural anthropology* (London: Allen Lane, 1968).

9 Walter J. Ong, *Interfaces of the word* (Ithaca: Cornell University Press, 1977).
10 Harold Innis, *Empire and communications* (Oxford University Press, 1950); Marshall McLuhan, *Understanding media* (London: Sphere, 1967).
11 Christopher Small, *The printed word* (Aberdeen: Aberdeen University Press, 1982).
12 J.M. Strawson: 'Future methods and techniques', in Philip Hills (ed.), *The future of the printed word* (London: Pinter, 1980), p. 15.
13 Susan R. Brooker-Gross, 'The changing concept of place in the news'. In Jacquelin Burgess and John R. Gold, *Geography, the media and popular culture* (London: Croom Helm, 1985), p. 63.
14 *Cf.* E. Relph, *Place and placelessness* (London: Pion, 1976). Joshua Meyrowitz, *No sense of place* (Oxford: Oxford University Press, 1985).
15 Especially Jean Baudrillard. See Mark Poster, *Jean Baudrillard* (Cambridge: Polity, 1989).
16 Yi-Fu Tuan, *Topophilia* (Englewood Cliffs: Prentice-Hall, 1974); Robert David Sack, *Conceptions of space in social thought* (London: Macmillan, 1980).

IV 22: The unique perspective of television and its effect: a pilot study (Lang and Lang)

Notes

1 'MacArthur Day in Chicago' includes the following occasions which were televised: arrival at Midway Airport, parade through the city including the dedication at the Bataan-Corregidor Bridge, and the evening speech at Soldiers Field.
2 This paper reports only one aspect of a larger study of MacArthur Day in Chicago. A report of the larger study is nearing completion. This writeup is limited to drawing together some of the implications concerning the role of television in public events, this particular study being considered as a pilot study for the framing of hypotheses and categories prerequisite for a more complete analysis of other such events in general. The present study could not test these categories, but was limited to an analysis of the television content in terms of the observed 'landslide effect' of the telecast. The authors wish to express their indebtedness to Dr. Tamatsu Shibutani (then of the Department of Sociology, University of Chicago) for lending his encouragement and giving us absolute freedom for a study which, due to the short notice of MacArthur's planned arrival in Chicago, had to be prepared and drawn up in three days, and for allowing his classes to be used for soliciting volunteers. No funds of any sort were at our disposal. Dr. Donald Horton was kind enough to supply us with television sets and tape recorders. In discussions of the general problems involved in the analysis of television content, he has indirectly been of invaluable aid. Finally, we are indebted to the other twenty-nine observers, without whose splendid cooperation the data could never have been gathered.
3 That this check together with our observation of the general impression left by MacArthur Day constitutes only a very limited validation is beyond question. Under the conditions of the study – carried on without financial support and as an adjunct to other research commitments – it was the best we could do.
4 Analysis of personal data sheets, filled out by participants prior to MacArthur Day, revealed that 'objectivity' in observation was not related to political opinion held, papers and periodicals subscribed to, and previous exposure to radio or TV coverage of MacArthur's homecoming. The significant factor in evaluating the reports for individual or deviant interpretation was found to reside in the degree to which individual observers were committed to scientific and objective procedures. Our observers were all advanced graduate students in the social sciences.
5 An analysis of televised interviews is omitted in this condensation. Interviews obtained for the study by observers posing as press representatives elicited responses similar to those given over TV. Without exception, those questioned referred to the magnitude, import, and other formal aspects of the event. These stand in contrast to results obtained through informal probes and most overheard conversation. One informant connected with television volunteered that television announcers had had specific instructions to emphasize that this was a 'dramatic event.' Another of Chicago's TV newsmen noted that throughout the telecast the commentary from each position made it sound as if the high points of the day's activity were about to occur or were occurring right on their own spot.
6 The day's activities at a nearby race track were not cancelled. At one point in the motorcade from the airport to the Loop, a traffic block resulted in a partially 'captive audience.' An irritated 'captive' remarked, 'I hope this doesn't make me late for the races.'

7 In a subsequent interview, a TV producer explained his conception of the MacArthur Day coverage as 'being the best in the country.' He especially recalled bracketing and then closing in on the General during the motorcade, the assumption being that he was the centre of attraction.
8 During the evening ceremonies, MacArthur's failure to show fatigue in spite of the strenuous experiences of the day received special notice. A report from a public viewing of the evening speech indicates the centering of discussion about this 'lack of fatigue' in relation to the General's advanced years (Observer 24).
9 It must be re-emphasized that there was no independent check – in the form of a validation – of the specific effect of TV. However, newspaper coverage emphasized the overwhelming enthusiasm. Informal interviews, moreover, even months later, showed that the event was still being interpreted as a display of mass hysteria.

IV 24: Visualizing the news (Richard Ericson, Patricia Baranek and Janet Chan)

References

Gans, H. 1980: *Deciding what's news*. New York: Vintage.
GUMG (Glasgow University Media Group) 1976: *Bad news*. London: Routledge.
GUMG (Glasgow University Media Group) 1980: *More bad news*. London: Routledge.
Schlesinger, P. 1978: *Putting 'reality' together: BBC news*. London: Constable.
Williams, R. 1974: *Television: technology and cultural form*. London: Fontana.

IV 25: Documentary meanings and the discourse of interpretation (John Corner and Kay Richardson)

Notes

1 Morley, D. *The 'Nationwide audience'*, (London: British Film Institute, 1980).
2 Richardson, K. and Corner, J. Reading reception: mediation and transparency in viewers' accounts of a TV programme', *Media, Culture and Society* 1980, 8, 3.

IV 26: *Dallas* between reality and fiction (Ien Ang)

Notes

1 For a foundation of this semiological approach to television programmes, see *inter alia* U. Eco, 'Towards a semiotic inquiry into the television message', *Working Papers in Cultural Studies* 2, 1972; and S. Hall, 'Encoding and decoding in the television discourse', C.C.C.S. Occasional Stencilled Papers, University of Birmingham, 1973.
2 See also R.C. Allen, 'On reading soaps: a semiotic primer', in E. Ann Kaplan (ed.), *Regarding television*, (Los Angeles: American Film Institute, 1983).
3 D. Morley, *The 'nationwide' audience*, (London: British Film Institute, 1980), 10.
4 J-M Piemme, *La propagande inavoué*, Paris: Union Generale d'Editions, 1975.
5 *ibid.*, 176.
6 *ibid.*, 114.
7 'Empiricist' because the basic premise is used that reality can be gathered from the manifestation of the world. *Cf.* C. MacCabe, 'Theory and film: principles of realism and pleasure', *Screen* 17.3, 1976, 9–11.
8 R. Williams, *Marxism and literature*, Oxford: OUP, 1977, 97.
9 Piemme, *La propagande inavoué*, 120–1.
10 See also Piemme, 170–1.
11 For a critique of the theory of the classic-realist text, see *inter alia* T. Lovell, *Pictures of Reality*, pp. 84–87; also D. Morley, 'Texts, readers, subjects', in S. Hall, D. Hobson, A. Lowe, P. Willis (eds.) *Culture, Media, Language*, Hutchinson, London, 1980, 163–73.
12 The distinction between denotation and connotation is made among others by Roland Barthes in

his *Elements of Semiology*, (London: Jonathan Cape, 1967). Subsequently various semiologists have contested this distinction, because it suggests a hierarchy between 'literal' and 'figurative' meaning, which does not in fact exist. However, in his *S/Z*, Hill and Wang, New York, 1974/Jonathan Cape, London 1975, Barthes defends this distinction if it is a matter of the analysis of what he calls 'the classical text' (as opposed to the 'modern text'). It is in any case important to regard the distinction between denotation and connotation as an analytical difference. See also S. Hall, 'Encoding/Decoding', in Hall *et al.* (eds.), *Culture, Media, Language*.

13 The concept 'structure of feeling' comes from Raymond Williams. See for example his *Marxism and Literature*, 128–35.

IV 27: Captured on videotape: camcorders and the personalization of television. (Lawrence J. Vale)

Notes

1 In some cases the camcorder is not an accessory to the VCR, but, instead, its substitute, because some users connect their camcorder directly to their television and do not own a separate VCR.

2 Video cameras and camcorders are, of course, used in many settings and for many purposes. They are used for diagnostic and documentary purposes in many professions, and are becoming an increasingly important component of television production. In addition, video, as a medium in itself, is an increasingly popular art form. My central concern here, however, is with the social uses and cultural meanings of video in the home context.

3 This figure for still camera use coincides exactly with the national average – 73% of households (Wolfman, 1988).

4 To date, the verb 'to film' seems to encompass the act of making a video as well. Although the verb 'to videotape' seems also to be used frequently, it is treated – by the nonprofessional at least – as virtually synonymous with the verb 'to film.' Conversely, there seems to be a tendency to refer to the finished product as only 'a video' rather than a 'film' or even a 'videotape.' This idea of 'videotape' as a noun seems to refer primarily to the blank videocassette before the camcorder user has acted on it. This points to the curious tripartite existence of the videocassette (shared with its predecessor the audiocassette): it is (a) a format for consuming prerecorded and packaged data that is borrowed, rented or purchased (b) a format for personalizing this prerecorded and packaged data not only by selecting it, but by time-shifting it, editing it, and storing it (c) a format for home-producing personalized equivalents, approximations or alternatives to pre-recorded fare. It will be interesting to see how long it will be before 'to video' and/or 'to videotape' gains a separate currency as a verb. Perhaps, though the underlying technologies of film and video are very different, there is such a great deal that is similar about the physical and social acts of holding and using the camera, that the new word may be slower to catch on than might otherwise be expected.

5 Video 'instant replay' has also had a controversial impact on professional sports, where play is interrupted until such time as what has just happened can be reviewed and assessed. The controversy occurs not simply because the audience has a chance to 'relive' the moment that has just passed but because game officials rely on this technology to overrule or reinterpret these events, which thereby transforms them. The assumption that the videotaped instant replay should be accepted as a more 'accurate' basis for judgment than the eye of the appointed human authority seems significant. Where there are clear 'rules' that the videocamera is asked to help adjudicate, there is some justification; as an example of a tendency to abandon the authority of one's own experiential vantage point, this seems part of a disturbing trend.

References

Armes, R. 1988: *On video*. London: Routledge.

Beloff, H. 1985: *Camera culture*. Oxford: Basil Blackwell.

Benjamin, W. 1969: The work of art in the age of mechanical reproduction. In H. Arendt (ed.) & H. Zohn (trans.), *Illuminations*, pp. 217–252. New York: Schocken. (Originally published 1936).

Chalfen, R. 1983: Exploiting the vernacular: Studies in snapshot photography. *Studies in Visual Communication*, 9(3), 70–84.

Chalfen, R. 1984: The sociovidistic wisdom of Abby and Ann: Toward an etiquette of home mode

photography. *Journal of American Culture*, 7(1–2), 22–31.
Chalfen, R. 1987: *Snapshot versions of life*. Bowling Green, OH: Bowling Green State University Press.
Collins, M. January 19, 1989: Latest newshounds: Video camera buffs. *USA Today*, p. 3D.
Dobrow, J.R. 1989: Away from the mainstream? VCRs and ethnic identity. In M. Levy (ed.), *The VCR age*, pp. 193–208. Newbury Park, CA: Sage.
Hattersley, R. June, 1971: Family photography as a sacrament. *Popular Photography*, pp. 106–108.
Hirsch, J. 1981: *Family photographs: Content, meaning and effect*. New York: Oxford University Press.
Hughey, A. March 17, 1982: Sales of home-movie equipment falling as firms abandon market. *Wall Street Journal*.
Jacobs, D. 1981: Domestic snapshots: Toward a grammar of motives. *Journal of American Culture*, 4(Wolf1), 101.
Kealy, J. July 12, 1981: Will videotape [sic] systems replace home movies? *New York Times*.
Lasica, J.D. January 2, 1989: Pictures *don't* always tell truth. *Boston Globe*, pp. 29–30.
Lesy, M. 1980: *Time frames: The meaning of family photographs*. New York: Pantheon.
Marbach, W. December 30, 1985: Video's new focus: A small camera makes a big difference, *Newsweek*, pp. 56–57.
Olshaker, M. 1978: *The instant image*. New York: Stein & Day.
Sontag, S. 1977: *On photography*. New York: Farrar, Straus & Giroux.
Stewart, D. 1979: Photo therapy: Theory and practice. *Art Psychotherapy*, 6(1), 42.
Stocker, C. December 8, 1988: Camcorders zooming in on family life. *Boston Globe*, pp. 49, 54–55.
Waters, H. December 30, 1985: The age of video. *Newsweek*, pp. 44–53.
Wolfman, A. (ed.) December, 1966: *Photo dealer: 1966 annual statistical report: The photographic industry in the United States*.
Wolfman, A. (ed.) 1980: *1979–1980 Wolfman report on the photographic industry in the United States*. New York: Modern Photography.
Wolfman, A. (ed.) 1981: *1980–1981 Wolfman report on the photographic industry in the United States*. New York: Modern Photography.
Wolfman, L.(ed.) 1988: *1987–1988 Wolfman report on the photographic and imaging industry in the United States*. New York: Diamandis Communications.

IV 28: Music, text and image in Coca-Cola commercials (Hroar Klempe)

References

Barthes, R. 1973: *Mythologies*. Translated by A. Lavers. London: Paladin.
Cassirer, E. 1955: *The philosophy of symbolic forms. Vol. 2: 'Mythical Thought'*. New Haven and London: Yale University Press.
Freeman, J.P. 1986: 'The Real Thing: "Lifestyle" and "Cultural" Appeals in Television Advertising for Coca-Cola, 1969–1976.' Unpublished PhD thesis, The University of North Carolina at Chapel Hill (University Microfilms International).
Johansen, J.D. 1990: 'Semiotics, rhetoric and advertising'. Unpublished lecture given at seminar 'Marketing and Semiotics', Oslo, November 3.
Lévi-Strauss C. 1981: *The naked man*. New York: Harper and Row.
Schenker, H. 1979: *Free composition*. Second edition. New York and London: Longman.

Subject index